INTRODUCTORY ALGEBRA
SECOND EDITION

John Gillam
Donald O. Norris
Ohio University—Athens, OH

Man M. Sharma
Clark College—Atlanta, GA

Kendall/Hunt Publishing Company
Dubuque, Iowa

Copyright © 1983 by John Gillam, Donald O. Norris and Man M. Sharma
Copyright © 1986 by Kendall Hunt Publishing Company

ISBN 0-8403-3915-1

All rights reserved. No part of this publication may be reproduced, stored in a retrieval system, or transmitted, in any form or by any means, electronic, mechanical, photocopying, recording, or otherwise, without the prior written permission of the copyright owner.

Printed in the United States of America
10 9 8 7 6 5 4 3

Contents

Preface — vii

1. **Arithmetic Refresher** — 1
    - 1.1 Real numbers, 1
    - 1.2 The order of operations, 5
    - 1.3 The number line and signed numbers, 8
    - 1.4 Fractions, 12
    - 1.5 Operations on decimal numbers, 15
    - 1.6 Basic operations on fractions, 19
    - 1.7 Percentage, 27
    - Chapter Summary, 30
    - Review Problems, 30
    - Chapter Test, 35

2. **Exponents, Square Roots and Algebraic Expressions** — 38
    - 2.1 An introduction to exponents, 38
    - 2.2 Natural number exponents, 41
    - 2.3 Integer exponents, 47
    - 2.4 Scientific notation, 51
    - 2.5 Square roots, 55
    - 2.6 Algebraic expressions, 60
    - 2.7 Distributive law, 62
    - 2.8 Evaluating algebraic expressions, 66
    - Chapter Summary, 69
    - Review Problems, 70
    - Chapter Test, 73

3. **Linear Equations and Inequalities** — 76
    - 3.1 Types of equations, 76
    - 3.2 Solutions of simple linear equations, 78
    - 3.3 Solution of equations having the unknown on both sides of the equation, 83
    - 3.4 Equations involving grouping symbols and fractions, 89

    3.5   Linear inequalities, 93
            Chapter Summary, 98
            Review Problems, 98
            Chapter Test, 101

4. Introduction to Word Problems   104

    4.1   From English to symbols, 104
    4.2   Translating word problems into equations, 106
    4.3   Ratio and proportion problems, 113
    4.4   Mixture problems, 118
    4.5   Distance problems, 122
    4.6   Problems on geometric figures, 127
    4.7   General applications of algebra, 130
            Review Problems, 132
            Chapter Test, 134

5. Basic Operations on Algebraic Expressions   137

    5.1   Review, 137
    5.2   Multiplication of multinomials, 140
    5.3   Division by monomials, 144
    5.4   Long division, 146
            Chapter Summary, 150
            Review Problems, 151
            Chapter Test, 153

6. Special Products and Factoring   156

    6.1   Review, 156
    6.2   Special products, 157
    6.3   Simple factors, 160
    6.4   Factors of a difference of two squares, 162
    6.5   Critical integers, 164
    6.6   Factoring trinomials of the type $x^2 + bx + c$, 165
    6.7   Factoring a trinomial of the type $ax^2 + bx + c$, 168
    6.8   More on factors, 170
    6.9   Factoring the sum or difference of two cubes, 172
    6.10   Factors by completing the squares, 174
            Chapter Summary, 177
            Review Problems, 178
            Chapter Test, 182

7. Quadratic Equations   184

    7.1   Some basic facts, 184
    7.2   Solution by factoring, 186
    7.3   Checking solutions of $ax^2 + bx + c = 0$, 189
    7.4   Solutions by use of the quadratic formula, 194
    7.5   Complex number solutions of $ax^2 + bx + c = 0$, 201
            Chapter Summary, 206
            Review Problems, 207
            Chapter Test, 212

8. Algebraic Fractions — 214

    8.1 Some basic facts, 214
    8.2 Multiplication and division, 219
    8.3 Addition and subtraction of fractions, 222
    8.4 Complex fractions, 232
    8.5 Rational equations, 235
    8.6 Equations with more than one unknown, 239
         Chapter Summary, 241
         Review Problems, 242
         Chapter Test, 246

9. Exponents and Radicals — 248

    9.1 Review, 248
    9.2 Fractional exponents and radicals, 250
    9.3 Expressions with rational exponents, 253
    9.4 Simplifying radicals, 257
    9.5 Basic operations with radicals, 263
    9.6 Complex numbers, 268
         Chapter Summary, 272
         Review Problems, 274
         Chapter Test, 277

10. Graphing and Systems of Linear Equations — 279

    10.1 The Cartesian Coordinate System, 279
    10.2 Graphing straight lines and simple quadratics, 284
    10.3 Graphical solution of two linear equations, 295
    10.4 Algebraic solution of two linear equations—substitution, 301
    10.5 Algebraic solution of two linear equations—elimination, 304
         Chapter Summary, 308
         Review Problems, 309
         Chapter Test, 313

11. Logarithms — 315

    11.1 More on exponents, 315
    11.2 Introduction to logarithms, 318
    11.3 Properties of logarithms, 324
    11.4 Logarithmic equations, 330
    11.5 Computing logarithms on the calculator, 334
    11.6 Base changing formula, 341
         Chapter Summary, 343
         Review Problems, 344
         Final Examination, 347

Answers — 351

Preface

This workbook is designed to serve those students who need to sharpen their skills in the use of elementary algebra. There are many textbooks in elementary algebra available in the market place today and the fact that we have chosen to add yet another one deserves some explanation.

For the most part the presentation of new concepts is done in the following four-step process:

(1) use examples to illustrate a new concept,
(2) abstract from the examples a formal rule, procedure or definition,
(3) illustrate with further examples the rule, procedure or definition presented, and
(4) reinforce the concepts with a large number of problems to be solved.

In addition, there are several other features of this book which deserve some comment.

Factoring Trinomials. The method for factoring presented in most books is essentially the method of trial and error. In this book we give a simple algorithm for factoring which is based on the fact that the sum and product of roots are related to the coefficients of the trinomial.

Checking Solutions of Quadratic Equations. The relationship between roots and coefficients is used to check solutions. This is much easier than checking by substitution. Furthermore, when roots are complex or irrational it is easy to get a partial check by summing the solutions.

Logarithms. In our opinion hand-held calculators have made the traditional material based on tables and interpolation obsolete. We emphasize developing the skill to estimate logarithms, the use of calculators to compute logarithms and the algebraic properties of logarithms.

Other features worth noting are:

- There are always several illustrative examples in each section with explanatory comments.
- Each problem set is divided into different sections with graded exercises. Each section of a problem set starts out with a worked exercise so that the student knows what is expected in the way of a procedure for solution. Exercises are often formatted so that the student can show the algorithm he or she is using.
- The topics in algebra are introduced as generalizations of the same topic in arithmetic. Detailed algorithms are provided for working out problems.
- The arrow sign ➡ is used to focus attention on notes of caution or important results.
- Important rules and methods are boxed for easy reference and review.
- A chapter summary is provided at the end of each chapter.
- The answers to all the problems are given at the back of the book.

We believe that any teaching method for a course in elementary algebra, where the emphasis is on developing basic skills, must involve supervised regular drill and practice. We have found that "teaching by testing" is quite effective in terms of student achievement. This method makes heavy use of the computer.

We have developed computer software which will ease the instructor's task in teaching by testing and managing the course. The main features of the package include:

(a) a question bank containing more than 2000 items organized by topic with categories and sub-categories,
(b) the ability to generate any number of tests in the format illustrated in this text, (See the cover or any of the Chapter Tests.)
(c) provisions for taking practice tests and make-up tests at the computer terminal.
(d) test grading, record keeping and the production of progress reports for each student.

All the chapter tests in this book have been produced by the computer package. These chapter tests are comprehensive, in the sense that each test includes those items from previous chapters which the student found most difficult. This way students get regular practice and review throughout the course.

We are indebted to the instructors who taught algebra at Ohio University during the 1982–83 academic year for their valuable contributions in class testing the material presented in this book. Dr. David Keck deserves special mention for his continued guidance and support in the development of this text. We are also indebted to the people who read the manuscript and offered useful comments. They include Amos Lee, Davender Malik, Danny Clark, and S. Nair. We appreciate the tireless effort of Stephanie Goldsberry in typing the manuscript.

John Gillam
Donald O. Norris
Man M. Sharma

Chapter 1

Arithmetic Refresher

Most of the concepts in algebra are generalizations of what you have already learned in arithmetic. It is for this reason that in this chapter we first review the essentials of arithmetic. You should understand the logic of the rules used for arithmetic computations. The same logic will apply for algebraic computations. It is not necessary to memorize these rules. Regular practice in computation will automatically help you to remember the rules as well.

In this chapter you will review

(a) different types of real numbers,
(b) the order of arithmetic operations,
(c) the number line and signed numbers,
(d) fractions,
(e) operations on fractions, and
(f) the computation of percents,

1.1 Real Numbers

The real numbers constitute a very broad category of numbers. They include positive integers, integers, rational numbers, and irrational numbers.

1. **Positive Integers:** All those numbers that are used in counting are called positive integers. Positive integers are also called *counting numbers* or *whole numbers*.

> Positive Integers
> 1, 2, 3 . . .

2. **Integers:** All positive whole numbers, negative whole numbers, and the number zero, together are called Integers.

$$\boxed{\begin{array}{l}\text{Integers}\\ \ldots -3,-2,-1,0,1,2,3,\ldots\end{array}}$$

There are some special types of integers that we shall be using quite often.
   (a) *Even Integers:* The integers divisible by 2, such as, 2, 4, and 6.
   (b) *Odd Integers:* The integers that are not even.
   (c) *Prime Numbers:* The integers that are greater than 1 and are not divisible by any other integer except by themselves, such as 2,3,5,7,11,13, ...
   (d) *Composite Numbers:* The integers that are not prime and can be expressed as product of two or more prime numbers, such as 4,6,9,15,30 . . .

$$4 = 2 \cdot 2 \qquad 6 = 2 \cdot 3$$
$$9 = 3 \cdot 3 \qquad 15 = 3 \cdot 5$$
$$30 = 2 \cdot 3 \cdot 5$$

3. **Rational Numbers:** All numbers of the type $\frac{a}{b}$ where a, b are any two integers and $b \neq 0$ are called Rational Numbers.

$$\boxed{\begin{array}{l}\text{Rational Numbers}\\ \frac{a}{b}, \text{ a, b are Integers and } b \neq 0\end{array}}$$

Examples: Rational numbers.

$\frac{5}{6}$ is a rational number.

$\frac{-5}{4}$ is a rational number.

$\frac{5}{0}$ is not a rational number, since b = 0.

9 or $\frac{9}{1}$ is a rational number.

-3 or $\frac{-3}{1}$ is a rational number.

Thus all integers and common fractions are rational numbers.

4. **Irrational Numbers:** All real numbers that are not rational are called irrational numbers. For example:
   (a) the length ($\sqrt{2}$) of the diagonal of a square each of whose sides is 1 unit is not a rational number

(b) the ratio of the length of the circumference of a circle to the diameter of a circle, π, (called "pie") is not a rational number.

> **Irrational Numbers**
> $\sqrt{2}, \sqrt{3}, \sqrt{5}, \pi, \ldots$

> The numbers defined in 1 thru 4, all put together, are called real numbers.

There are some special laws for performing arithmetic operations with real numbers which will prove very useful in later chapters. Basically these laws state that the order in which we add or multiply real numbers is immaterial.

For example,
$$4 + 7 = 11$$
and $7 + 4 = 11$
Thus, $4 + 7 = 7 + 4$.
This is an example of the **Commutative Law of Addition.**

Also, we know
$$5 \cdot 4 = 20$$
and $4 \cdot 5 = 20$
Thus, $5 \cdot 4 = 4 \cdot 5$
This is an example of the **Commutative Law of Multiplication.**

In general, if x and y are any two numbers, then

$$x + y = y + x \quad \text{Commutative law of addition}$$
and $$x \cdot y = y \cdot x \quad \text{Commutative law of multiplication}$$

➡ The Commutative Law does not hold for subtraction or division
That is, $5 - 7 \neq 7 - 5$
and $10 \div 5 \neq 5 \div 10$

Whenever you add more than two numbers, you always perform the operation on them two at a time. In the real number system it does not matter which pair of numbers you choose to add first. A similar comment applies when you multiply more than two numbers.

For example,
$$(3 + 4) + 5 = 7 + 5 = 12$$
and $3 + (4 + 5) = 3 + 9 = 12$
Thus, $(3 + 4) + 5 = 3 + (4 + 5)$
This is an example of the **Associative Law of Addition.**

Also we know that
$$(2 \cdot 4) \cdot 5 = 8 \cdot 5 = 40$$
and $2 \cdot (4 \cdot 5) = 2 \cdot 20 = 40$
Thus, $(2 \cdot 4) \cdot 5 = 2 \cdot (4 \cdot 5)$
This is an example of the **Associative Law of Multiplication.**

In general, if x, y and z are three real numbers, then

$$(x + y) + z = x + (y + z) \quad \text{Associative law of addition}$$
$$\text{and} \quad (x \cdot y) \cdot z = x \cdot (y \cdot z) \quad \text{Associative law of multiplication.}$$

Problem Set 1.1

In Problems 1–15, mark the correct response.

1. -39 is an integer.
   (a) True   (b) False

2. $\frac{3}{4}$ is an irrational number.
   (a) True   (b) False

3. Integers are real numbers.
   (a) True   (b) False

4. Every prime number, except 2, is an odd number.
   (a) True   (b) False

5. The sum of two odd numbers is always an odd number.
   (a) True   (b) False

6. If we multiply an odd number by an even number, we get an odd number.
   (a) True   (b) False

7. If we add two odd numbers we get an even number.
   (a) True   (b) False

8. $\frac{9}{0}$ is a rational number.
   (a) True   (b) False

9. All irrational numbers are real numbers.
   (a) True   (b) False

10. The length of the circumference of a circle with diameter one unit is a rational number.
    (a) True   (b) False

11. Consider the following statements P and Q:
    P: $5 + 4 = 4 + 5$ Q: $5 - 4 = 4 - 5$
    Which of these statements is true?
    (a) P only   (b) Q only   (c) P and Q   (d) None of these

12. Consider the following statements P and Q:
    P: $5 \cdot 4 = 4 \cdot 5$ Q: $5 \div 4 = 4 \div 5$
    Which of these statements is true?
    (a) P only   (b) Q only   (c) P and Q   (d) None of these

13. Consider the following statements P and Q:
    P: $(5 + 4) + 7 = 5 + (4 + 7)$ Q: $(3 \cdot 4) \cdot 7 = 3 \cdot (4 \cdot 7)$
    Which of these statements is true?
    (a) P only   (b) Q only   (c) P and Q   (d) None of these

14. Consider the following statements P and Q:
    P: $x \cdot 6 = 6 \cdot x$ Q: $2 \cdot x + y = y + 2 \cdot x$
    Which of these statements is true?
    (a) P only   (b) Q only   (c) P and Q   (d) None of these

15. Consider the following statements P and Q:
    P: If a and b are two nonzero integers other than 1 or -1, then $a \cdot b$ is a composite number.
    Q: If x is any prime number then $x \cdot x$ is also a prime number.
    Which of these statements is true?
    (a) P only   (b) Q only   (c) P and Q   (d) None of these

1.2   The Order of Operations

There are two categories of basic operations in Arithmetic:

   (i)   addition or subtraction, and

   (ii)   multiplication or division.

Whenever both of these types of operations appear in the same expression, it is a convention to perform multiplication and division before performing addition and subtraction. We agree to follow this convention.

Example 1: $5 + 4 \times 2 = 5 + 8$
$$= 13$$

➡ $5 + 4 \times 2 \neq 9 \times 2$ because of our agreement to perform multiplication first.

Sometimes the priority of performing certain operations is indicated by using grouping symbols. If in a certain problem we are required to multiply the sum of 5 and 4 by 2, then we can express the same in symbols as $(5 + 4)2$.

$$(5 + 4)2 = (5 + 4) \times 2 = 9 \times 2 = 18$$

The fraction bar in $\dfrac{16 - 7}{3}$ is also a grouping symbol. It indicates that the numerator as one group is to be divided by another group, the denominator.

$$\frac{16 - 7}{3} = (16 - 7) \div 3$$

➡ $\dfrac{16 - 7}{3} \neq 16 - 7 \div 3$

The following rule describes universally accepted procedures for simplifying expressions.

Rule 1.1 Order of Operations

Step 1. Simplify all expressions within grouping symbols.
Step 2. Moving from left to right, perform multiplication or division, in the order of their appearance in the expression.
Step 3. Moving from left to right, perform addition or subtraction, in the order of their appearance in the expression.

➡ Note: Some calculators do not follow this order of operations, and perform operations in the same order as the expression is entered in the calculator. We shall refer to such calculators as nonscientific calculators. Those calculators which do follow the prescribed convention will be called scientific calculators.

Example 2. Simplify $15 - 10 + 21 \div 7 \times 3$
Since there is no grouping symbol, Step 1 is skipped.

$$15 - 10 + 21 \div 7 \times 3 = 15 - 10 + 3 \times 3 = 15 - 10 + 9 \qquad \text{Step 2}$$
$$= 5 + 9 = 14 \qquad \text{Step 3}$$

Example 3. Simplify $12 - 8 + 3(18 - 9)$

$$\begin{aligned} 12 - 8 + 3(18 - 9) &= 12 - 8 + 3 \cdot 9 & \text{Step 1} \\ &= 12 - 8 + 27 & \text{Step 2} \\ &= 4 + 27 = 31 & \text{Step 3} \end{aligned}$$

Example 4. Simplify $\dfrac{10 + 4 - 21 \div 7}{5 \times 2 - 7}$

$$\begin{aligned} \text{Numerator} &= 10 + 4 - 21 \div 7 \\ &= 10 + 4 - 3 & \text{Step 2} \\ &= 14 - 3 = 11 & \text{Step 3} \end{aligned}$$

$$\begin{aligned} \text{Denominator} &= 5 \times 2 - 7 \\ &= 10 - 7 & \text{Step 2} \\ &= 3 & \text{Step 3} \end{aligned}$$

Thus, $\dfrac{10 + 4 - 21 \div 7}{5 \times 2 - 7} = \dfrac{11}{3}$

Problem Set 1.2

Simplify the expressions in Problems 1–24.

	Expression	Answer		Expression	Answer
1.	$5 + 2 \times 3$ $= 5 + 6 = 11$	11	2.	$15 - 4 \div 2$ $=$ $=$	____
3.	$7 - 15 \div 5$ $=$ $=$	____	4.	$5 + 7 - 3 \times 4$ $=$ $=$	____

6

5. $19 - 3 + 24 \div 3$
   =
   =   =

6. $25 - 7 - 35 \div 5$
   =
   =   =

7. $27 - 15 \times 2 \div 10$
   =
   =   =

8. $42 + 12 \times 2 \div 8$
   =
   =   =

9. $12 - 10 \div 2 + 5$
   =
   =   =

10. $17 + 5 - 12 \div 6 \times 2$
    =
    =   =

11. $29 - 10 + 16 \div 4 \times 2$
    =
    =   =

12. $72 - 2 \times 9 + 10 \div 5$
    =
    =   =

13. $84 - 21 \div 3 + 5 \times 3$
    =
    =   =

14. $12 + 7 \times 5 - 8 \div 4$
    =
    =   =

15. $5 + 3(7 - 5)$
    $= 5 + 3(2)$
    $= 5 + 6$ =   11

16. $17 + 4(18 - 12)$
    =
    =   =

17. $25 - 4 \times 3 + 5(17 - 9)$
    =
    =   =

18. $32 - 12 \div (4 - 2)$
    =
    =   =

19. $\dfrac{4 + 3 \times 8}{4}$
    $= \dfrac{4 + 24}{4} = \dfrac{28}{4} = 7$ 7

20. $\dfrac{12 - 3 \times 3}{5 - 3}$
    =
    =   =

21. $\dfrac{7 - 2(6 - 4)}{7 - 4}$
    =
    =   =

22. $\dfrac{12 + 3(8 - 6)}{6}$
    =
    =   =

23. $\dfrac{12 - 3(9 - 6)}{12 - 3}$
    =
    =   =

24. $\dfrac{17 - 5(13 - 10)}{2(9 - 6)}$
    =
    =   =

Mark the correct response in Problems 25–30.

25. Compute: $5 + 3 \times 4 = 32$
    (a) True   (b) False

26. Compute: $15 - 10 + 4 \times 2$
    (a) 38   (b) -3 (c) 18   (d) 13   (e) None of these

27. Compute: $25 - 10 + 15 \div 5 \times 3$
    (a) 14 (b) -14 (c) 6 (d) 18 (e) None of these

28. Simplify: $10 + 2(7 - 5)$
    (a) 24 (b) 19 (c) 14 (d) 79 (e) None of these

29. Simplify: $\dfrac{6 + 2 \times 3}{6}$
    (a) 7 (b) $\dfrac{11}{6}$ (c) 6 (d) 4 (e) None of these

30. Simplify: $\dfrac{12 - 3(8 - 6)}{12 - 6}$
    (a) 3 (b) -4 (c) -3 (d) 1 (e) None of these

1.3 The Number Line and Signed Numbers

It is convenient to represent points on a line as numbers. This identification of points and numbers enables us to use our geometric intuition to visualize relationship between numbers.

We start out with a line of unending length and choose any point to stand for the number zero. In Figure 1.1 below the number zero is made to correspond to the point labeled Z. We call Z the *origin* of the number line.

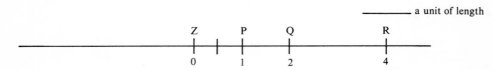

Figure 1.1 The number line

Next we choose any length to represent one unit of length. We have marked off one, two, and four units of lengths to the *right of Z* and have identified these markings by points P, Q and R respectively. The points P, Q and R then correspond to the numbers 1, 2 and 4 (Figure 1.1), respectively.

Just as we identify points on the number line to the right of the origin with positive numbers, we identify points to the left of the origin with negative numbers. In Figure 1.2, we have marked off one, two and three units of lengths to the *left of Z* and have identified these markings by points A, B and C, respectively.

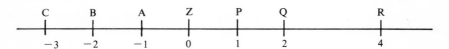

Figure 1.2 The number line

The points A, B and C then corresponds to the negative integers -1, -2 and -3. In a similar way, it is not difficult to see that any other integer number can be made to correspond to a point on the number line.

It is very helpful to use the number line to visualize when one number is larger or smaller than another. The fact that the number 5 is less than 7 is often expressed symbolically as

$5 < 7$

The fact that the number 8 greater than 6 is often expressed as

$8 > 6$

The following rule is very convenient for determining the order relation between any two numbers.

> **Rule 1.2 Order Relation**
> Given two distinct numbers on a number line, the number to the right is always the greater of the two and the number to the left is always the smaller of the two.

Examples: *Order Relations*

1. $5 < 9$ 5 is to the left of 9 on the number line.
2. $-2 > -4$ -2 is to the right of -4 on the number line.
3. $0 > -4$ 0 is to the right of -4 on the number line.
4. $-3 < 2$ -3 is to the left of 2 on the number line.

The *absolute value* of a number is the distance of the number from the origin. Since we usually think of distance as being a positive number or zero, the absolute value of a number is positive or zero. We use vertical bars around a number to denote its absolute value.

Examples: *Absolute Values*

5. The absolute value of $9 = |9| = 9$
6. The absolute value of $-9 = |-9| = 9$

Addition of Signed Numbers

When we encounter a subtraction problem such as $5 - 2$, we have learned to solve the problem almost automatically without giving it much thought. However, if we are given the problem $2 - 5$ or the problem $-2 - 5$ to solve we may feel less sure about how to proceed. This is probably due to the fact that we have been accustomed to thinking of addition and subtraction as being two different operations. In reality, *subtraction is nothing more than addition of a negative number.*

Examples: *Subtraction as addition*

7. $9 - 7 = 9 + (-7)$
8. $5 - 7 = 5 + (-7)$
9. $-7 - 9 = (-7) + (-9)$

Thus each problem of subtraction can be converted to a problem of addition of two numbers. These two numbers can be of like signs, both positive or both negative, or of unlike signs. The following rule is very convenient in adding such numbers.

> **Rule 1.3 Addition of Signed Numbers**
>
> **Like Signs**
> Find the sum of their absolute values and prefix the common sign. The addition is positive if both numbers are positive, and negative if both numbers are negative.
>
> **Unlike Signs**
> Compute the absolute value of each number and subtract the smaller absolute value from the larger absolute value. Prefix the sign of the number having the larger absolute value.

Examples: *Addition of Signed Numbers*

10. $10 - 15 = 10 + (-15)$ unlike signs
    $|10| = 10, \quad |-15| = 15$
    Since $15 - 10 = 5$ and $|-15| > |10|$, we prefix the sign of -15 before the 5.
    Thus, $10 - 15 = -5$

11. $-13 + 9 = (-13) + (9)$ unlike signs
    $|-13| = 13, \quad |9| = 9$
    Since $13 - 9 = 4$ and $|-13| > |9|$, we prefix the sign of -13 before the 4.
    Thus, $-13 + 9 = -4$

12. $-8 - 7 = (-8) + (-7)$ like signs
    $|-8| = 8, \quad |-7| = 7$
    Since $8 + 7 = 15$ and both numbers are negative $-8 - 7 = -15$

Multiplication of Signed Numbers

The number obtained by multiplying two numbers is called the *product* of two numbers. The two numbers are called the *factors* of the product. Thus, in

$$6 \times 3 = 18$$

18 is the product of the two numbers of 6 and 3, whereas the numbers 6 and 3 are factors of 18. The following rule is quite convenient for computing the product of two signed numbers.

Rule 1.4 Multiplication of Two Signed Numbers   SAME Applies DIV.

Both Factors Have Like Signs
Multiply the absolute values of the signed numbers and prefix a plus sign.

The Factors Have Different Signs
Multiply the absolute values of the signed numbers and prefix a negative sign.

Examples: *Multiplication of signed numbers*

13. $(-7)(-9) = |-7| \times |-9| = 7 \times 9 = 63$ or $+63$
14. $(-7)(9) = -(|-7|)(9) = -(7 \times 9) = -63$
15. $5 \times (-2) \times (-3) = (-10)(-3) = 30$
16. $15 - 10(7 - 9) = 15 + (-10)(-2)$ Rule 1.1
                       $= 15 + 20$ Rule 1.4
                       $= 35$

    You know $(-1)(-9) = 9$
    Also, $(-1)(-9) = -(1)(-9)$
                       $= -(-9)$ Since $(1)(-9) = -9$

    Thus, we see $\boxed{-(-9) = 9}$

$$\boxed{\text{The negative of a negative number is positive}}$$

Examples: *The negative of a negative number is positive*

17. $-(-5) = 5$
18. $5 - (3 - 7) = 5 - (-4) = 5 + 4 = 9$

The rule for division of signed numbers is similar to Rule 1.4. If we divide two numbers with like signs, we divide their absolute values and prefix a positive sign to the result. If we divide two numbers with unlike signs, then we prefix a negative sign to the result obtained by dividing their absolute values.

Examples: Division of signed numbers

19. $(-15) \div (-5)$ or $\frac{-15}{-5} = 3$

20. $(-10) \div 2$ or $\frac{-10}{2} = -5$

21. $10 \div (-2)$ or $\frac{10}{-2} = -5$

➡ $\quad \frac{-10}{2} = \frac{10}{-2} = -5$

Problem Set 1.3

In Problems 1–8, insert "<" or ">" between the two numbers so that the inequality is true.

1. 5 __<__ 7
2. 0 _____ 4
3. 0 _____ −2
4. −2 _____ −1
5. −3 _____ −4
6. 1 _____ −2
7. 5 _____ −2
8. −2 _____ 4

Compute the expressions in Problems 9–30.

9. $15 + 3 \times 2 = 15 + 6 = 21$ 9. __21__

10. $-9 + 4$ 10. _____

11. $-8 - 7$ 11. _____

12. $9 + 7 - 4 - 9 = 9 + 7 + (-4) + (-9)$
    $= 16 + (-13) = 3$ 12. __3__

13. $3 - 2 - 7 - 9$ 13. _____

14. $12 - 17 - 8 + 5$ 14. _____

15. $2(-4)$ 15. _____

16. $(-2)(-3)$ 16. _____

17. $(-2)(-4)(-5)$ 17. _____

18. $5(-3)(-4)(4)$ 18. _____

19. $7 + 2(-4)$ 19. _____

20. $-7 + 2(-4)$ 20. _____

21. $21 - 15 \div 5$ 21. _____

22. $8 - 2(7 - 9)$ 22. _____

23. $-8 - 4(12 - 8)$ 23. _____

24. $-2 + 3 - 12 \div (-4)$ 24. _____

25. $20 - 30 \div (-15)$ 25. _____

26. $\dfrac{4 - 2(-3)}{-8 + 3} = \dfrac{4 + (-2)(-3)}{(-8) + 3} = \dfrac{4 + 6}{-5} = \dfrac{10}{-5} = -2$ 26. $\underline{-2}$

27. $\dfrac{5 - 4(3 - 4)}{-3 - 6}$ 27. _____

28. $\dfrac{21 \div (-7) - (-2)}{3 - 2(7 - 5)}$ 28. _____

29. $15 - 10 + 2(-4) - 25 \div 5$ 29. _____

30. $(30 - 20 \div 5) \div (-24 - 2)$ 30. _____

1.4 Fractions

A fraction, as you know, is a ratio of two integers such as $\dfrac{a}{b}$ where a and b are any two integers and $b \neq 0$. The number a is called the *numerator* and the number b is called the *denominator*. For example, in the fraction $\dfrac{3}{4}$, 3 is the numerator and 4 is the denominator.

Let a and b be two positive integers. Then a fraction $\dfrac{a}{b}$ is called a *proper* fraction if a is less than b, and an *improper* fraction if a is greater than or equal to b.

Examples: *Fractions*

1. $\dfrac{5}{7}$ is a proper fraction since $5 < 7$.

2. $\dfrac{9}{4}$ is an improper fraction since $9 > 4$.

3. $\dfrac{5}{0}$ is not a fraction since $b = 0$.

➡ $\dfrac{5}{0}$ is not defined

$\dfrac{0}{5} = 0$ since 0 divided by any nonzero number is 0.

An improper fraction is either a whole number or the sum of a whole number and a proper fraction. The fraction $\frac{9}{4}$ is an improper fraction. As you know $\frac{9}{4}$ means $9 \div 4$. If you divide 9 by 4, you get 2 as the quotient and 1 as the remainder.

$$4\overline{)9} \quad \begin{array}{r} 2 \\ \underline{8} \\ 1 \end{array}$$

That is, $\frac{9}{4} = 2 + \frac{1}{4} = 2\frac{1}{4}$. Numbers of the type $2\frac{1}{4}$ are called *mixed numbers*. Thus, we can write any improper fraction as a mixed number.

Examples: *Improper fractions as mixed numbers*

4. $\frac{17}{5} = 3 + \frac{2}{5} = 3\frac{2}{5}$

5. $\frac{25}{8} = 3 + \frac{1}{8} = 3\frac{1}{8}$

[handwritten note: a number is divisible by three if the sum of its numbers are 3.]

Equivalent Fractions

Sometimes fractions look different, but in fact have the same value.

Consider, for example, the fraction $\frac{1}{2}$

$$\frac{1}{2} = \frac{1}{2} \cdot 1 = \frac{1}{2} \cdot \frac{2}{2} = \frac{2}{4}$$
$$\text{or} = \frac{1}{2} \cdot \frac{3}{3} = \frac{3}{6}$$
$$\text{or} = \frac{1}{2} \cdot \frac{4}{4} = \frac{4}{8}$$

When you multiply fractions, you multiply their numerators and their denominators.

Thus the fractions $\frac{2}{4}, \frac{3}{6}, \frac{4}{8}$ etc. are essentially the same as $\frac{1}{2}$. Such fractions are called *equivalent fractions*.

Examples: *Equivalent fractions*

6. $\frac{1}{2}, \frac{1 \times 2}{2 \times 2}$ or $\frac{2}{4}, \frac{1 \times 3}{2 \times 3}$ or $\frac{3}{6}, \ldots$ are equivalent fractions since they are all equal to .5
(Check with your calculator)

7. $\frac{16}{24}, \frac{16 \div 2}{24 \div 2}$ or $\frac{8}{12}, \frac{16 \div 4}{24 \div 4}$ or $\frac{4}{6}, \frac{16 \div 8}{24 \div 8}$ or $\frac{2}{3}$
are equivalent fractions since they are all equal to .666666 . . .
(Check with your calculator. Your calculator will round the answer.)

You can easily see from Example 6 that given a fraction you can find an equivalent fraction whose denominator (numerator) is some multiple of the numerator (denominator) of the given fraction.

Example 8. Find the fraction equivalent to $\frac{3}{5}$ whose denominator is 20.

$$\frac{3}{5} = \frac{?}{20} = \frac{?}{5 \times 4} \qquad \frac{3}{5} = \frac{3 \times 4}{5 \times 4} = \frac{12}{20}$$

Example 9. Find the fraction equivalent to $\frac{3}{5}$ whose numerator is 21.

$$\frac{3}{5} = \frac{21}{?} = \frac{3 \times 7}{?} \qquad \frac{3}{5} = \frac{3 \times 7}{5 \times 7} = \frac{21}{35}$$

Reducing Fractions

In Example 7, you have seen how $\frac{16}{24}$ was made equivalent to $\frac{2}{3}$ by dividing the numerator and the denominator by 8 (the greatest common factor in 16 and 24). This process of dividing the numerator and the denominator by a common factor is called *reducing the fraction*.

➡ While performing computations with fractions you should always leave the final answer, if a fraction, in completely reduced form. That is, the numerator and the denominator of the resulting fraction should have no common factor.

Completely reducing a fraction is also sometimes called *reducing a fraction to its lowest terms*.

Examples: *Reducing a fraction to lowest terms*

10. $\frac{12}{16} = \frac{(2 \cdot 2) \cdot 3}{(2 \cdot 2) \cdot 2 \cdot 2} = \frac{3}{4}$ . . . divide the numerator and the denominator by 4.

    or $\frac{12}{16} = \frac{\not{2} \cdot \not{2} \cdot 3}{\not{2} \cdot \not{2} \cdot 2 \cdot 2}$ . . . cancel the common factors.

11. $\frac{30}{42} = \frac{2 \cdot 3 \cdot 5}{2 \cdot 3 \cdot 7} = \frac{5}{7}$ . . . divide the numerator and the denominator by 6.

    or $\frac{30}{42} = \frac{\not{2} \cdot \not{3} \cdot 5}{\not{2} \cdot \not{3} \cdot 7} = \frac{5}{7}$ . . . cancel the common factors.

Problem Set 1.4

Complete the statements in Problems 1–6.

1. In the fraction $\frac{4}{7}$, the numerator is _____ and the denominator is _____ .

2. The fraction $\frac{3}{7}$ is called _____ fraction.
   (proper or improper)

3. The fraction $\frac{10}{3}$ is called _____ fraction.
   (proper or improper)

4. All whole numbers are equivalent to fractions with denominators equal to _____ .

5. A fraction with zero numerator is always equal to _____ .

6. A fraction with zero denominator is _____ .

In Problems 7–14, write the improper fraction as a mixed number.

	Im. Fraction		Mixed No.		Im. Fraction	Mixed No.
7.	$\dfrac{13}{5}$	$5\overline{)13}$ $\underline{10}$ 3 with quotient 2	$2\dfrac{3}{5}$	8.	$\dfrac{12}{5}$	_____
9.	$\dfrac{17}{3}$		_____	10.	$\dfrac{25}{4}$	_____
11.	$\dfrac{21}{8}$		_____	12.	$\dfrac{134}{7}$	_____
13.	$\dfrac{315}{43}$		_____	14.	$\dfrac{479}{55}$	_____

In Problems 15–20, find the missing numerator or the missing denominator.

15. $\dfrac{1}{3} = \dfrac{?}{9}$, $\dfrac{1}{3} = \dfrac{(1 \times 3)}{(3 \times 3)} = \dfrac{3}{9}$ 16. $\dfrac{2}{5} = \dfrac{?}{20}$

17. $\dfrac{4}{7} = \dfrac{?}{42}$ 18. $\dfrac{3}{5} = \dfrac{12}{?}$

19. $\dfrac{10}{9} = \dfrac{70}{?}$ 20. $\dfrac{15}{22} = \dfrac{45}{?}$

Reduce the fractions to lowest terms in Problems 21–28.

	Fraction	Reduced Form		Fraction	Reduced Form
21.	$\dfrac{4}{6} = \dfrac{2 \cdot 2}{2 \cdot 3}$	$\dfrac{2}{3}$	22.	$\dfrac{15}{25}$	_____
23.	$\dfrac{16}{40}$	_____	24.	$\dfrac{72}{60}$	_____
25.	$\dfrac{4 \times 5 \times 9}{5 \times 4 \times 3}$	_____	26.	$\dfrac{120}{315}$	_____
27.	$\dfrac{21 \times 10}{15 \times 14}$	_____	28.	$\dfrac{4 \times 9 \times 7}{6 \times 21}$	_____

1.5 Operations on Decimal Numbers

Addition of Decimal Numbers

The addition or subtraction of decimal numbers is very similar to the addition of integers, except that a little extra care is needed to align the decimal points and digits with the same place value. The following examples should remind you of the process.

Example 1. Add 43.57 and 9.69.

$$\begin{array}{r} 43.57 \\ +9.69 \\ \hline 53.26 \end{array}$$

Example 2. Add 47.931 and 2.02.

$$\begin{array}{r} 47.931 \\ +2.020 \\ \hline 49.951 \end{array}$$

Note that in this example we appended a zero to 2.02 to make it 2.020 so that we had a digit in each place to add. Of course, 2.02 and 2.020 are the same number, so the change we made was for the purpose of making our calculation easier.

Example 3. Subtract 3.17 from 21.96.

$$\begin{array}{r} 21.96 \\ -3.17 \\ \hline 18.79 \end{array}$$

Example 4. Subtract 16.09 from 98.317.

$$\begin{array}{r} 98.317 \\ -16.090 \\ \hline 82.227 \end{array}$$

Here we have adjoined a zero to the subtrahend in order to facilitate our calculation. As you can see from these examples, it is only necessary to make sure that the decimals are lined up and that the number of places to the right of the decimal point are the same in both addition and subtraction.

Multiplication of Decimal Numbers

The multiplication of two decimal numbers is also performed in almost the same way that we multiply integers. However, extra care must be taken to correctly place the decimal point in the product. Here are two ways of accomplishing that.

1. Count the number of digits to the right of the decimal point in the first number. Call it r. Count the number of digits to the right of the decimal point in the second number. Call it s. Now multiply the two numbers as though they were integers and place the decimal point in the product so that the number of places to the right of the decimal point is r + s.

Example 5. Multiply 4.57 and 3.8.

$$\begin{array}{r} 4.57 \\ \times3.8 \\ \hline 3656 \\ 1371 \\ \hline 17.366 \end{array} \quad \begin{array}{l} r = 2 \\ s = 1 \\ \\ r + s = 3 \end{array}$$

2. We can also place the decimal point in the product by use of estimation.

Example 6. Multiply 4.57 and 3.8.

    4.57 lies between 4 and 5.
    3.8 lies between 3 and 4.

Therefore, the product, 4.57×3.8, must lie between $4 \times 3 = 12$ and $5 \times 4 = 20$. That is, the decimal point must be placed in the product to ensure that the product is a number between 12 and 20. Thus,

    $4.57 \times 3.8 = 17.366$.

Example 7. Compute 3.75×14.2.

```
      3.75           r = 2         3 < 3.75 < 4
   ×  14.2           s = 1         14 < 14.2 < 15
      750
     1500
      375
     53.250         r + s = 3      42 < 3.75 × 14.2 < 60
```

Division of Decimal Numbers

The division of decimal numbers is made simpler by converting the division problem into an equivalent one for which the divisor is an integer. This is because we can divide in the usual manner and we will have a simple rule for placing the decimal point in the quotient.

The fractions $\frac{2.55}{1.5}$ and $\frac{25.5}{15}$ are equivalent because

$$\frac{2.55}{1.5} = \frac{2.55}{1.5} \times \frac{10}{10} = \frac{25.5}{15}$$

Therefore, we can compute $2.55 \div 1.5$ by computing $25.5 \div 15$. The procedure we use for computing $25.5 \div 15$ is to perform the division as we do with whole numbers and place the decimal point in the quotient directly above the decimal point in the dividend.

    You should recall that multiplication of a decimal number by 10 may also be done by shifting the decimal point to the right by one digit; multiplication by 100 may also be done by shifting the decimal point to the right by two digits; etc. So, you can create the equivalent division problem by merely shifting the decimal point in the numerator and the denominator the same number of places so that the denominator is an integer.

Example 8. Compute $2.55 \div 1.5$. As noted above, we convert this to the equivalent problem $25.5 \div 15$ and use our usual long division procedure.

```
        1.7
   15)25.5           Note the placement of the decimal point in the quotient.
      15
      10 5
      10 5
         0
```

Example 9. Compute $6.25 \div 1.25$.

$$\frac{6.25}{1.25} = \frac{6.25}{1.25} \times \frac{100}{100} = \frac{625}{125}$$ Multiply numerator and denominator by 100

$$\frac{625.}{125.} = \frac{625}{125}$$ Shift the decimal points two places to the right.

$$125 \overline{)625} \atop \underline{625} \atop 0 \quad 5$$

Example 10. Compute $7.35 + 4 - 2.75 \div 1.1$. We apply Rule 1.1 and calculate $2.75 \div 1.1$ first.

$$\frac{2.75}{1.1} = \frac{27.5}{11} = 2.5$$

To further refresh your memory, we have done the long division on the right.

$$11 \overline{)27.5} \atop \underline{22} \atop 5\ 5 \atop \underline{5\ 5} \atop 0 \quad 2.5$$

Therefore, $7.35 + 4 - 2.75 \div 1.1 = 7.35 + 4 - 2.5 = 11.35 - 2.5 = 8.85$.

Problem Set 1.5

In Problems 1–3, write the number in symbols.

1. Twenty nine and five tenths. _____

2. Twelve million two hundred forty seven and twenty five thousandths. _____

3. Eight and eight thousandths. _____

In Problems 4–19, perform the indicated operations and check your answer with a calculator.

4. $\quad 19.75$
   $+ \ 2.59$

5. $\quad 897.2$
   $+ \ 75.49$

6. $\quad 2497.04$
   $+ \ 799.87$

7. $\quad 89.892$
   $- \ 5.72$

8. $\quad 249.72$
   $- 189.54$

9. $\quad 3454.09$
   $- \ 278.42$

10. $4.79 \times .02 =$ _____

11. $285.4 \times 2.54 =$ _____

12. $345.57 \times 2.5 =$ _____

13. $4.79 \div .02 =$ _____

14. $285.4 \div 2.54 =$ _____

15. $345.57 \div 2.5 =$ _____

16. $2.45 + 47.2 - 7.1 =$ _____

17. $3.42 - 7.5 + 15 =$ _____

18. $2.79 + 2.5 \times 1.2 =$ _____

19. $57.2 + 2.45 \div 3.5 =$ _____

In Problems 20–23, find the answer.

20. $39.72 more than $28.50 equals _____ .

21. The cost of one calculator is $8.92. What is the cost of 25 calculators? _____

22. You pay 15 cents income tax on each dollar that you earn. What is your income tax if you earn $21,784.50? _____

23. What is the total carpet required to cover a room 10.5 feet by 14.5 feet? _____

1.6 Basic Operations on Fractions

In this Section you will review the concepts of performing the basic operations $(+, -, \times, \div)$ on fractions.

Addition of Fractions

If you add fractions with the same denominators, the result is a fraction whose numerator is the sum of the numerators of the given fractions and the denominator is the common denominator.

Examples: *Addition of fractions with the same denominator*

1. $\frac{2}{5} + \frac{1}{5} = \frac{2+1}{5} = \frac{3}{5}$

2. $\frac{4}{15} + \frac{3}{15} + \frac{2}{15} = \frac{4+3+2}{15} = \frac{9}{15}$

   $\frac{9}{15} = \frac{3 \cdot 3}{3 \cdot 5} = \frac{3}{5}$

   Thus, $\frac{4}{15} + \frac{3}{15} + \frac{2}{15} = \frac{3}{5}$

3. $\frac{2}{7} + \frac{4}{7} - \frac{3}{7} = \frac{2}{7} + \frac{4}{7} + \left(-\frac{3}{7}\right)$

   $= \frac{2}{7} + \frac{4}{7} + \frac{-3}{7}$

   $= \frac{2 + 4 + (-3)}{7} = \frac{3}{7}$

   Or you may compute the sum directly by taking the sign in between the fractions to the numerator of the resulting fraction.

   $\frac{2}{7} + \frac{4}{7} - \frac{3}{7} = \frac{2 + 4 - 3}{7} = \frac{3}{7}$

4. $\frac{5}{9} - \frac{2}{9} + \frac{6}{9} = \frac{5 - 2 + 6}{9} = \frac{9}{9} = 1$

➡ $\frac{2}{3} + \frac{1}{3} \neq \frac{2+1}{3+3}$ or $\frac{3}{6}$

When the denominator of the fractions to be added are not the same, we change each of the fractions to an equivalent fraction with a common denominator and perform addition as in Examples 1–4. The most desirable common denominator is the least common multiple of the denominators, called the *least common denominator* or simply the LCD. The following rule is quite convenient for finding the LCD.

Rule 1.5 Finding the LCD

Step 1. Write each denominator as a product of prime factors.
Step 2. Write all the distinct prime factors. If a certain prime factor occurs more than once in a single denominator then take it the maximum number of times it occurs in any of the denominators.
Step 3. LCD = the product of all the prime numbers in Step 2.

Example 5. Find the LCD of the fractions $\frac{2}{9}, \frac{3}{15}$.

The denominators are 9 and 15

Step 1. $9 = 3 \cdot 3$
$15 = 3 \cdot 5$

Step 2. Since 3 occurs twice in 9 the prime number factors in the LCD are 3, 3, 5

Step 3. LCD $= 3 \cdot 3 \cdot 5 = 45$

Example 6. Find the LCD of the fractions $\frac{5}{12}, \frac{7}{40}, \frac{11}{18}$.

The denominators are 12, 40 and 18

Step 1. $12 = 2 \cdot 2 \cdot 3$
$40 = 2 \cdot 2 \cdot 2 \cdot 5$
$18 = 2 \cdot 3 \cdot 3$

Step 2. 2 occurs three times in 40,
3 occurs two times in 18, and
5 occurs only once in 40.
Thus, the prime number factors in the LCD are 2, 2, 2, 3, 3, 5

Step 3. LCD $= 2 \cdot 2 \cdot 2 \cdot 3 \cdot 3 \cdot 5 = 360$

The following three step rule for adding fractions when the denominators are not the same is useful.

Rule 1.6 Addition of Fractions

Step 1. Find the LCD
Step 2. Change each fraction to an equivalent fraction having the LCD as its denominator.
Step 3. Add the fractions obtained in Step 2 (add the numerators and place it over the common denominator).

Example 7. Add $\frac{2}{9} + \frac{3}{15}$

Step 1. LCD = 45   See Example 5

Step 2. $\frac{2}{9} = \frac{?}{45}, \quad \frac{2}{9} = \frac{2 \cdot 5}{9 \cdot 5} = \frac{10}{45}$

$\frac{3}{15} = \frac{?}{45}, \quad \frac{3}{15} = \frac{3 \cdot 3}{15 \cdot 3} = \frac{9}{45}$

Step 3. $\dfrac{10}{45} + \dfrac{9}{45} = \dfrac{10+9}{45} = \dfrac{19}{45}$.

Thus, $\dfrac{2}{9} + \dfrac{3}{15} = \dfrac{10}{45} + \dfrac{9}{45} = \dfrac{19}{45}$

Example 8. Perform the indicated operations

$$\dfrac{8}{15} + \dfrac{9}{20} - \dfrac{7}{30}$$

Step 1. $15 = 3 \cdot 5$
$20 = 2 \cdot 2 \cdot 5 \qquad \text{LCD} = 2 \cdot 2 \cdot 3 \cdot 5 = 60$
$30 = 2 \cdot 3 \cdot 5$

Step 2. $\dfrac{8}{15} = \dfrac{?}{60}, \quad \dfrac{8}{15} = \dfrac{8 \times 4}{15 \times 4} = \dfrac{32}{60}$

$\dfrac{9}{20} = \dfrac{?}{60}, \quad \dfrac{9}{20} = \dfrac{9 \times 3}{20 \times 3} = \dfrac{27}{60}$

$\dfrac{7}{30} = \dfrac{?}{60}, \quad \dfrac{7}{30} = \dfrac{7 \times 2}{30 \times 2} = \dfrac{14}{60}$

Step 3. $\dfrac{8}{15} + \dfrac{9}{20} - \dfrac{7}{30} = \dfrac{32}{60} + \dfrac{27}{60} - \dfrac{14}{60}$

$= \dfrac{32 + 27 - 14}{60}$

$= \dfrac{45}{60} = \dfrac{3 \cdot 3 \cdot 5}{2 \cdot 2 \cdot 3 \cdot 5} = \dfrac{3}{4}$

Thus, $\dfrac{8}{15} + \dfrac{9}{20} - \dfrac{7}{30} = \dfrac{32}{60} + \dfrac{27}{60} - \dfrac{14}{60} = \dfrac{45}{60} = \dfrac{3}{4}$

➡ 1. Never leave the answer as $\dfrac{45}{60}$. It should be reduced to lowest terms: $\dfrac{3}{4}$ in this case.

2. It is very important that you understand Rule 1.5 and Rule 1.6, since we will be using the same rules later for the addition of algebraic fractions.

Examples: *Addition of fractions*

9. $\dfrac{2}{3} + \dfrac{5}{6}$ $\qquad\qquad\qquad\qquad$ LCD = 6

$\dfrac{2}{3} + \dfrac{5}{6} = \dfrac{2 \cdot 2}{6} + \dfrac{5}{6}$

$= \dfrac{4}{6} + \dfrac{5}{6} = \dfrac{9}{6} = \dfrac{3}{2}$

10. $\dfrac{3}{4} + \dfrac{5}{12} - \dfrac{7}{18}$ $\qquad\qquad\qquad\qquad$ LCD = 36

$\dfrac{3}{4} + \dfrac{5}{12} - \dfrac{7}{18} = \dfrac{3 \cdot 9}{36} + \dfrac{5 \cdot 3}{36} - \dfrac{7 \cdot 2}{36}$

$= \dfrac{27}{36} + \dfrac{15}{36} - \dfrac{14}{36} = \dfrac{27 + 15 - 14}{36} = \dfrac{28}{36} = \dfrac{7}{9}$

11. $3\frac{4}{5} = 3 + \frac{4}{5} = \frac{3}{1} + \frac{4}{5}$ LCD = 5

    $= \frac{15}{5} + \frac{4}{5}$

    $= \frac{15 + 4}{5} = \frac{19}{5}$

12. $2\frac{7}{4} = 2 + \frac{7}{4} = \frac{8}{4} + \frac{7}{4} = \frac{15}{4}$

13. $3\frac{4}{5} - 2\frac{7}{4}$

    Change the mixed numbers $3\frac{4}{5}$ and $2\frac{7}{4}$ to improper fractions as in Examples 11–12.

    $3\frac{4}{5} - 2\frac{7}{4} = \frac{19}{5} - \frac{15}{4}$ LCD = 20

    $\frac{19}{5} - \frac{15}{4} = \frac{19 \times 4}{20} - \frac{15 \times 5}{20}$

    $= \frac{76}{20} - \frac{75}{20} = \frac{1}{20}$

Multiplication and Division of Fractions

Multiplication of fractions is a much simpler operation to perform than the addition of fractions. *Multiply the numerators and the denominators and reduce the resulting fraction.*

Examples: *Multiplication of fractions*

14. $\frac{3}{4} \cdot \frac{8}{9} = \frac{3 \cdot 8}{4 \cdot 9} = \frac{24}{36} = \frac{2}{3}$

15. $\frac{4}{15} \cdot \frac{10}{3} \cdot \frac{9}{5} = \frac{4 \cdot 10 \cdot 9}{15 \cdot 3 \cdot 5} = \frac{360}{225} = \frac{2 \cdot 2 \cdot 2 \cdot 3 \cdot 3 \cdot 5}{3 \cdot 3 \cdot 5 \cdot 5} = \frac{8}{5}$

To divide two fractions, we make use of the notion of the reciprocal of a fraction. The *reciprocal* of a given fraction is the fraction obtained by interchanging the position of the numerator and the denominator.

The reciprocal of 6 or $\frac{6}{1}$ is $\frac{1}{6}$.

The reciprocal of $\frac{4}{5}$ is $\frac{5}{4}$.

The reciprocal of $\frac{7}{8}$ is $\frac{8}{7}$.

You may note that the product of a non zero number and its reciprocal is always 1.

$6 \cdot \frac{1}{6} = \frac{6}{6} = 1, \quad \frac{4}{5} \cdot \frac{5}{4} = \frac{20}{20} = 1, \quad \frac{7}{8} \cdot \frac{8}{7} = \frac{56}{56} = 1$

▶ The *reciprocal of "0"* does not exist. There is no number which when multiplied by 0 gives the product 1.

Let 'a' and 'b' be any two numbers. You already know that

$$a \div b = \frac{a}{b} = \frac{a}{1} \times \frac{1}{b} = a \times \frac{1}{b}$$

Therefore, $a \div b = a \times \frac{1}{b}$

That is,

$$\boxed{a \div b = a \times \text{reciprocal of b}}$$

Examples: *Division of two fractions*

16. $\frac{6}{25} \div \frac{12}{5} = \frac{6}{25} \times \frac{5}{12} = \frac{6 \cdot 5}{25 \cdot 12} = \frac{30}{300} = \frac{1}{10}$

17. $\frac{8}{15} \div \frac{16}{9} = \frac{8}{15} \times \frac{9}{16} = \frac{3}{10}$

Mixed Operations on Fractions

In the following examples, we will illustrate methods of simplifying expressions that involve more than one type of operation.

Example 18. Simplify:

$$\frac{2}{3} + \frac{4}{5} \div \frac{12}{25}$$

Recall Rule 1.1 for the order of performing operations

$$\frac{2}{3} + \frac{4}{5} \div \frac{12}{25} = \frac{2}{3} + \frac{4}{5} \times \frac{25}{12} = \frac{2}{3} + \frac{100}{60} = \frac{2}{3} + \frac{5}{3} = \frac{7}{3}$$

Example 19. Simplify:

$$\frac{\frac{5}{6}}{\frac{15}{8}}$$

$$\frac{\frac{5}{6}}{\frac{15}{8}} = \frac{5}{6} \div \frac{15}{8} = \frac{5}{6} \times \frac{8}{15} = \frac{4}{9}$$

Example 20. Simplify:

$$\frac{2 + \frac{4}{5}}{2 - \frac{5}{6}}$$

This type of fraction is called a complex fraction. In order to simplify this fraction we convert the numerator "$2 + \frac{4}{5}$", and the denominator "$2 - \frac{5}{6}$" to single fractions and then solve the problem as in Example 19. (This is an important example because the procedure we use here is the same procedure we will use to simplify complex algebraic fractions.)

$$\frac{2 + \frac{4}{5}}{2 - \frac{5}{6}} = \frac{\frac{2}{1} + \frac{4}{5}}{\frac{2}{1} - \frac{5}{6}} = \frac{\frac{10}{5} + \frac{4}{5}}{\frac{12}{6} - \frac{5}{6}} = \frac{\frac{14}{5}}{\frac{7}{6}}$$

$$= \frac{14}{5} \div \frac{7}{6} = \frac{14}{5} \times \frac{6}{7} = \frac{12}{5}$$

Example 21. Simplify:

$$\frac{2}{3} + \frac{1}{5} \times \frac{20}{9}\left(\frac{2}{3} + \frac{1}{4}\right).$$

We shall use Rule 1.1 to perform the simplification.

$$\frac{2}{3} + \frac{1}{5} \times \frac{20}{9}\left(\frac{2}{3} + \frac{1}{4}\right)$$

$$= \frac{2}{3} + \frac{1}{5} \times \frac{20}{9}\left(\frac{11}{12}\right) \qquad \text{Simplify grouping symbols}$$

$$= \frac{2}{3} + \frac{11}{27} \qquad \text{Simplify multiplication}$$

$$= \frac{18 + 11}{27} = \frac{29}{27} \qquad \text{Add fractions}$$

Problem Set 1.6

In Problems 1–10, perform the indicated operations.

1. $\frac{4}{15} + \frac{7}{15} - \frac{1}{15}$ $\frac{2}{3}$ 2. $\frac{2}{3} + \frac{1}{3} =$ _____

  $= \frac{4 + 7 - 1}{15} = \frac{10}{15} = \frac{2}{3}$

3. $\frac{5}{11} - \frac{3}{11} =$ _____   4. $\frac{7}{12} - \frac{2}{12} =$ _____

5. $\frac{4}{9} - \frac{7}{9} =$ _____   6. $\frac{6}{7} - \frac{4}{7} =$ _____

7. $\frac{3}{5} - \frac{1}{5} + \frac{13}{5} =$ _____   8. $\frac{2}{8} - \frac{1}{8} + \frac{7}{8} =$ _____

9. $\frac{7}{12} - \frac{5}{12} + \frac{11}{12} =$ _____   10. $\frac{9}{32} + \frac{8}{32} - \frac{5}{32} =$ _____

In Problems 11–20, find the LCD of the given fractions.

11. $\frac{2}{5}, \frac{3}{20}, \frac{7}{15}$ 60   12. $\frac{3}{20}, \frac{7}{10}$ _____

     $5 = 1 \cdot 5$
     $20 = 2 \cdot 2 \cdot 5$
     $15 = 3 \cdot 5$
     $\text{LCD} = 1 \cdot 5 \cdot 2 \cdot 2 \cdot 3 = 60$

13. $\frac{3}{8}, \frac{5}{4}$ _____ 14. $\frac{5}{12}, \frac{3}{20}$ _____

15. $\frac{5}{12}, \frac{1}{18}$ _____ 16. $\frac{8}{15}, \frac{9}{25}$ ___75___

17. $\frac{2}{5}, \frac{3}{10}, \frac{7}{15}$ _____ 18. $\frac{3}{4}, \frac{5}{18}, \frac{7}{12}$ ___36___

19. $\frac{3}{16}, \frac{7}{20}, \frac{9}{10}$ _____ 20. $\frac{5}{12}, \frac{9}{16}, \frac{7}{60}$ _____

In Problems 21–35, perform the indicated operations.

21. $\frac{2}{5} + \frac{3}{20} + \frac{7}{15}$ LCD = 60   Ans. $\frac{61}{60}$

$\frac{2}{5} + \frac{3}{20} + \frac{7}{15} = \frac{24}{60} + \frac{9}{60} + \frac{28}{60} = \frac{61}{60}$

22. $\frac{2}{15} + \frac{9}{20}$ LCD = _____   Ans. _____

23. $\frac{3}{10} + \frac{4}{15} - \frac{2}{5}$ LCD = _____   Ans. _____

24. $2\frac{1}{4} - 1\frac{7}{8}$ LCD = _____   Ans. _____

25. $4 - \frac{3}{5} - 2\frac{7}{10}$ LCD = _____   Ans. _____

26. $\frac{12}{25} \times \frac{15}{16}$ Ans. _____

27. $\frac{12}{25} \div \frac{16}{15}$ Ans. _____

28. $\frac{32}{75} \div \frac{16}{125}$ Ans. _____

29. $\frac{2}{3} + \frac{1}{3} \times \frac{6}{5}$ Ans. _____

30. $\frac{4}{5} - \frac{2}{3} \div \frac{4}{9}$ Ans. _____

31. $\frac{2}{3} + \frac{1}{2}\left(\frac{2}{3} + \frac{8}{3}\right)$ Ans. _____

32. $\frac{4}{5} \div \left(\frac{16}{25} \times \frac{15}{8}\right)$ Ans. _____

33. $\left(\frac{21}{20} - \frac{5}{12}\right) \div \frac{10}{16}$ Ans. _____

34. $\frac{5}{12} - \left(\frac{7}{10} + \frac{3}{8}\right) \div \frac{9}{16}$ Ans. _____

35. $\frac{5}{12} - \frac{3}{8} \times \left(\frac{4}{9} \div \frac{8}{27}\right)$ Ans. _____

36. $\dfrac{2 + \frac{4}{5}}{1 - \frac{2}{3}}$ Ans. _____

37. $\dfrac{\frac{1}{3} - \frac{2}{5}}{\frac{3}{10} + \frac{2}{3}}$ Ans. _____

In Problems 38–46, mark the correct response.

38. Compute: $\frac{2}{3} + \frac{4}{5}$
    (a) $\frac{3}{4}$ (b) $\frac{14}{15}$ (c) $\frac{22}{15}$ (d) $\frac{26}{15}$ (e) None of these

39. Compute: $2\frac{1}{4} - 1\frac{1}{4}$
    (a) 1   (b) $1\frac{1}{4}$ (c) 0   (d) $\frac{4}{0}$ (e) None of these

40. Consider the following statements P and Q:
    P: $\frac{5}{0} = 0$ Q: $\frac{0}{5} = 0$
    Which of the statements above is true?
    (a) P only   (b) Q only   (c) P and Q   (d) None of these

41. The LCD of the fractions $\frac{3}{4}$ and $\frac{5}{12}$ is 48.
    (a) True   (b) False

42. Consider the following statements P and Q:
    P: Reciprocal of $\frac{2}{3}$ is $\frac{3}{2}$
    Q: The product of a nonzero number and its reciprocal is equal to 1.
    Which of these statements is true?
    (a) P only   (b) Q only   (c) P and Q   (d) None of these

43. Compute: $5 - \frac{3}{4}$
    (a) $\frac{1}{2}$ (b) $\frac{23}{4}$ (c) $\frac{17}{4}$ (d) None of these

44. Compute: $\frac{5}{6} + \frac{1}{6} \times \frac{4}{3}$
    (a) $\frac{4}{3}$ (b) $\frac{3}{4}$ (c) $\frac{19}{18}$ (d) $\frac{3}{7}$ (e) None of these

45. Compute: $\dfrac{4 - \dfrac{2}{3}}{\dfrac{2}{3}}$

    (a) 3   (b) 5   (c) 1   (d) 4   (e) None of these

46. Simplify: $\dfrac{3}{4} \div \dfrac{9}{16}\left(\dfrac{3}{4} + \dfrac{1}{2}\right)$

    (a) $\dfrac{135}{256}$ (b) $\dfrac{5}{3}$ (c) $\dfrac{15}{16}$ (d) $\dfrac{3}{2}$ (e) None of these

1.7 Percentage

A *percent* is just a fraction whose denominator is 100. For example, twenty-five percent means twenty-five out of 100 or $\dfrac{25}{100}$.

That is, $25\% = \dfrac{25}{100} = \dfrac{1}{4}$

Thus, for writing percents as fractions you divide the number by 100 and reduce the fraction thus obtained.

Examples: *Change percent to fraction*

1. $30\% = \dfrac{30}{100} = \dfrac{3}{10}$
2. $12\% = \dfrac{12}{100} = \dfrac{3}{25}$

Converting a percent to a decimal or a decimal to a percent is quite easy. *For changing a percent to a decimal, shift the decimal point in the percent by two places to the left.* For example:

$30\% = 30.0\% = .30$
$75.8\% = .758$
$12\% = 12.0\% = .12$
$6\% = 6.0\% = .06$

For changing a decimal to a percent, shift the decimal point by two digits to the right. For example:

$.30 = 30.\% = 30\%$
$.12 = 12.\% = 12\%$
$.045 = 4.5\%$
$2.38 = 238.\% = 238\%$

Changing fractions to percent is a two step process. Change the fraction to a decimal and then the decimal to a percent. For example,

$\dfrac{3}{4} = .75 = 75\%$

$\dfrac{2}{5} = .4 = .40 = 40\%$

Or, we can also change a fraction to a percent by multiplying the fraction by 100. Thus,

$\frac{3}{4}$ is equivalent to $\left(\frac{3}{4} \times 100\right)\%$ or 75% and

$\frac{2}{5}$ is equivalent to $\left(\frac{2}{5} \times 100\right)\%$ or 40%.

To find the percent of a number you multiply the percent, either as a fraction or as a decimal, by the number. For example:

$$20\% \text{ of } 120 = \frac{20}{100} \times 120 = 24$$
$$\text{or } 20\% \text{ of } 120 = .20 \times 120 = 24$$

In all such cases, "of" always means multiplication.

Example: *Computing the percent of a number*

3. 15% of $48 = .15 \times 48 = 7.20$

4. $2\frac{1}{2}\%$ of $30 = 2.5\%$ of $30 = .025 \times 30 = .750 = .75$

Example 5.
In a 25% off sale, what do you pay if the original price of an article is $52?

$$\text{Discount} = 25\% \text{ of } \$52$$
$$= .25 \times 52 = \$13$$
$$\text{The amount you pay} = \text{Original price} - \text{Discount}$$
$$= \$52 - \$13 = \$39$$

Example 6.
If you deposit $500 in a savings account that gives 5% interest compounded annually, how much will you have in your savings account after two years?

Interest after one year $= 5\%$ of $\$500 = .05 \times 500 = \$25$

Principal amount after one year $= \$500 + \$25 = \$525$

Interest for the second year $= 5\%$ of $\$525 = .05 \times 525$
$$= \$26.25$$

Principal amount after two years $= \$525 + \$26.25$
$$= \$551.25$$

Problem Set 1.7

In Problems 1–6, change percents to decimals and fractions reduced to lowest terms.

	Percent	Decimal	Fraction
1.	40%	.40	$\frac{40}{100} = \frac{2}{5}$
2.	2.5%	_____	_____
3.	.5%	_____	_____
4.	1%	_____	_____

5. .04% _____ _____

6. 215% _____ _____

In Problems 7–12, change decimals or fractions to percent.

Percent	Decimal	Fraction
7. ___5%___	.05	$\frac{5}{100} = \frac{1}{20}$
8. _____	.15	_____
9. _____	1.30	_____
10. _____	_____	$\frac{1}{2}$
11. _____	_____	$\frac{4}{5}$
12. _____	_____	$\frac{7}{4}$

In Problems 13–17, compute the percent of the number.

                                                                                                        Answer

13. 4% of 32   .04 × 32 = 1.28   13. ___1.28___

14. 33% of 45   14. _____

15. 2.5% of 200   15. _____

16. .5% of $21.50   16. _____

17. 125% of 240   17. _____

18. Out of 250 students taking an algebra course, 8% got an A grade on the final. How many students got an A grade?

                                                                                      Ans. _____

19. Johnson spent 15% of his monthly salary on food and 25% on rent. If his net monthly salary is $1500, what is the total amount he spent on these two items?

                                                                                      Ans. _____

20. Linda got a 10% discount on an item costing $32. If the sales tax is 5%, what is the total amount she paid?

                                                                                      Ans. _____

Chapter Summary

1. **Positive integers** 1, 2, 3, . . .
2. **Integers** . . . $-3, -2, -1, 0, 1, 2, 3,$ . . .
3. **Rational numbers** All integers and fractions.
4. **Irrational numbers** Numbers of the type $\sqrt{2}, \sqrt{3}, \sqrt{5}, \ldots$ or π (pie), i.e. numbers which are not rational.
5. **Real numbers** All rational and irrational numbers, such as $2, -5, \frac{4}{5}, \sqrt{7}$.
6. **Even numbers** Integers divisible by 2, such as 2, 4, 6, 8, . . .
7. **Odd numbers** Integers not divisible by 2, such as 1, 3, 5, 7, 9, . . .
8. **Prime numbers** Integers greater than one and not divisible by any integer except by themselves, such as 2, 3, 5, 7, 11, . . .
9. **Composite numbers** Integers that can be expressed as product of prime numbers, such as 4, 6, 9, 12, 15, . . .
10. **Order of operation** Perform multiplication and division before performing addition and subtraction. $5 + 2 \times 7 = 5 + 14 = 19$
11. **Equivalent fractions** Fractions having the same value when expressed in decimals.
12. **Addition of signed numbers** Refer to Rule 1.3.
13. **Multiplication (division) of signed numbers** Perform multiplication (division) of their absolute values and prefix a positive sign if the numbers are of like sign and prefix a negative sign if the numbers are of opposite signs.
14. **Addition or subtraction of decimals** Line up the decimal points and perform the operations.
15. **Multiplication and division of decimal numbers** Multiply or divide without the decimal point and place the decimal at an appropriate place.
16. **Least common denominator** Refer to Rule 1.5.
17. **Addition of fractions** Refer to Rule 1.6.
18. **Multiplication of fraction:** Multiply the numerators and the denominators and reduce the fraction.
19. **Division of fractions** Multiply the numerator by the reciprocal of the denominator.
20. **Division by zero** Has no meaning.
21. **Percent of a number** Percent, changed to fraction or decimal, times the number.

Review Problems

In Problems 1–10, write the numbers as a product of prime factors. The number of bars indicates the number of prime factors.

1. $12 = \underline{\ 2\ } \cdot \underline{\ 2\ } \cdot \underline{\ 3\ }$
2. $20 = \underline{\quad}\ \underline{\quad}\ \underline{\quad}$
3. $30 = \underline{\quad}\ \underline{\quad}\ \underline{\quad}$
4. $36 = \underline{\quad}\ \underline{\quad}\ \underline{\quad}\ \underline{\quad}$

5. 48 = ___ ___ ___ ___ ___   6. 90 = ___ ___ ___ ___

7. 68 = ___ ___ ___   8. 195 = ___ ___ ___

9. 240 = ___ ___ ___ ___ ___ ___   10. 924 = ___ ___ ___ ___ ___

In Problems 11–26, evaluate the given expression.

11. $5 + 4 \times 3 = 5 + 12 = 17$ __17__   12. $7 + 4 \times 2$ _____

13. $12 - 10 \div 2 =$ _____   14. $10 - 5 \div 5$ _____

15. $7 + 3 \times 5 - 10 =$ _____   16. $9 + 4 \times 7 - 3 =$ _____

17. $(3 + 4) \times 7 =$ _____   18. $9 + 4(3 + 2) =$ _____

19. $15 - 10 \div 5 \times 4 =$ _____   20. $24 - 16 \div 4 \times 2 =$ _____

21. $\dfrac{6 - 2}{2} =$ _____   22. $\dfrac{12 - 4}{4} =$ _____

23. $\dfrac{12 \times 3 - 15 \div 3}{12 + 19} =$ _____   24. $\dfrac{9 \times 3 - 14 \div 7}{9 + 16} =$ _____

25. $\dfrac{14 - 2(7 - 4)}{4} =$ _____   26. $\dfrac{26 - 3(8 - 2)}{8} =$ _____

In Problems 27–33, insert '>' or '<' between the two numbers to make a true statement.

27. 12 _____ 9   28. 9 _____ 11

29. 0 _____ 2   30. −2 _____ 0

31. −3 _____ 2   32. −5 _____ −7

33. −11 _____ −9

In Problems 34–53, evaluate the given expression.

34. $14 - 21 = 14 + (-21) =$ __−7__   35. $9 - 11 =$ _____

36. $-5 - 6 =$ _____   37. $-10 - 22 =$ _____

38. $-4 - 2(3 - 1) =$ _____   39. $-5 - 2(7 - 3) =$ _____

40. $5 - 4(4 - 8) =$ _____   41. $7 + 2(4 - 12) =$ _____

42. $-5 - 4(3 - 9) =$ _____   43. $9 - 9(7 - 12) =$ _____

44. $18 - 19 + 12 - 7 =$ _____   45. $12 - 9 + 4 - 13 =$ _____

46. $21 - 10 \div 2 =$ _____   47. $-35 + 18 \div 2 =$ _____

48. $-24 + 40 \div (-10) =$ _____ 49. $-12 - 16 \div (-4) =$ _____

50. $\dfrac{-10 - 2(-4)}{-2} =$ _____ 51. $\dfrac{8 - 4(-2)}{-4} =$ _____

52. $13 - 2(4 + 2 \times 3) =$ _____ 53. $25 - 3(4 - 2 \times 8) =$ _____

In Problems 54–61, write the improper fraction as a mixed number.

54. $\dfrac{12}{5}$ $5\overline{)12}$ $\begin{array}{r}2\\\underline{10}\\2\end{array}$ $2\dfrac{2}{5}$ 55. $\dfrac{17}{5}$ _____

56. $\dfrac{17}{4} =$ _____ 57. $\dfrac{25}{6} =$ _____

58. $\dfrac{27}{11} =$ _____ 59. $\dfrac{49}{13} =$ _____

60. $\dfrac{127}{15} =$ _____ 61. $\dfrac{285}{31} =$ _____

In Problems 62–67, find the missing numerator or missing denominator.

62. $\dfrac{2}{3} = \dfrac{}{12}$ _____8_____ 63. $\dfrac{4}{5} = \dfrac{}{25}$ _____

The answer is 8 because $3 \times 4 = 12$, so $2 \times 4 = 8$

64. $\dfrac{7}{9} = \dfrac{}{27}$ _____ 65. $\dfrac{12}{13} = \dfrac{24}{}$ _____

66. $\dfrac{5}{8} = \dfrac{30}{}$ _____ 67. $\dfrac{5}{12} = \dfrac{}{144}$ _____

In Problems 68–75, reduce the fraction to lowest terms.

68. $\dfrac{12}{30} = \dfrac{2 \cdot 2 \cdot 3}{2 \cdot 3 \cdot 5} = \dfrac{2}{5}$ $\dfrac{2}{5}$ 69. $\dfrac{12}{36} =$ _____

70. $\dfrac{20}{45} =$ _____ 71. $\dfrac{15}{25} =$ _____

72. $\dfrac{36}{42} =$ _____ 73. $\dfrac{72}{90} =$ _____

74. $\dfrac{4 \times 5 \times 7}{10 \times 14} =$ _____ 75. $\dfrac{5 \times 7 \times 8}{14 \times 25} =$ _____

In Problems 76–88, perform the indicated operations. Check your answer with a calculator.

76. $2.87 + 1.37 =$ _____

77. $29.35 + 14.24 =$ _____

78. $5 + 12.79 - 8.86 =$ _____

79. $1.2 + 298.45 - 72 =$ _____

80. $3.45 \times .02 =$ _____

81. $45.23 \times 0.5 =$ _____

82. $2.85 \div .05 =$ _____

83. $12.25 \div 3.5 =$ _____

84. $12.45 - 3.2 \times 1.5 =$ _____

85. $42.15 - 2.7 \div .09 =$ _____

86. $2.5 - 1.2(4.8 - 3.5) =$ _____

87. $\dfrac{4.85 + 2.9}{2.45} =$ _____

88. $\dfrac{7.2 - 2.45 \div .5}{.023} =$ _____

In Problems 89–92 find the LCD of the given fractions.

	Fractions		LCD
89.	$\dfrac{3}{10}, \dfrac{7}{15}$	$10 = 2 \cdot 5$ $15 = 3 \cdot 5$	$2 \cdot 3 \cdot 5 = 30$
90.	$\dfrac{2}{9}, \dfrac{5}{6}$	$9 =$ $6 =$	18
91.	$\dfrac{4}{15}, \dfrac{7}{20}, \dfrac{9}{10}$	$15 =$ $20 =$ $10 =$	60
92.	$\dfrac{3}{28}, \dfrac{7}{40}, \dfrac{11}{70}$	$28 =$ $40 =$ $70 =$	_____

In Problems 93–102, perform the indicated operations.

93. $\dfrac{4}{5} + \dfrac{3}{5} = \dfrac{4+3}{5}$ $\dfrac{7}{5}$

94. $\dfrac{2}{7} - \dfrac{2}{7} + \dfrac{4}{7}$ _____

95. $\dfrac{7}{5} + \dfrac{3}{10}$ LCD =   _____

96. $\dfrac{4}{21} + \dfrac{5}{14}$ LCD =   _____

97. $\dfrac{4}{15} + \dfrac{7}{20} + \dfrac{9}{10}$ LCD =   _____

98. $\dfrac{15}{24} \cdot \dfrac{30}{25}$ _____

99. $\dfrac{12}{25} \div \dfrac{84}{105}$ _____

100. $\dfrac{4}{5} - \dfrac{10}{21} \div \dfrac{15}{14}$ _____

101. $5 - \dfrac{7}{10}$ _____

102. $\dfrac{\frac{1}{2} + \frac{2}{5}}{3 \cdot \frac{7}{10}}$ _____

In Problems 103–107, complete the table.

	Percent	Decimal	Fraction
103.	25%	_____	_____
104.	$3\frac{1}{2}\%$	_____	_____
105.	_____	.45	_____
106.	_____	1.05	_____
107.	_____	_____	$\dfrac{3}{8}$

In Problems 108–112, determine if the number in Column A is less than, equal to, or greater than the number in Column B.

	Column A	Column B

108. $1 + \frac{1}{2}$ $\frac{2}{3}$
   (a) <   (b) =   (c) >   (d) Cannot be determined

109. Given that
$$\begin{array}{r} 20M \\ \times 6 \\ \hline 125N \end{array}$$
   (M and N are digits.)

   Column A   Column B
   M   N
   (a) <   (b) =   (c) >   (d) Cannot be determined

110. Column A   Column B
   20% of 40 $3.2 \div .4$
   (a) <   (b) =   (c) >   (d) Cannot be determined

111. Column A   Column B
   $\frac{4}{5}$ 40%
   (a) <   (b) =   (c) >   (d) Cannot be determined

112. Column A   Column B
   one-third of 84   30% of 90
   (a) <   (b) =   (c) >   (d) Cannot be determined

Chapter Test

1. Determine if the number in column A is less than, greater than, or equal to the number in column B.
   Column A   Column B
   5,632 $[5(1000) + 6(1001) + 3(10) + 2(10)]$
   (a) <   (b) >   (c) =   (d) Cannot be determined

2. Consider the following statements P and Q:
   P: 11 is an odd integer.
   Q: 21 is a composite number.
   Which of the above statements are true?
   (a) P only   (b) Q only   (c) P and Q   (d) None

3. Find the quotient using long division: $\frac{1325}{53} =$
   (a) 250   (b) 205   (c) 25   (d) None of these

4. $\begin{array}{r}20\,M\\ \times\ 6\\ \hline 1{,}25N\end{array}$ M and N are integers

   Determine if the number in column A is less than, greater than, or equal to the number in column B.

   Column A   Column B
     M   [   ]   N

   (a) <   (b) >   (c) =   (d) Cannot be determined

5. On the number line, the number -4 lies to the right of the number -6.
   (a) True   (b) False

6. Consider the following statements P and Q:
   P: $0 \times (-21) = 0$.
   Q: $\frac{-21}{0}$ is not defined.
   Which of the above statements are true?
   (a) P only   (b) Q only   (c) P and Q   (d) None

7. Compute: $5 + 2(-4) =$
   (a) -28 (b) 13   (c) 3   (d) -3 (e) None of these

8. Simplify: $5Y - 2[-3X + 2(Y - 2X) - 4Y]$
   (a) $Y - 14X$ (b) $-9Y - 14X$ (c) $9Y + 14X$ (d) None of these

9. Insert the proper symbol to make a true statement: $\frac{3}{4}$ [   ] $\frac{4}{3}$
   (a) <   (b) >   (c) =   (d) Cannot be determined

10. The product of 0.977×40.35 is closest to: (Use estimation)
    (a) 0.4   (b) 40   (c) 4   (d) 400

11. Round off to the nearest centimeter: 44.8 cm
    (a) 44.9 cm   (b) 45 cm   (c) 44 cm   (d) 4 cm

12. Insert the proper symbol to make a true statement: .50 [   ] $\frac{1}{2}$
    (a) <   (b) >   (c) =   (d) Cannot be determined

13. Change $\frac{3}{8}$ to a decimal
    (a) 3.75   (b) .375   (c) 375   (d) 2.66   (e) None of these

14. Change .04 to a fraction
    (a) $\frac{.04}{100}$ (b) $\frac{4}{100}$ (c) $\frac{2}{50}$ (d) None of these

15. Complete the following statement:
    $\frac{27.8}{0.05} = \frac{?}{5}$
    (a) 0.278   (b) 2780   (c) 278   (d) 27.8   (e) None of these

16. Out of a total of 2000 dollars, Mr. Jones spends 400 dollars on food. What fraction of his income does he spend on food?
    (a) $\frac{1}{4}$ (b) $\frac{5}{1}$ (c) $\frac{1}{2}$ (d) $\frac{1}{5}$ (e) None of these

17. At 4% sales tax, the total price of a color TV whose marked price is 380 dollars is:
    (a) 364.80 (b) 395.20 (c) 384 (d) None of these

18. Compute: $5 - \frac{-3}{7} =$
    (a) $\frac{32}{7}$ (b) $\frac{38}{5}$ (c) $\frac{38}{7}$ (d) $\frac{-38}{7}$ (e) None of these

19. Determine if the entry in Column A is less than, greater than, or equal to the entry in Column B.
    Column A [ ] Column B
    $\frac{5x}{6}$ $\frac{x}{\frac{6}{5}}$

    take the reciprocal of the denominator, and x' it by the numerator

    (a) = (b) > (c) < (d) Cannot be determined

20. Compute: $\frac{1}{2} - \frac{3}{4} \times \frac{8}{9} =$
    (a) $\frac{1}{6}$ (b) $-\frac{1}{6}$ (c) $-\frac{1}{2}$ (d) $\frac{5}{36}$ (e) None of these

Chapter 2

Exponents, Square Roots and Algebraic Expressions

In this chapter we will discuss the following:

- (a) An introduction to exponents
- (b) Rules for natural number exponents
- (c) Integer exponents
- (d) Scientific notation
- (e) Square roots
- (f) Algebraic expressions
- (g) The distributive law
- (h) Evaluation of algebraic expressions

2.1 An Introduction to Exponents

We often encounter situations where the same number is repeated as a factor in a product several times. For example,

$$16 = 2 \cdot 2 \cdot 2 \cdot 2.$$

The number 2 is repeated four times in the product 16. A shorter way of writing

$$2 \cdot 2 \cdot 2 \cdot 2 \text{ is } 2^4.$$

The number 4 in the expression 2^4 is called the *exponent* and indicates that the number 2 is repeated 4 times. The number 2 is called the *base*.

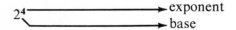

We read 2^4 as "two to the fourth power" or "two to the exponent four". In general, if x stands for a number that is repeated n times in a product, we have

$$\underbrace{x \cdot x \cdot x \ldots x}_{n \text{ factors}} = x^n$$

where n is the exponent and x is the base.

Example 1: *Exponents*

(a) $5 \cdot 5 \cdot 5 = 5^3$

(b) $5 = 5^1$ If a number appears only once as a factor then the exponent for that number is 1.

➡ Note that if there is no exponent over a number then it is understood that the exponent is 1.

(c) $5^4 = 5 \cdot 5 \cdot 5 \cdot 5$

(d) $(2x)^4 = (2x)(2x)(2x)(2x)$
$= (2 \cdot 2 \cdot 2 \cdot 2)(x \cdot x \cdot x \cdot x)$
$= 2^4 \cdot x^4$

In the expression $(2x)^4$, the exponent 4 has 2x as its base.

(e) $2x^4 = 2 \cdot x \cdot x \cdot x \cdot x$

In the expression $2x^4$, the exponent 4 has x as its base.

➡ $(2x)^4 \neq 2x^4$

An exponent applies to the number to its immediate left unless modified by the use of parentheses.

Example 2:

(a) $-2^4 = -(2^4) = -(2 \cdot 2 \cdot 2 \cdot 2) = -16$
The base is 2

(b) $(-2)^4 = (-2)(-2)(-2)(-2) = 16$
The base is -2

➡ $-2^4 \neq 16$
$-2^2 \neq 4$

(c) In the expression $5x^4$, the exponent 4 has x as its base.

(d) In the expression $(3x)^5$, the exponent 5 has 3x as its base.

(e) In the expression $2(5x + 1)^2$, the exponent 2 has $5x + 1$ as its base.

If in an expression the base is negative and the exponent is an even number then the product will be positive. That is

Rule 2.1

If $b > 0$ and n is even then $(-b)^n = b^n$

Example 3: *Negative base and even exponent*

(a) $(-2)^2 = (-2)(-2) = 4 = 2 \cdot 2 = 2^2$
Thus, $(-2)^2 = 2^2$
(b) $(-4)^{10} = 4^{10}$
(c) $(-x)^{10} = x^{10}$ For all x
(d) $-(-4)^{10} = -4^{10}$ Why?

If in an expression the base is negative and the exponent is an odd number then the product will be negative. That is

Rule 2.2

If $b > 0$ and n is odd then $(-b)^n = -(b)^n = -b^n$

Example 4: *Negative base and odd exponent*

(a) $(-2)^3 = (-2)(-2)(-2) = -8 = -(8) = -2^3$
Thus, $(-2)^3 = -2^3$
(b) $(-3)^{11} = -3^{11}$
(c) $(-x)^{21} = -x^{21}$
(d) $-(-5)^{11} = 5^{11}$ Why?

Problem Set 2.1

Write the expressions in Problems 1–10 in factored form.

1. $x^5 = $ \qquad $x \cdot x \cdot x \cdot x \cdot x$
2. $4^5 = $ \qquad
3. $(2x)^4 = $ \qquad
4. $2x^4 = $ \qquad
5. $-4^2 = $ \qquad
6. $(-4)^2 = $ \qquad
7. $-3x^4 = $ \qquad
8. $-(3x)^4 = $ \qquad
9. $(-3x)^4 = $ \qquad
10. $5(2x)^3 = $ \qquad

In Problems 11–20, identify the exponent other than 1 and its corresponding base.

		Base	Exponent			Base	Exponent
11.	x^9	x	9	12.	$(-4)^3$	___	_____
13.	-4^3	___	_____	14.	$2x^3$	___	_____
15.	$-3x^4$	___	_____	16.	$-(3x)^4$	___	_____
17.	$(-4x)^5$	___	_____	18.	$7(-3x)^5$	___	_____
19.	$4xy^3$	___	_____	20.	$3(ab)^7$	___	_____

Mark True or False in Problems 21–30. Assume x > 0.

21. $(-2)^4 = 2^4$ (a) True   (b) False

22. $(-3)^{21} = 3^{21}$ (a) True   (b) False

23. $2(-2)^2 = 8$ (a) True   (b) False

24. $-(-x)^4 = x^4$ (a) True   (b) False

25. $-(-5)^{11} = 5^{11}$ (a) True   (b) False

26. $-2^2 = 2^2$ (a) True   (b) False

27. $-3(-2)^{15} = 3 \cdot 2^{15}$ (a) True   (b) False

28. $3(-2)^2 = 12$ (a) True   (b) False

29. $y(-x)^{24} = y \cdot x^{24}$ (a) True   (b) False

30. $(-5x)^{39} = -5x^{39}$ (a) True   (b) False

2.2 Natural Number Exponents

Consider the product $x^2 \cdot x^3$ where x is any number.

$$x^2 \cdot x^3 = (x \cdot x)(x \cdot x \cdot x)$$
$$= x \cdot x \cdot x \cdot x \cdot x$$
$$= x^5$$

That is, $x^2 \cdot x^3 = x^{2+3} = x^5$

In general when we multiply factors having a like base, the exponent of the product is the sum of the exponents of the factors and the base of the product is the same as the base of the factors.

Rule 2.3

$$x^m \cdot x^n = x^{m+n}$$

Example 1: *Rules 2.1–2.3*

(a) $2^3 \cdot 2^5 = 2^{3+5} = 2^8$

(b) $x^9 \cdot x^{21} = x^{9+21} = x^{30}$

(c) $x^2 \cdot x^3 \cdot x^7 = x^5 \cdot x^7 = x^{12}$

(d) $(-2)^4 2^5$. Here the base is not the same in both the factors.
However, you know that $(-2)^4 = 2^4$ Rule 2.1.
Therefore, $(-2)^4 2^5 = 2^4 \cdot 2^5 = 2^9$

(e) $(-2)^5 2^4$. Here the base is not the same in both the factors.
However, you know that $(-2)^5 = -2^5$ Rule 2.2.
Therefore, $(-2)^5 2^4 = -2^5 \cdot 2^4 = -2^9$

(f) $(x+y)^3 (x+y)^4 = (x+y)^{3+4} = (x+y)^7$

(g) $(-2)^4(-2)^5 = (-2)^{4+5} = (-2)^9 = -2^9$
or since $(-2)^4 = 2^4$ Rule 2.1
and $(-2)^5 = -2^5$ Rule 2.2
therefore $(-2)^4(-2)^5 = 2^4(-2)^5$
$= -2^4 \cdot 2^5 = -2^9$

Now, consider the expression $(a \cdot b)^4$. Here the base is a product of two factors "a" and "b". By definition

$(a \cdot b)^4 = (a \cdot b)(a \cdot b)(a \cdot b)(a \cdot b)$
$= (a \cdot a \cdot a \cdot a)(b \cdot b \cdot b \cdot b)$
$= a^4 \cdot b^4$

That is, $(a \cdot b)^4 = a^4 \cdot b^4$ (1)

In a similar way consider $\left(\dfrac{a}{b}\right)^4$, where $b \neq 0$.

$\left(\dfrac{a}{b}\right)^4 = \dfrac{a}{b} \cdot \dfrac{a}{b} \cdot \dfrac{a}{b} \cdot \dfrac{a}{b}$
$= \dfrac{a \cdot a \cdot a \cdot a}{b \cdot b \cdot b \cdot b} = \dfrac{a^4}{b^4}$

That is, $\left(\dfrac{a}{b}\right)^4 = \dfrac{a^4}{b^4}$ (2)

In general, from (1) and (2) we have

Rule 2.4

$(x \cdot y)^n = x^n y^n$

$\left(\dfrac{x}{y}\right)^n = \dfrac{x^n}{y^n}$

Example 2: *Rules 2.1–2.4*

(a) $(2 \cdot 3)^2 = 2^2 \cdot 3^2 = 4 \cdot 9 = 36$
Also, $(2 \cdot 3)^2 = 6^2 = 36$

(b) $(2x)^2 = 2^2 \cdot x^2 = 4x^2$

➡ $(2x)^2 \neq 2x^2$

(c) $(-x)^4 = x^4$
We have $-x = (-1)x$.
Therefore, $(-x)^4 = (-1)^4 x^4 = 1 \cdot x^4 = x^4$ Rule 2.4

(d) $(-2y)^7 = -2^7 y^7$
We have $(-2y) = (-1) \cdot 2 \cdot y$
Therefore, $(-2y)^7 = (-1)^7 2^7 y^7 = -2^7 y^7$ Rule 2.4

We can now restate Rules 2.1 and 2.2.

> **Rule 2.2A**
> (a) if n is even, $(-b)^n = b^n$
> (b) if n is odd, $(-b)^n = -(b^n) = -b^n$

(e) $(-3x)^4 = (3x)^4$ Rule 2.2A
$= 3^4 x^4 = 81 x^4$

(f) $(-2xy)^3 = -(2xy)^3$ Rule 2.2A
$= -2^3 x^3 y^3 = -8 x^3 y^3$

(g) $\left(\dfrac{2}{3}\right)^2 = \dfrac{2^2}{3^2} = \dfrac{4}{9}$

(h) $\left(\dfrac{x}{4}\right)^3 = \dfrac{x^3}{4^3} = \dfrac{x^3}{64}$

(i) $\left(\dfrac{2x}{3y}\right)^2 = \dfrac{(2x)^2}{(3y)^2} = \dfrac{2^2 x^2}{3^2 y^2} = \dfrac{4x^2}{9y^2}$

Consider the expression $(a^3)^4$. Here 4 is the exponent and a^3 is the base. By definition

$(a^3)^4 = a^3 \cdot a^3 \cdot a^3 \cdot a^3$
$= a^{3+3+3+3}$ Rule 2.3
$= a^{12}$

That is, $(a^3)^4 = a^{12} = a^{3 \cdot 4}$

In general, we have

> **Rule 2.5**
> $(x^m)^n = x^{m \cdot n}$

Example 3: *Rules 2.1–2.5*

(a) $(2^2)^3 = 2^{2 \cdot 3} = 2^6 = 64.$ Rule 2.5
Also, $(2^2)^3 = 4^3 = 64$

(b) $(x^3)^5 = x^{3 \cdot 5} = x^{15}$ Rule 2.5

(c) $(2x^3)^4 = 2^4 \cdot (x^3)^4$ Rule 2.4
$= 2^4 \cdot x^{12}$ Rule 2.5
$= 16 x^{12}$

(d) $\left(\dfrac{2x^2}{y}\right)^3 = \dfrac{(2x^2)^3}{y^3}$ Rule 2.4

$= \dfrac{2^3(x^2)^3}{y^3}$ Rule 2.4

$= \dfrac{2^3 x^6}{y^3} = \dfrac{8x^6}{y^3}$ Rule 2.5

(e) $(-3x^2)^2(-2x)^3$

$= (3x^2)^2(-1)(2x)^3$ Rule 2.2A

$= -3^2(x^2)^2 \cdot 2^3 x^3$ Rule 2.4

$= -3^2 \cdot 2^3 \cdot x^4 \cdot x^3$ Rule 2.5

$= -9 \cdot 8 \cdot x^7 = -72x^7$ Rule 2.3

Finally, consider the expression $\dfrac{x^5}{x^2}$, where x is any nonzero number. By definition

$$\dfrac{x^5}{x^2} = \dfrac{x \cdot x \cdot x \cdot x \cdot x}{x \cdot x} = \dfrac{x}{x} \cdot \dfrac{x}{x} \cdot x \cdot x \cdot x$$

$$= 1 \cdot 1 \cdot x^3 = x^3$$

That is, $\dfrac{x^5}{x^2} = x^3 = x^{5-2}$

Again, consider the expression $\dfrac{x^2}{x^5}$, where x is any nonzero number. By definition

$$\dfrac{x^2}{x^5} = \dfrac{x \cdot x}{x \cdot x \cdot x \cdot x \cdot x} = \dfrac{x \cdot x \cdot 1 \cdot 1 \cdot 1}{x \cdot x \cdot x \cdot x \cdot x} = 1 \cdot \dfrac{1}{x^3}$$

$$= \dfrac{1}{x^3} = \dfrac{1}{x^{5-2}}$$

In general, we have

Rule 2.6

For $x \neq 0$,

$$\dfrac{x^m}{x^n} = \begin{cases} x^{m-n} & \text{if } m > n \\ \dfrac{1}{x^{n-m}} & \text{if } n > m. \end{cases}$$

It should be noticed that in Rule 2.6 the base of the expressions in the numerator and the denominator is the same nonzero number.

Example 4: *Rules 2.1–2.6*

(a) $\dfrac{2^5}{2^3} = 2^{5-3} = 2^2 = 4$ Rule 2.6

Also, $\dfrac{2^5}{2^3} = \dfrac{32}{8} = 4$

(b) $\dfrac{x^9}{x^4} = x^{9-4} = x^5$ Rule 2.6

(c) $\dfrac{(-2x^2)^3}{4x^4} = \dfrac{-(2x^2)^3}{4x^4}$ Rule 2.2A

$= \dfrac{-2^3(x^2)^3}{4x^4}$ Rule 2.4

$= \dfrac{-2^3 x^6}{4x^4}$ Rule 2.5

$= \dfrac{-8}{4} \cdot \dfrac{x^6}{x^4} = -2x^2$ Rule 2.6

(d) $\dfrac{(2x^2)^3 x^4}{4x^7} = \dfrac{2^3(x^2)^3 x^4}{4x^7}$ Rule 2.4

$= \dfrac{2^3 x^6 x^4}{4x^7}$ Rule 2.5

$= \dfrac{8x^{10}}{4x^7}$ Rule 2.3

$= \dfrac{8}{4} \cdot \dfrac{x^{10}}{x^7} = 2x^3$ Rule 2.6

Problem Set 2.2

In Problems 1–10, identify the base for the given exponent.

		Exponent	Base			Exponent	Base
1.	$4x^3$	3	x	2.	$(5x)^2$	2	
3.	$2x^3 x^5$	3		4.	$-4x^5$	5	
5.	$\left(\dfrac{4}{x}\right)^3$	3		6.	$(x^2)^3$	3	
7.	$x^3 y^2$	2		8.	$-(3x)^3$	3	
9.	$\left(\dfrac{2x}{5y}\right)^{11}$	11		10.	$-4z^5$	5	

In Problems 11–16, identify the exponent for the given base.

		Base	Exponent			Base	Exponent
11.	$-4x^5$	4	1	12.	$2^2 x^8$	x	
13.	$3^4 x^5$	3		14.	$5x^4 y^5$	y	
15.	$2^4 y^5$	2		16.	$\left(\dfrac{x}{2y}\right)^5$	$\dfrac{x}{2y}$	

In Problems 17–40, simplify the given expressions by use of Exponent Rules 2.1–2.6. Assume that the variables involved in these problems are all nonzero.

	Expression	Show work here		Simplified form (Answer)
17.	$x^2 \cdot x^5 =$	$x^{2+5} = x^7$	17.	x^7
18.	$x^5 \cdot x^9 \cdot x^{12} =$		18.	_____
19.	$(2x)^2 =$		19.	_____
20.	$(2x)^4(-2x)^3 =$		20.	_____
21.	$(x^2)^4 =$		21.	_____
22.	$(3a^5)^3 =$		22.	_____
23.	$(-2x^7)^4 x^9 =$		23.	_____
24.	$(-x^2)^3(-x)^4 =$		24.	_____
25.	$(a^b)^c =$		25.	_____
26.	$(2x^4)^3(x^3)^4 =$		26.	_____
27.	$(2x+y)^3(2x+y)^5 =$		27.	_____
28.	$\dfrac{x^9}{x^7} =$		28.	_____
29.	$\dfrac{x^5}{x^7} =$		29.	_____
30.	$\dfrac{a^5 \cdot a^9}{a^{20}} = \dfrac{a^{14}}{a^{20}} =$		30.	$\dfrac{1}{a^6}$
31.	$\dfrac{(2x^2)^3}{x^4} =$		31.	_____
32.	$\dfrac{-x^5}{-x^2} =$		32.	_____
33.	$\left(\dfrac{2x}{y}\right)^3 =$		33.	_____
34.	$\left(\dfrac{2x^2y^3}{3z}\right)^2 =$		34.	_____
35.	$\dfrac{2x^3}{4x^4} =$		35.	_____
36.	$\dfrac{2x^4y}{2x^3y} =$		36.	_____

37. $\dfrac{(-3x^2y^3)^3}{9(-xy^2)^2} = \dfrac{-3^3 x^6 y^9}{9x^2y^4} = \dfrac{-27 x^6 y^9}{9x^2y^4} = -\dfrac{3}{1} \cdot \dfrac{x^6}{x^2} \cdot \dfrac{y^9}{y^4} =$ 37. $\underline{\quad -3x^4y^5 \quad}$

38. $\dfrac{(5x^3y^4)^3}{(-2xy^2)^4} =$ 38. _____

39. $\dfrac{5x^3y^4(z^3)^5}{(-2x)^3(4y^3)} =$ 39. _____

40. $\dfrac{(-2x)^4(3y)^5}{(2x^2y)^6} =$ 40. _____

In Problems 41–45, evaluate the given expressions by use of the exponent rules.

                        Show work here   Answer

41. $\dfrac{(12)^4}{(6)^3} =$ $\dfrac{(2 \cdot 6)^4}{(6)^3} = \dfrac{2^4 \cdot 6^4}{6^3} = 2^4 \cdot 6^1 = 16 \cdot 6 =$ 41. $\underline{\quad 96 \quad}$

42. $\dfrac{9 \cdot 4^3}{4^2} =$ 42. _____

43. $\dfrac{12 \cdot 5^7}{4 \cdot 5^6} =$ 43. _____

44. $\dfrac{6^6}{3^5} =$ 44. _____

45. $\dfrac{5(4)^9}{15(2)^7} =$ 45. _____

46. An employer agreed to pay an employee one cent for the first day's labor, two cents for the second day's labor, four cents for the third day's labor, etc. How much does the employer have to pay on the 30th day? (You can earn a fortune by using this trick on a multimillionaire who does not have the knowledge of exponents.)

2.3 Integer Exponents

Zero Exponents

In the last section we considered expressions with natural number exponents and we summarized how to deal with these expressions in Rules 2.1–2.6. We now extend these rules to apply to expressions which involve zero exponents or negative exponents. We will give meaning to expressions such as

    5^0 and 4^{-2}

and in the process we will define the zero and negative exponents.

    If Rule 2.3 is to hold even for a zero exponent, then we must have, for example,

    $x^0 x^3 = x^{0+3} = x^3.$

This suggests that x^0 should be 1, in which case we have $x^0 x^3 = 1 \cdot x^3 = x^3$ holds for all nonzero x. For reasons which are beyond the scope of this book, we must leave 0^0 undefined. Thus, in general, we have:

> **Rule 2.7**
>
> For $x \neq 0$, $x^0 = 1$, and 0^0 is not defined.

Example 1:

(a) $3^0 = 1$

(b) $3x^0 = 3 \cdot 1 = 3$
Note: $3x^0 \neq 1$ because the exponent 0 has only x as its base.

(c) $(3x)^0 = 1$

(d) $2(x^2)^0 y = 2 \cdot 1 \cdot y = 2y$

(e) $\left(\dfrac{2x^2}{6}\right)^0 = 1$

Negative Exponents

In order to define a negative integer exponent, we again must make certain that Rules 2.1–2.6 hold. In particular, we want Rule 2.3 to hold.

Therefore, as an example, for $x \neq 0$ we want

$$x^{-5} x^5 = x^{-5+5} = x^0 = 1.$$

But, from $x^{-5} \cdot x^5 = 1$ we can see that x^{-5} and x^5 are reciprocals of each other. That is,

$$x^{-5} = \frac{1}{x^5}$$

$$\text{and } x^5 = \frac{1}{x^{-5}}.$$

In general, we have the following rule:

> **Rule 2.8**
>
> For $x \neq 0$, $x^{-n} = \dfrac{1}{x^n}$ and $x^n = \dfrac{1}{x^{-n}}$

Example 2:

(a) $5^{-2} = \dfrac{1}{5^2} = \dfrac{1}{25}$

(b) $\dfrac{1}{4^{-2}} = 4^2 = 16$

(c) $\dfrac{x^5}{x^{-3}} = x^5 \cdot \dfrac{1}{x^{-3}} = x^5 \cdot x^3 = x^8$

(d) $\dfrac{y^4}{y^{-7}} = y^4 \cdot \dfrac{1}{y^{-7}} = y^4 \cdot y^7 = y^{11}$

Now that we have defined zero and negative integer exponents, Rule 2.6 may be restated in a less restrictive form.

Rule 2.6 (Modified)

For any $x \neq 0$,
$$\frac{x^m}{x^n} = x^{m-n}$$

Example 3:

(a) $\dfrac{x^4}{x^2} = x^{4-2} = x^2$ (Rule 2.6)

(b) $\dfrac{x^2}{x^4} = \dfrac{1}{x^{4-2}} = \dfrac{1}{x^2}$ (Rule 2.6)

or $\dfrac{x^2}{x^4} = x^{2-4} = x^{-2}$ (Rule 2.6 modified)

$\qquad = \dfrac{1}{x^2}$ (Rule 2.8)

(c) $\dfrac{x^2 \cdot x^{-3}}{x^9} = \dfrac{x^{2-3}}{x^9} = \dfrac{x^{-1}}{x^9} = \dfrac{1}{x} \cdot \dfrac{1}{x^9} = \dfrac{1}{x^{10}}$

➡ Note: As a general rule, when we simplify an expression involving exponents we try to obtain an answer in which only positive exponents appear.

(d) $\dfrac{2x^0 x^{-5}}{-x^{-4}} = -\dfrac{2 \cdot 1 \cdot x^{-5}}{x^{-4}} = -2 \cdot \dfrac{x^{-5}}{x^{-4}} = -2 \cdot \dfrac{x^4}{x^5} = -2 \cdot \dfrac{1}{x} = -\dfrac{2}{x}$

(e) $\dfrac{(-2x^2)^3 \, x^{-4}}{4x^{-2}x^4} = \dfrac{-2^3 x^6 x^{-4}}{4x^{-2}x^4} = \dfrac{-8x^6 \cdot x^{-4}}{4x^4} \cdot \dfrac{1}{x^{-2}} = -\dfrac{8x^6}{4x^4} \cdot \dfrac{1}{x^4} \cdot x^2$

$\qquad = -\dfrac{8x^6 x^2}{4x^4 x^4} = -2\dfrac{x^8}{x^8} = -2x^0 = -2$

(f) $\dfrac{4x^{-2}y^2}{9x^2 y^{-3}} = \dfrac{4y^2 y^3}{9x^2 x^2} = \dfrac{4}{9} \cdot \dfrac{y^5}{x^4}$

(g) $\left(\dfrac{2x}{3y}\right)^{-3} = \dfrac{(2x)^{-3}}{(3y)^{-3}} = \dfrac{(3y)^3}{(2x)^3}$

$\qquad = \dfrac{3^3 y^3}{2^3 x^3} = \dfrac{27y^3}{8x^3}$

Problem Set 2.3

In Problems 1–36, simplify the given expression so that only positive exponents appear in your answer.

1. $x^0 =$ 1

2. $(3x)^0 =$ _____

3. $\left(\dfrac{4}{x}\right)^0 =$ _____

4. $2x^0 =$ $2 \cdot 1 = 2$

5. $3(2x)^0 =$ _____

6. $7x(5x)^0 =$ _____

7. $\dfrac{1}{x^{-4}} =$ x^4

8. $\dfrac{1}{2^{-9}} =$ _____

9. $\dfrac{3}{3^{-2}} =$ _____

10. $\dfrac{a^{-3}}{a^2} =$ $\dfrac{1}{a^2} \cdot \dfrac{1}{a^3} = \dfrac{1}{a^5}$

11. $\dfrac{a^{-2}}{a^3} =$ _____

12. $\dfrac{6a^{-2}}{18a} =$ _____

13. $\dfrac{a^{-5}}{a^{-4}} =$ _____ $\dfrac{a^4}{a^5} = \dfrac{1}{a}$ 14. $\dfrac{x^2 y^{-4}}{xy^{-2}} =$ _____   15. $\dfrac{4 \cdot 5^{-2}}{5^{-3}} =$ _____

16. $(10^2)^{-2} =$ _____ $10^{-4} = \dfrac{1}{10^4}$ 17. $(x^{-3})^2 =$ _____   18. $(2x^3)^{-5} =$ _____

19. $(2x^{-5})^{-3} =$ _____   20. $(5x^2)^{-4} =$ _____

21. $(3x^{-4})^{-2} =$ _____   22. $(5xy^{-2})^{-2} =$ _____

23. $\dfrac{5x^0 x^{-6} y^2}{10xy^{-3}} =$ _____   24. $\dfrac{-50x^{-3}(5x)^0}{(-5x)^2} =$ _____

25. $(2ab^2 c^{-2})^{-3} =$ _____   26. $\left(\dfrac{x^2}{x^{-2}}\right)^4 =$ _____

27. $\dfrac{27 \cdot (10)^{-4}}{9 \cdot (10)^{-6}} =$ _____   28. $\dfrac{10^3 \cdot (10)^{-4}}{10^2 \cdot (10)^{-5}} =$ _____

29. $\dfrac{2 \cdot 4(10)^{-5}}{2 \cdot 2(10)^2} =$ _____   30. $(10^9 \cdot 10^{-8})^{-2} =$ _____

31. $\left(\dfrac{3x^{-1}}{4y^{-2}}\right)^{-2} =$ _____   32. $\dfrac{1}{(-2)^{-3}} =$ _____

33. $\dfrac{6x^{-4} y^{-3} z}{3x^{-2} y^3 z^{-2}} =$ _____   34. $\dfrac{2x^0 y^{-1} z}{4x^{-3} y^2 z^{-5}} =$ _____

35. $\left(\dfrac{x^2 y^{-3}}{x^{-1}}\right)^{-4} =$ _____   36. $(-3x^{-3} y^2)^{-3} =$ _____

In Problems 37–41, mark the correct response.

37. Simplify: $\dfrac{1}{(-4)^{-3}} =$
    (a) 64   (b) 12   (c) -12 (d) -64 (e) $-\dfrac{1}{64}$

38. Simplify: $(2x^{-2} y^3)^{-3} =$
    (a) $\dfrac{y^9}{8x^6}$ (b) $\dfrac{6y^5}{x^6}$ (c) $\dfrac{x^6}{8y^9}$ (d) $-\dfrac{8x^5}{y^6}$ (d) none of these

39. Simplify: $\dfrac{2x^{-5} y^{-4} z^2}{4x^2 yz^{-3}} =$
    (a) $\dfrac{x^7 y^5 z^5}{2}$ (b) $\dfrac{z^5}{2x^7 y^5}$ (c) $\dfrac{x^3 y^3 z^{-5}}{2}$ (d) $\dfrac{z^5}{2x^3 y^3}$ (e) none of these

40. Simplify: $\left(\dfrac{x^2 y^{-3}}{x^{-1}}\right)^{-4} =$
    (a) $x^{12} y^{12}$ (b) $\dfrac{x^{12}}{y^{12}}$ (c) $\dfrac{y^{12}}{x^{12}}$ (d) $\dfrac{1}{x^4 y^{12}}$ (e) $x^4 y^{12}$

41. Consider the following statements P and Q:
    P: $(-4x^4)^5 = -20x^{20}$
    Q: $(3x^2)^2 = 3x^4$
    Which of the statements above is true?
    (a) P only   (b) Q only   (c) P and Q   (d) none

2.4 Scientific Notation

If you use your calculator to compute $1{,}200{,}000 \times 8{,}000$ you will not see 9,600,000,000 in the display window; instead, you will see 9.6   09.

Your calculator cannot display 10 digits and a decimal point. When the result of a calculation is too large (or too small) to be displayed, the calculator resorts to an abbreviated form of what we call *scientific notation*. You must interpret 9.6   09 to mean 9.6×10^9. We know that multiplication of 9.6 by 10^9 can be achieved by simply moving the decimal point 9 places to the right. And, when we move the decimal point in 9.6 nine places to the right we obtain 9,600,000,000.

The purpose of scientific notation is to enable us to write very large or very small numbers in a concise form. For example, there are 6.02×10^{23} molecules in 32 grams of oxygen. If you were to see 6.02×10^{23} written as 602 followed by 21 zeros you would probably feel uncomfortable with the number. It would be too hard to read. The scientific notation for 6.02×10^{23} is far more compact and conveys all the information you need to compute with the number.

We want to learn, in this Section, how to convert numbers to scientific notation and how to multiply and divide them.

Let us first learn how to represent numbers in scientific notation. We always write a given number as a number between 1 and 10 multiplied by a power of 10.

Illustrations

1. $8000 = 8.0 \times 10^3$

    In 8000 we know the decimal point is assumed to be after the last zero. If we move the decimal point three places to the left we will have accomplished the same thing as division by $1000 = 10^3$. In order not to change the value of 8000 we must also multiply by $1000 = 10^3$. Thus, we have

    $$8000. \times \frac{10^3}{10^3} = \frac{8000.}{1000.} \times 10^3 = 8.0 \times 10^3.$$

2. $73{,}400 = 7.34 \times 10^4$

    Here we shift the decimal point 4 places to the left, which is equivalent to division by $10^4 = 10{,}000$. Thus, we must also multiply by 10^4 to leave the value of 73,400 unchanged. We have

    $$73{,}400. \times \frac{10^4}{10^4} = \frac{73{,}400}{10{,}000} \times 10^4 = 7.34 \times 10^4.$$

3. $.0012 = 1.2 \times 10^{-3}$

    In this problem we must shift the decimal point three places to the right in order to place it after the 1. This is equivalent to multiplication by 10^3. In order not to change the value of .0012, we must also divide by 10^3. Thus,

    $$.0012 = \frac{10^3}{10^3}$$
    $$= .0012 \times 10^3 \times \frac{1}{10^3}$$
    $$= 1.2 \times \frac{1}{10^3} = 1.2 \times 10^{-3}.$$

4. $.000002 = 2.0 \times 10^{-6}$
To position the decimal point to the right of the 2 requires us to shift it to the right by 6 places. This is equivalent to multiplication by 10^6. We must also divide by 10^6. Therefore,

$$.000002 = .000002 \times \frac{10^6}{10^6}$$
$$= .000002 \times 10^6 \times \frac{1}{10^6}$$
$$= 2.0 \times \frac{1}{10^6} = 2.0 \times 10^{-6}.$$

5. Write .01345 in scientific notation.
To position the decimal point to the right of 1 requires us to shift it to the right by 2 places. This is equivalent to multiplying by 10^2. We must also divide by 10^2. Therefore, as in Illustrations 3 and 4 above, we get

$$.01345 = 1.345 \times 10^{-2}.$$

We can now summarize our computations above in the following rule.

Rule 2.9 Conversion to Scientific Notation

Step 1. Shift the decimal left or right to make the number a number between 1 and 10.
Step 2. (a) If the shift in Step 1 was left, multiply by 10 to the number of places shifted.
(b) If the shift in Step 1 was right, multiply by 10 to the *negative* of the number of places shifted.

Example 1:

(a) $8000 = 8.0 \times 10^3$
Step 1. 8.000, shifted three places left
Step 2. $8000 = 8.0 \times 10^3$

(b) $73,4000 = 7.34 \times 10^4$
Step 1. 7.3400, shifted four places left
Step 2. $73,400 = 7.34 \times 10^4$

(c) $.0012 = 1.2 \times 10^{-3}$
Step 1. 001.2, shifted three places right
Step 2. $.0012 = 1.2 \times 10^{-3}$

Now we can apply the rules for multiplication and division of powers to compute products and quotients in scientific notation.

Example 2:

(a) Compute $1,200,000 \times 8000$.
First we convert these numbers to scientific notation.
$1,200,000 = 1.2 \times 10^6$
$8000 = 8.0 \times 10^3$

Now we compute the product.
$$1{,}200{,}000 \times 8000 = 1.2 \times 10^6 \times 8.0 \times 10^3$$
$$= (1.2 \times 8.0) \times (10^6 \times 10^3)$$
$$= 9.6 \times 10^9$$

(b) Multiply 1.73×10^8 by 8.41×10^4
$$1.73 \times 10^8 \times 8.41 \times 10^4 = (1.73 \times 8.41) \times (10^8 \times 10^4)$$
$$= 14.5493 \times 10^{12}$$
$$= (1.45493 \times 10^1) \times 10^{12}$$
$$= 1.45493 \times 10^{13}$$

(c) Compute $7.21 \times 10^3 \div 4.0 \times 10^8$
$$(7.21 \times 10^3) \div (4.0 \times 10^8) = \frac{7.21 \times 10^3}{4.0 \times 10^8}$$
$$= \frac{7.21}{4.0} \times \frac{10^3}{10^8}$$
$$= 1.8025 \times 10^{-5}$$

(d) Divide 1.21 by 1.1×10^4
$$1.21 \div (1.1 \times 10^4) = \frac{1.21 \times 10^0}{1.1 \times 10^4}$$
$$= 1.1 \times 10^{-4}$$

Calculator Checks

On most scientific calculators there is a key labelled EE↓. This key enables us to enter a number in scientific notation. We illustrate this now.

(a) Multiply 1.73×10^8 by 8.41×10^4

| 1 | . | 73 | EE↓ | 8 | x | 8 | . | 41 | EE↓ | 4 | = | 1.4549 13 |

Recall that 1.4549   13 is short-hand for 1.4549×10^{13}.

(b) Compute $7.21 \times 10^3 \div 4.0 \times 10^8$.

| 7 | . | 21 | EE↓ | 3 | ÷ | 4 | . | 0 | EE↓ | 8 | = | 1.8025 − 05 |

Problem Set 2.4

In Problems 1–14, write the given number in scientific notation.

1. 75 = 7.5×10^1 2. 150 = _____

3. 29.7 = _____   4. 7459 = _____

5. 15000 = _____   6. .4 = _____

7. 0.04 = _____   8. 0.004 = _____

9. 429.75 = _____   10. 0.023 = _____

11. 7.345 = _____   12. .43429 = _____

13. 0.000042 = _____   14. 7934.425 = _____

In Problems 15–22, you are given numbers as they might appear in your calculator. Interpret these results in scientific notation.

	Calc. Display	Sci. Notat.		Calc. Display	Sci. Notat.
15.	1.2 08	1.2×10^8	16.	2.7 09	_____
17.	2.789 12	_____	18.	1.3 23	_____
19.	2.52 − 12	2.52×10^{-12}	20.	7.53 − 09	_____
21.	3.4 − 08	_____	22.	9.005 − 05	_____

In Problems 23–30, write the given number in decimal form.

23. $1.5 \times 10^4 =$ _____15000_____   24. $4.53 \times 10^2 =$ _____

25. $2.2 \times 10^3 =$ _____   26. $9.003 \times 10^7 =$ _____

27. $1.9 \times 10^{-2} =$ _____.019_____   28. $8.4 \times 10^{-5} =$ _____

29. $7.52 \times 10^{-7} =$ _____   30. $5.4 \times 10^{-6} =$ _____

In Problems 31–40, compute the given product or quotient without the use of a calculator and express your answer in scientific notation.

31. $(3 \times 10^4)(2.1 \times 10^{-8}) = (3 \times 2.1)10^4 10^{-8} =$ _____6.3×10^{-4}_____

32. $(2.2 \times 10^9)(5 \times 10^{-30}) =$ _____

33. $(8 \times 10^{-9})(.4 \times 10^5) =$ _____

34. $(5 \times 10^8)(2.1 \times 10^{-9}) =$ _____

35. $(2 \times 10^5)(3 \times 10^{-9})(4 \times 10^{-12}) =$ _____

36. $\dfrac{2.4 \times 10^8}{1.2 \times 10^{-9}} = \dfrac{2.4}{1.2} \cdot 10^8 \cdot 10^9 = 2 \times 10^{8+9} =$ _____2×10^{17}_____

37. $\dfrac{1.5 \times 10^{18}}{0.5 \times 10^{-3}} =$ _____

38. $\dfrac{4.5 \times 10^{-15}}{0.9 \times 10^9} =$ _____

39. $\dfrac{(2.1 \times 10^8)(0.2 \times 10^6)}{1.4 \times 10^{-9}} =$ _____

40. $\dfrac{(4 \times 10^8)(1.2 \times 10^{-9})}{(2 \times 10^{-8})(3 \times 10^{12})} =$ _____

In Problems 41–46, compute the product or quotient on your calculator and write the calculator answer in scientific notation.

41. $(2.73 \times 10^7)(3.45 \times 10^5) =$ 9.4185×10^{12}

42. $(1.89 \times 10^{-7})(2.79 \times 10^{12}) =$

43. $(283 \times 10^9)(.03 \times 10^7) =$

44. $\dfrac{2.793}{.0083} =$

45. $\dfrac{.00457}{278999} =$

46. $\dfrac{2.753 \times 10^7}{1.39 \times 10^{15}} =$

2.5 Square Roots

We define a square root of a number x to be a number y such that $y^2 = x$.

    y is a square root of x if $y^2 = x$.

For example,

    3 is a square root of 9, because $3^2 = 9$.
    -3 is a square root of 9, because $(-3)^2 = 9$.
    4 is a square root of 16, because $4^2 = 16$.
    -4 is a square root of 16, because $(-4)^2 = 16$.

Note that:

1. The square of a real number is always positive or zero. Thus, it is not possible to square a real number and produce a negative number. For the present, we will restrict ourselves to computing square roots of nonnegative real numbers. Later on, in Chapter 9, we will extend the definition of square root so that we can compute the square roots of a negative number.
2. There are always two square roots of a positive real number, each being the negative of the other.

Example 1:

(a) The square roots of 16 are 4 and -4, because

    $4^2 = 16$ and $(-4)^2 = 16$.

(b) The square roots of 25 are 5 and -5, because

    $5^2 = 25$ and $(-5)^2 = 25$.

The symbol $\sqrt{}$ is called a *radical*. A number or expression under a radical is called the *radicand*. Thus, in the expression $\sqrt{25}$, 25 is the radicand. We reserve the radical symbol for the positive square root of a number. If we want to indicate the negative square root of a number, we put a negative sign in front of the radical.

The square roots of 16 are $\sqrt{16}$ and $-\sqrt{16}$, or 4 and -4.
$\sqrt{25} = 5$ is the positive square root of 25.
$-\sqrt{25} = -5$ is the negative square root of 25.

Example 2:

(a) $\sqrt{4} = 2$ (b) $-\sqrt{4} = -2$

(c) $\sqrt{36} = 6$ (d) $-\sqrt{36} = -6$

(e) $\sqrt{0} = 0$

(f) $\sqrt{-4}$ is not defined in the real numbers, because there is no real number whose square is -4.

Consider the following examples:
$$\sqrt{4 \cdot 9} = \sqrt{36} = 6$$
On the other hand
$$\sqrt{4} \cdot \sqrt{9} = 2 \cdot 3 = 6.$$
Thus, we can conclude that
$$\sqrt{4 \cdot 9} = \sqrt{4} \cdot \sqrt{9}.$$

Now observe that
$$\frac{\sqrt{36}}{\sqrt{9}} = \frac{6}{3} = 2$$
and that
$$\sqrt{\frac{36}{9}} = \sqrt{4} = 2.$$
Thus,
$$\sqrt{\frac{36}{9}} = \frac{\sqrt{36}}{\sqrt{9}}.$$
These two examples are merely special cases of a general rule which we now formulate.

Rule 2.10

For $x \geq 0$ and $y \geq 0$
$$\sqrt{x \cdot y} = \sqrt{x}\,\sqrt{y}$$
$$\sqrt{\frac{x}{y}} = \frac{\sqrt{x}}{\sqrt{y}},\ y \neq 0$$

Example 3:

(a) $\sqrt{24} = \sqrt{4 \cdot 6} = \sqrt{4}\,\sqrt{6} = 2\sqrt{6}$

(b) $\sqrt{8}\,\sqrt{2} = \sqrt{8 \cdot 2} = \sqrt{16} = 4$

(c) $\dfrac{\sqrt{12}}{\sqrt{3}} = \sqrt{\dfrac{12}{3}} = \sqrt{4} = 2$

(d) $\sqrt{\dfrac{11}{4}} = \dfrac{\sqrt{11}}{\sqrt{4}} = \dfrac{\sqrt{11}}{2}$

Expressions involving radicals may be written in several equivalent forms. However, in the future when we say that a radical expression is in 'simplified' form, we shall mean that each of the following conditions is satisfied.

(i) *The radicand* (expression under the radical sign) *has no repeated factors.*
For example,
$$\sqrt{8} = \sqrt{2 \cdot 2 \cdot 2} = \sqrt{4}\sqrt{2} = 2\sqrt{2}.$$
Thus, $2\sqrt{2}$ is the simplified form of $\sqrt{8}$.

(ii) *There is no radical sign in the denominator of a fraction.*
For example,
$$\dfrac{4}{\sqrt{5}} \text{ is not in simplified form.}$$

To produce an equivalent expression which is in simplified form we proceed as follows:
$$\dfrac{4}{\sqrt{5}} = \dfrac{4}{\sqrt{5}} \cdot \dfrac{\sqrt{5}}{\sqrt{5}} = \dfrac{4\sqrt{5}}{\sqrt{25}} = \dfrac{4\sqrt{5}}{5}.$$
Thus, $\dfrac{4\sqrt{5}}{5}$ is the simplified form of $\dfrac{4}{\sqrt{5}}$.

(iii) *There is no fraction under a radical.*
For example,
$$\sqrt{\dfrac{3}{2}} \text{ is not in simplified form.}$$
However,
$$\sqrt{\dfrac{3}{2}} = \dfrac{\sqrt{3}}{\sqrt{2}} = \dfrac{\sqrt{3}}{\sqrt{2}} \cdot \dfrac{\sqrt{2}}{\sqrt{2}} = \dfrac{\sqrt{6}}{\sqrt{4}} = \dfrac{\sqrt{6}}{2}$$
and $\dfrac{\sqrt{6}}{2}$ is the simplified form of $\sqrt{\dfrac{3}{2}}$.

The following examples provide further demonstrations of how to simplify radical expressions.

Example 4: Simplify the following radicals.

(a) $\sqrt{50} = \sqrt{5 \cdot 5 \cdot 2}$ 5 is a repeated factor
$= \sqrt{25 \cdot 2} = \sqrt{25} \cdot \sqrt{2} = 5\sqrt{2}.$

(b) $\sqrt{40} = \sqrt{2 \cdot 2 \cdot 2 \cdot 5}$ 2 is a repeated factor
$= \sqrt{4} \cdot \sqrt{10} = \sqrt{4} \cdot \sqrt{10} = 2\sqrt{10}.$

(c) $\sqrt{x^3} = \sqrt{x \cdot x \cdot x}$ x is a repeated factor
$= \sqrt{x^2 \cdot x} = x\sqrt{x}.$

In this example we must assume that $x \geq 0$, because $\sqrt{x}$ has not yet been defined for negative x.

Example 5: Simplify the following radicals.

(a) $\dfrac{4}{\sqrt{3}}$ This has a radical in the denominator.

$$\dfrac{4}{\sqrt{3}} = \dfrac{4}{\sqrt{3}} \cdot \dfrac{\sqrt{3}}{\sqrt{3}} = \dfrac{4\sqrt{3}}{3}$$

(b) $\dfrac{7}{\sqrt{2}} = \dfrac{7}{\sqrt{2}} \cdot \dfrac{\sqrt{2}}{\sqrt{2}} = \dfrac{7\sqrt{2}}{2}$

Example 6: Simplify the following radicals.

(a) $\sqrt{\dfrac{4}{5}}$ This has a fraction under the radical.

$$\sqrt{\dfrac{4}{5}} = \dfrac{\sqrt{4}}{\sqrt{5}} = \dfrac{\sqrt{4}}{\sqrt{5}} \cdot \dfrac{\sqrt{5}}{\sqrt{5}} = \dfrac{2\sqrt{5}}{5}$$

(b) $\sqrt{\dfrac{5}{7}} = \dfrac{\sqrt{5}}{\sqrt{7}} = \dfrac{\sqrt{5}}{\sqrt{7}} \cdot \dfrac{\sqrt{7}}{\sqrt{7}} = \dfrac{\sqrt{35}}{7}$

Problem Set 2.5

In Problems 1–5, complete the statement.

1. $\sqrt{25}$ = _____ because the square of _____ is _____ .

2. $\sqrt{36}$ = _____ because the square of _____ is _____ .

3. $12^2 = 144$. Therefore, $\sqrt{\underline{}}$ = _____

4. $(\sqrt{7})^2 = \sqrt{\underline{}}$ = _____

5. $\sqrt{(x \cdot y)^2}$ = _____ , where x and y are nonnegative

In Problems 6–39, simplify the radicals. Assume all variables are nonnegative.

6. $\sqrt{20} = \sqrt{4 \cdot 5} = \sqrt{4}\sqrt{5}$ __$2\sqrt{5}$__   7. $\sqrt{32}$ = _____

8. $\sqrt{200}$ = _____   9. $\sqrt{8}$ = _____

10. $\sqrt{12}$ = _____   11. $\sqrt{28}$ = _____

12. $\sqrt{18}$ = _____   13. $\sqrt{27}$ = _____

14. $\sqrt{45}$ = _____   15. $\sqrt{75}$ = _____

16. $\sqrt{125}$ = _____   17. $\sqrt{98}$ = _____

18. $\sqrt{y^3}$ = _____   19. $\sqrt{4x^3} = \sqrt{4 \cdot x^2 \cdot x}$ __$2x\sqrt{x}$__

$\phantom{19.\ \sqrt{4x^3}} = \sqrt{4}\sqrt{x^2}\sqrt{x}$

$\phantom{19.\ \sqrt{4x^3}} = 2x\sqrt{x}$

20. $\sqrt{16a^3} =$ _____   21. $\sqrt{12a^3} =$ _____

22. $\sqrt{8x^3y^3} =$ _____   23. $\sqrt{20x^3y^4} =$ _____

24. $\sqrt{x^2y^2} =$ _____   25. $\sqrt{\dfrac{5}{12}} =$ _____

26. $\dfrac{7}{\sqrt{5}} =$ _____   27. $\dfrac{10}{\sqrt{3}} =$ _____

28. $\dfrac{\sqrt{8}}{\sqrt{3}} =$ _____

29. $\dfrac{5}{\sqrt{8}} = \dfrac{5\sqrt{8}}{\sqrt{8}\sqrt{8}} = \dfrac{5\sqrt{4\cdot 2}}{\sqrt{64}} = \dfrac{10\sqrt{2}}{8} = \dfrac{5\sqrt{2}}{4}$ $\dfrac{5\sqrt{2}}{4}$

30. $\dfrac{7}{\sqrt{12}} =$ _____ $=$ _____ $=$ _____ $=$ _____

31. $\dfrac{5}{\sqrt{24}} =$ _____ $=$ _____ $=$ _____ $=$ _____

32. $\dfrac{5x}{\sqrt{y}} =$ _____   33. $\dfrac{2}{\sqrt{11}} =$ _____

34. $\dfrac{2x}{\sqrt{3y}} =$ _____   35. $\dfrac{5}{\sqrt{4x^3}} =$ _____

36. $\sqrt{\dfrac{5}{7}} =$ _____   37. $\sqrt{\dfrac{5a}{x}} =$ _____

38. $\sqrt{\dfrac{x^3}{y}} =$ _____   39. $\sqrt{\dfrac{4x^3}{3y}} =$ _____

In Problems 40–45, circle the correct response.

40. $\sqrt{x^2} = x$ for all real x. (Hint: Check for $x = -4$.)
    (a) True   (b) False

41. $\sqrt{x^2} = |x|$ for all real x.
    (a) True   (b) False

42. $\sqrt{16 + 9} = \sqrt{16} + \sqrt{9}$. (Hint: Compute both sides.)
    (a) True   (b) False

43. $\sqrt{x + y} = \sqrt{x} + \sqrt{y}$.
    (a) True   (b) False

44. $\sqrt{xy} = \sqrt{x}\sqrt{y}$ where x and y are nonnegative real numbers.
    (a) True   (b) False

45. If $x^2 = y^2$, then $x = y$ for all real numbers x and y.
    (a) True   (b) False

2.6 Algebraic Expressions

Expressions such as $5x$, $2x + 5xy - 7y^2$, or $7x^2 - 3x$ which involve one or more unknowns or constants combined through arithmetic operations are called *algebraic expressions*. The expression $7x^2 - 3x$ has two *terms*, $7x^2$ and $-3x$. The term $7x^2$ has two parts: the constant 7, called the coefficient, and the variable part x^2, called the *literal part*.

Example 1: $2x^2 - 3xy$ is an algebraic expression. It has two terms, $2x^2$ and $-3xy$. The coefficient of the term $2x^2$ is 2 and its literal part is x^2. The coefficient of the term $-3xy$ is -3 and its literal part is xy.

Example 2: $10x^3 - 4x^2 + 3xy - 7$ is an algebraic expression with four terms: $10x^3$, $-4x^2$, $3xy$, and -7. The coefficient of the term $-4x^2$ is -4 and its literal part is x^2. The term -7 has no literal part. It is the constant term -7.

Sometimes an algebraic expression may have two or more terms with the same literal part. For example:

$$5x^2 - 7x + 9x^2 - 3x^2$$

has three terms with the same literal part. These terms are $5x^2$, $9x^2$, $-3x^2$.

> *Like Terms*
> In an algebraic expression, terms with the same literal part are called *like terms*.

When an algebraic expression has two or more like terms, it can be simplified by combining the like terms into one term. Thus, the expression

$$5x^2 - 7x + 9x^2 - 3x^2,$$
$$\text{or } 5x^2 + (-7x) + 9x^2 + (-3x^2)$$

may be written as

$$5x^2 + 9x^2 + (-3x^2) + (-7x),$$

and this may be further simplified to obtain $(5 + 9 - 3)x^2 - 7x$ or $11x^2 - 7x$, because the commutative law allows us to add terms in any order. When we rearrange terms of an expression and put like terms together we refer to this as *grouping like terms*. Like terms can be combined to a single term by use of the following rule.

> **Rule 2.11 Combining Like Terms**
> Step 1. Find the sum of the coefficients of the like terms.
> Step 2. Affix the common literal part to the sum obtained in Step 1.

Example 3: Simplify $12x - 4x + 7y - x + 4y - 5x$

$12x - 4x + 7y - x + 4y - 5x$
$= (12x - 4x - x - 5x) + (7y + 4y)$ Grouping like terms
$= (12 - 4 - 1 - 5)x + (7 + 4)y$
$= 2x + 11y$ Rule 2.11

Example 4: Simplify $9x^2 + 5x + x^2 - 7x$

$$\begin{aligned}
9x^2 + 5x + x^2 - 7x &= 9x^2 + x^2 + 5x - 7x &&\text{Grouping like terms}\\
&= (9+1)x^2 + (5-7)x\\
&= 10x^2 + (-2)x &&\text{Rule 2.11}\\
&= 10x^2 - 2x
\end{aligned}$$

The process of the addition of two algebraic expressions is similar to the simplification of an expression.

Example 5: Add: $(4x - 7) + (9 - 7x)$

$$\begin{aligned}
&= 4x - 7 + 9 - 7x\\
&= 4x - 7x - 7 + 9\\
&= (4 - 7)x + 2 &&\text{Grouping like terms}\\
&= -3x + 2 &&\text{Combining like terms}
\end{aligned}$$

Example 6: Add: $(2x^2 + 4x - 7) + (5x^2 - 7x - 9)$

$$\begin{aligned}
&= 2x^2 + 4x - 7 + 5x^2 - 7x - 9\\
&= (2x^2 + 5x^2) + 4x - 7x - 7 - 9 &&\text{Grouping like terms}\\
&= (2 + 5)x^2 + (4 - 7)x - 16 &&\text{Combining like terms}\\
&= 7x^2 - 3x - 16
\end{aligned}$$

Problem Set 2.6

Complete the statements in Problems 1–5.

1. The expression $3x^2 - 7x + 9$ has _____ terms.

2. The expression $4x^2 + 7x - 9x^2 - x^2$ has _____ like terms.

3. The coefficient of x^2 in $5x^3 - 9x^2 + 8x - 7$ is _____ .

4. The literal part in the second term of the expression $4x^2 - 9xy + 8y^2$ is _____ .

5. The literal part of the like terms in $5x^2 - 7x + 9x^3 - 8x$ is _____ .

In Problems 6–19, simplify the expressions by combining the like terms.

Expression		Simplified Form
6. $6x + 4 - 4x = 6x - 4x + 4$ $= (6 - 4)x + 4 =$	6.	$2x + 4$
7. $4x - 3 + 2x - 7$	7.	_____
8. $2x^2 - 3x + 4x^2$	8.	_____
9. $3x^2 - 4x - 2x^2 - 6x$	9.	_____
10. $4x^2 - 2x^2 - x^2 + 4x$	10.	_____

11. $x^3 - 3x^2 + 2x^2 - 3x^3$ 11. _____

12. $4 - x - 5 + 2x - x^2$ 12. _____

13. $a^2 - 2ab + a^2 - 3ab$ 13. _____

14. $T - 9 + 2T - 7T + 7$ 14. _____

15. $a^4 - 3a^3 + 5a^4 - 6a^3 - 7a$ 15. _____

16. $3x - 4 + 2x - 7 - x^2 + 9$ 16. _____

17. $2x^2 - xy - x^2 - 4xy - 9$ 17. _____

18. $A^3 - 2B - A^2B - 7B + 9A^2B$ 18. _____

19. $7 - x - x^2 + 9 - 9x - 11y + 2x^2$ 19. _____

Perform the indicated operations in Problems 20–23.

Problem	Show work here		Answer
20. $(2x^2 + 9x - 5) + (2x - 7)$	$= 2x^2 + 9x - 5 + 2x - 7$ $= 2x^2 + 9x + 2x - 5 - 7$ $= 2x^2 + 11x - 12$	20.	$2x^2 + 11x - 12$

21. $(7 - 2x) + (x^2 - 7x + 2)$ 21. _____

22. $(y^2 - 2y + 1) + (2y^2 - 8)$ 22. _____

23. $(4 - x^2 + 2x) + (x^2 - 7x - 4)$ 23. _____

2.7 Distributive Law

In Section 1.1 we discussed the commutative and associative laws of addition and multiplication of real numbers. Simply stated these laws described that if we add or multiply two or more real numbers then the order in which we perform these operations is immaterial. Another law which is very important in the study of algebra is what is commonly called *the distributive law*. This law states that if a, b and c are three real numbers then

$$a(b + c) = a \cdot b + a \cdot c$$

That is, when an expression is multiplied by a number then each term of that expression is multiplied by that number.

For example,
$$2(3 + 5) = 2 \cdot 3 + 2 \cdot 5 = 6 + 10 = 16.$$

Also, you know that
$$2(3 + 5) = 2 \cdot 8 = 16.$$

In a similar way
$$2(x + 2) = 2 \cdot x + 2 \cdot 2 = 2x + 4$$
$$-2(2x + 4) = (-2)(2x) + (-2)4 = -4x - 8$$

➡ $2(x + 2) \neq 2x + 2$ Why?
   $-2(x - 2) \neq -2x - 4$ Why?

Example 1: *Use of distributive law*

(a) $a(x + y) = ax + ay$

(b) $2(2x + 3) = 2 \cdot 2x + 2 \cdot 3 = 4x + 6$

(c) $-2(4 - 2x) = (-2)(4 - 2x) = (-2)[4 + (-2x)]$
$\qquad = (-2)(4) + (-2)(-2x) = -8 + 4x$

(d) $2x^2(x^3 - 2x) = 2x^2 \cdot x^3 - 2x^2 \cdot 2x$
$\qquad = 2 \cdot x^5 - 2 \cdot 2 \cdot x^2 \cdot x$
$\qquad = 2x^5 - 4x^3$

(e) $3y(-3y^2 + 7y - 5) = 3y(-3y^2) + 3y(7y) + 3y(-5)$
$\qquad = -3y \cdot 3y^2 + 21 \cdot y \cdot y - 15y$
$\qquad = -9y^3 + 21y^2 - 15y$

(f) $-2x(5x - 4) = (-2x)[5x + (-4)]$
$\qquad = (-2x)(5x) + (-2x)(-4)$
$\qquad = -10x^2 + 8x$

When you subtract two expressions you have to be a little more careful. Recall that
$$5 - 4 = 5 + (-4)$$

That is, subtracting a number is the same thing as adding the negative of that number. Thus, subtracting one algebraic expression from another means adding the negative of one to the other. What is the negative of an expression? Read the following example carefully.

Example 2: *The negative of an expression*

(a) $-5 = (-1)5$

(b) $-(2x + 3) = (-1)(2x + 3)$
$\qquad = (-1)(2x) + (-1)3$ Distributive law
$\qquad = -2x - 3$

(c) $-(3 - 2x) = (-1)(3 - 2x)$
$\qquad = (-1)[3 + (-2x)]$
$\qquad = (-1)3 + (-1)(-2x)$ Distributive law
$\qquad = -3 + 2x$

➤ $\qquad -(3 - 2x) \neq -3 - 2x$

You may notice that the negative of an expression is obtained by changing the signs of all the terms of that expression. Thus, in short,

$$-(2x + 3) = -2x - 3$$
$$-(3 - 2x) = -3 + 2x$$

Example 3: *Subtraction of two expressions*

(a) $(2x + 3) - (3x - 5)$
$= 2x + 3 - 3x + 5$
$= 2x - 3x + 3 + 5$
$= (2 - 3)x + 3 + 5$
$= -x + 8$

(b) $(x^2 + 2x - 3) - (2x^2 - x + 3)$
$= x^2 + 2x - 3 - 2x^2 + x - 3$
$= 1 \cdot x^2 - 2x^2 + 2x + x - 3 - 3$
$= (1 - 2)x^2 + (2 + 1)x - 6$
$= -x^2 + 3x - 6$

Example 4: Simplify $x^2 - 2x - x(2x - 3)$

$x^2 - 2x - x(2x - 3) = x^2 - 2x + (-x)(2x - 3)$
$= x^2 - 2x + (-x)(2x) + (-x)(-3)$
$= x^2 - 2x - 2x^2 + 3x$
$= x^2 - 2x^2 - 2x + 3x$
$= -x^2 + x$

Example 5: Simplify $2x(x - 3) + 3(2x - 1)$

$2x(x - 3) + 3(2x - 1) = 2x \cdot x - 2x \cdot 3 + 3 \cdot 2x - 3 \cdot 1$
$= 2x^2 - 6x + 6x - 3$
$= 2x^2 - 3$

Example 6: Subtract $2x - 3$ from $5 - 2x$

That is, from $5 - 2x$ subtract $2x - 3$
or $(5 - 2x) - (2x - 3)$
$= (5 - 2x) - 1(2x - 3)$
$= 5 - 2x - 2x + 3 = -2x - 2x + 5 + 3$
$= -4x + 8$

Problem Set 2.7

You are given algebraic statements in Problems 1–10. Read these statements carefully. If a statement is correct, write "yes" under the column "correct". Otherwise, write "no" and write the correct form of the statement under the column "corrected statement".

Given Statement	Correct	Corrected Statement
1. $2(x + 4) = 2x + 4$	1. no	$2(x + 4) = 2x + 8$
2. $2(3 + 4) = 14$	2. yes	

3. $-(x-2) = -x-2$　　　　　　　3. _____ _____

4. $-(x^2-3) = -x^2+3$　　　　　 4. _____ _____

5. $x(x^2-4) = x^3-4x$　　　　　 5. _____ _____

6. $2x(x-4) = 2x^2-4$　　　　　　6. _____ _____

7. $x^2(x-2x) = x^3-2x$　　　　　7. _____ _____

8. $-2(x-3) = -2x-6$　　　　　　8. _____ _____

9. $-x(2x+4) = -2x^2+4x$　　　　9. _____ _____

10. $-3x^2(-x+4) = 3x^3-12x^2$　 10. _____ _____

Simplify the expressions in Problems 11–20.

　　Given Expression　　　　　Show work here　　　　　　Simplified Form (Answer)

11. $x^2 + 5x + 2(x-3) = x^2 + 5x + 2x - 6$　　　　11. ____$x^2 + 7x - 6$____
　　　　　　　　　　　$= x^2 + 7x - 6$

12. $x(x+1) + 2(x+1)$　　　　　　　　　　　　　　12. _____

13. $3x + 7 - (2x+1) = 3x + 7 + (-2x-1)$　　　　　13. _____$x + 6$_____
　　　　　　　　　　　$= 3x + 7 - 2x - 1$
　　　　　　　　　　　$= 3x - 2x + 7 - 1$

14. $2x - 5 - (3x+4)$　　　　　　　　　　　　　　14. _____

15. $3(2x-4) - (4x-5)$　　　　　　　　　　　　　15. _____

16. $x(2x-1) + 3(2x-1)$　　　　　　　　　　　　　16. _____

17. $2x(3x+1) - 1(3x+1)$　　　　　　　　　　　　17. _____

18. $2x(x^2 - 7x - 5)$　　　　　　　　　　　　　　18. _____

19. $7x - 2 - 2x(7-x)$　　　　　　　　　　　　　　19. _____

20. $y(2y-1) - 3(y^2 - 2y + 1)$　　　　　　　　　 20. _____

2.8 Evaluating Algebraic Expressions

Algebraic expressions, as discussed earlier, can have any number of values depending upon the value assigned to the unknowns involved in that expression. We can find the value of the expression by replacing the unknowns by their assigned values. For example, when we substitute $x = -2$ in the expression $x^2 - 2x$ we obtain

$$(-2)^2 - 2(-2) = 4 - (-4) = 4 + 4 = 8.$$

Notice that x is replaced by (-2) and we use parentheses around -2, we do not replace x by -2 without the use of parentheses. You can avoid errors by following this procedure.

➡ x^2 for $x = -2$ is not $-2^2 = -4$ Why?
You can avoid this type of error if you replace x by (-2).
That is,
x^2 for $x = -2$ is $(-2)^2 = 4$.

Example 1: *Evaluating algebraic expressions*

(a) $4x - 9$ for $x = -5$
$4(-5) - 9 = -20 - 9 = -29$

(b) $3x^2 + 5x$ for $x = 2$
$3(2)^2 + 5(2) = 3 \cdot 4 + 10 = 12 + 10 = 22.$

(c) $\dfrac{x^2 - 1}{2x}$ for $x = -2$
$\dfrac{(-2)^2 - 1}{2(-2)} = \dfrac{4 - 1}{-4} = \dfrac{3}{-4} = -\dfrac{3}{4}$

(d) $\dfrac{x^3}{3} - \dfrac{2 - x}{x}$ for $x = -1$
$\dfrac{(-1)^3}{3} - \dfrac{2 - (-1)}{(-1)} = \dfrac{-1}{3} - \dfrac{2 + 1}{-1} = \dfrac{-1}{3} - \dfrac{3}{-1}$
$= -\dfrac{1}{3} + \dfrac{3}{1} = \dfrac{-1}{3} + \dfrac{9}{3} = \dfrac{8}{3}$

(e) $p \cdot r \cdot t$ for $p = 500$, $r = .08$, $t = 2$
$500(.08)(2) = 80$

Example 2: The formula $C = \dfrac{5}{9}(F - 32)$ is used for converting a Fahrenheit temperature to a Celsius temperature C. Find the Celsius temperature if the temperature in Fahrenheit is

(a) 32 (b) 95 (c) 113.

(a) For $F = 32$, $C = \dfrac{5}{9}(32 - 32)$
$= \dfrac{5}{9}(0) = 0.$

(b) For $F = 95$, $C = \dfrac{5}{9}(95 - 32)$
$= \dfrac{5}{9}(63) = 35.$

(c) For F = 113, C = $\frac{5}{9}(113 - 32)$
$= \frac{5}{9}(81) = 45.$

Example 3: Suppose you deposit "p" dollars in a savings account at a rate of interest "r" compounded annually. Then your savings "s" after "n" number of years is given by the formula

$$s = p\left(1 + \frac{r}{100}\right)^n.$$

If p = \$2000, r = 10 (rate of interest = 10%) and n = 3 then
$s = 2000\left(1 + \frac{10}{100}\right)^3$
$= 2000\left(\frac{110}{100}\right)^3 = 2000\left(\frac{11}{10}\right)^3$
$= 2000 \times \frac{11}{10} \times \frac{11}{10} \times \frac{11}{10}$
$= 2 \times 11 \times 11 \times 11 = \$2662.$

Problem Set 2.8

In Problems 1–13, evaluate the given expression for the indicated value of the unknown.

	Expression	Show work here		Answer
1.	$3x^2 - 4x^3$ for $x = -2$	$3(-2)^2 - 4(-2)^3 = 12 + 32$	1.	44
2.	a^2 for $a = -4$		2.	
3.	$2x - 4x^2$ for $x = 3$		3.	
4.	$2a^2 - 3$ for $a = 1$		4.	
5.	$3y - 4y^3$ for $y = -2$		5.	
6.	$\frac{4}{x} - \frac{x}{2}$ for $x = -2$		6.	
7.	$\frac{x}{x^2 + 1}$ for $x = -2$		7.	
8.	$2\sqrt{x} - 4x^2$ for $x = 4$		8.	
9.	$\frac{x^3}{5} - \frac{7}{2x}$ for $x = 4$		9.	
10.	$(2x + 3)(x^2 + 4)$ for $x = -2$		10.	
11.	$\frac{x^2 - x}{4} - \frac{5}{x^3}$ for $x = -1$		11.	
12.	x^2y for $x = 2, y = 3$		12.	
13.	$x(2x - y)$ for $x = -1, y = -2$		13.	

In Problems 14–17, evaluate the given expression for the indicated value of the unknown, and express your answer in decimal form rounded to the thousandths position.

14. πr^2 for $\pi = \frac{22}{7}$, $r = 2$ 14. _____

15. $2\pi r$ for $\pi = \frac{22}{7}$, $r = 4$ 15. _____

16. $\pi r^2 h$ for $\pi = \frac{22}{7}$, $r = 2$, $h = 5$ 16. _____

17. $\frac{4}{3}\pi r^3$ for $\pi = \frac{22}{7}$, $r = 3$ 17. _____

In Problems 18–25, evaluate $b^2 - 4ac$ for the values assigned to a, b, and c.

18. $a = 1$ $b = 2$ $c = 1$ $b^2 - 4ac =$ 18. _____

19. $a = 1$ $b = -1$ $c = -6$ $b^2 - 4ac =$ 19. _____

20. $a = 1$ $b = -2$ $c = -8$ $b^2 - 4ac =$ 20. _____

21. $a = 1$ $b = 3$ $c = 5$ $b^2 - 4ac =$ 21. _____

22. $a = 2$ $b = 6$ $c = 4$ $b^2 - 4ac =$ 22. _____

23. $a = 4$ $b = 7$ $c = -3$ $b^2 - 4ac =$ 23. _____

24. $a = -3$ $b = 4$ $c = -2$ $b^2 - 4ac =$ 24. _____

25. $a = 7$ $b = -7$ $c = 3$ $b^2 - 4ac =$ 25. _____

In Problems 26–31, use the formula

$$F = \frac{9C}{5} + 32$$

to calculate the value of F for the given value of C. Round your answer to the nearest whole number. (This is the formula used to convert temperature in Celsius to temperature in Fahrenheit.)

26. $C = 0$ $F = \frac{9 \cdot 0}{5} + 32 =$ 26.   32

27. $C = 20$ $F =$ 27. _____

28. $C = 80$ $F =$ 28. _____

29. $C = -12$ $F =$ 29. _____

30. $C = -4$ $F =$ 30. _____

31. $C = -9$ $F =$ 31. _____

In Problems 32–36, use the formula

$$S = P\left(1 + \frac{r}{100}\right)^n$$

to calculate the total money in an account which starts with a principal P, an annual rate of interest r, and is compounded annually for n years.

32. P = 2000   r = 10   n = 2
    $S = 2000\left(1 + \frac{10}{100}\right)^2 = 2000\left(\frac{11}{10}\right)^2 = 2000 \cdot \frac{11}{10} \cdot \frac{11}{10} =$ 32. __2420__

33. P = 1500   r = 10   n = 1
    S =   33. _____

34. P = 20,000   r = 5   n = 2
    S =   34. _____

35. P = 500   r = 15   n = 2
    S =   35. _____

36. P = 1000   r = 10   n = 3
    S =   36. _____

Chapter Summary

1. $x \, x \, x \ldots x = x^n$ **(n factors)**   x is the base and n is the exponent.

2. **Identification of the base**
   The base in $2x^5$ is x
   The base in -2^2 is 2
   The base in $(-2x)^4$ is $-2x$

3. **Exponents with a base of the form $(-x)$**
   $(-x)^n = x^n$ if n is even
   $(-x)^n = -(x)^n = -x^n$ if n is odd

4. **Exponent Laws**
   (i) $x^m \cdot x^n = x^{m+n}$

   (ii) $(xy)^n = x^n y^n$

   (iii) $\left(\frac{x}{y}\right)^n = \frac{x^n}{y^n}, y \neq 0$

   (iv) $(x^m)^n = x^{m \cdot n}$

   (v) $x^{-n} = \frac{1}{x^n}$ and $\frac{1}{x^{-n}} = x^n$ $(x \neq 0)$

   (vi) $x^0 = 1$ $(x \neq 0)$

5. **Scientific Notation** Any real number may be expressed in the form $r \times 10^n$ where r is a number between 1 and 10, including 1 but not 10.

    $23000 = 2.3 \times 10^4$
    $.00045 = 4.5 \times 10^{-4}$

6. **Square root** A square root of a number is a number whose square is the given number. That is, x is a square root of y if $x^2 = y$. If x and y are positive numbers then

    $$\sqrt{xy} = \sqrt{x}\sqrt{y}$$
    $$\sqrt{\frac{x}{y}} = \frac{\sqrt{x}}{\sqrt{y}}$$

7. **An algebraic expression** An expression that involves unknowns such as $3x - 5$, and

    $x^2 - 4x + 5$.

    The expression $x^2 - 4x + 5$ has three terms.

8. **The coefficient and the literal part of a term** The term $2x^2$ has two parts; 2 called the coefficient and x^2, called the literal part.

9. **The negative of an expression** An expression obtained by changing the sign of each term.

10. **Like terms** Terms in an expression having the same literal part.

11. **Simplifying expressions** Combine the like terms in an expression. We can combine like terms by adding their coefficients and affixing the common literal part to this sum.

12. **Distributive law** $a \cdot (b + c) = a \cdot b + a \cdot c$

Review Problems

In Problems 1–10, identify the base for the given exponent.

		Exponent	Base			Exponent	Base
1.	$2x^5$	5	x	2.	-2^4	4	
3.	x^9	9		4.	$(-2)^4$	4	
5.	$3x^4$	4		6.	$(3x)^4$	4	
7.	$(2x)^4$	4		8.	$-4x^5$	5	
9.	$2x^4$	1		10.	$3(2x)^4$	4	

Rewrite the expressions in Problems 11–16 without a negative sign in the base.

		Positive base			Positive base
11.	$(-2x)^{10}$	$(2x)^{10}$	12.	$2(-x)^3$	$-2x^3$
13.	$(-x)^4$		14.	$(-y)^3$	
15.	$(-2x)^4$		16.	$(-3x)^3$	

In Problems 17–34, simplify the expressions using the exponent laws.

	Expression	Show work here		Simplified Form (Answer)

17. $\dfrac{(x^2)^3 x^5}{x^7} = \dfrac{x^6 x^5}{x^7} = \dfrac{x^{11}}{x^7} = x^{11-7} = x^4$ 17. $\underline{\quad x^4 \quad}$

18. $x^4 x^9$ 18. _____

19. $\dfrac{x^{19}}{x^{12}}$ 19. _____

20. $\dfrac{x^{12}}{x^{16}}$ 20. _____

21. $(x^2)^3$ 21. _____

22. $(2x^3)^3$ 22. _____

23. $(3x^2 y^3)^2$ 23. _____

24. $(-x^2)^3 x^4$ 24. _____

25. $(-2x^3)^4 (3x)^2$ 25. _____

26. $(-a^2)(-a)^3$ 26. _____

27. $\dfrac{(2a^2)^3(-2a)}{(-4a^2)^2}$ 27. _____

28. $\dfrac{x^{-2} y^4}{y^{-2} x^2}$ 28. _____

29. $3x^0$ 29. _____

30. $\dfrac{(2x^0)(x)^{-5}}{(3x^2)^3}$ 30. _____

31. $\dfrac{2A^2(3B)^3}{(-2AB)^4}$ 31. _____

32. $\dfrac{s^3 t^{-5}}{2t^0 s^{-4}}$ 32. _____

33. $\dfrac{12x^{-5}y^3 z^{-2}}{6x^2 y^{-4} z^5}$ 33. _____

34. $\left(\dfrac{16x^{-2}y}{12xy^{-3}}\right)^{-3}$ 34. _____

In Problems 35–40, write the given number in scientific notation.

35. $150000 =$ _____1.5×10^5_____   36. $278000 =$ _____

37. $278.42 =$ _____   38. $0.02 =$ _____

39. $0.0004 =$ _____   40. $0.0000087 =$ _____

In Problems 41–44, write the product and quotients in scientific notation.

41. $275000 \times .0005 = (2.75 \times 10^5)(5.0 \times 10^{-4})$
    $= (2.75 \times 5.0)(10^5 \cdot 10^{-4})$
    $= 13.75 \times 10^1$ $= \underline{\ 1.375 \times 10^2\ }$

42. $182.75 \times .0002 =$
    $=$
    $=$ $=$ _____

43. $\dfrac{(2.1 \times 10^8)(.003)}{7.0 \times 10^{-4}} =$
    $=$
    $=$ $=$ _____

44. $\dfrac{(2.5 \times 10^{-8})(1.2 \times 10^{-7})}{3.0 \times 10^{12}} =$
    $=$
    $=$ $=$ _____

Complete the statements in Problems 45–48.

45. The expression $4x^2 - 5x + 3$ has _____ terms.

46. The coefficient of the second term in $3x^2 - 7x + 8$ is _____ .

47. The literal part of the term $-3y^3$ is _____ .

48. The expression $3x^2 - 4x + 2x^2 - 7x^3$ has _____ like terms.

Simplify the expressions in Problems 49–54.

Given Expression | Simplified Form (Answer)

49. $4x^2 - 7x + 4x^2 - 2x + 9 = 4x^2 + 4x^2 - 7x - 2x + 9$ $\underline{8x^2 - 9x + 9}$

50. $-2(2x + 3)$ _____

51. $3(x^2 + 4) + 2x(x + 1)$ _____

52. $x^2 - 3x - 3(2x - 3)$ _____

53. $2x(x^3 - x) - 2(x^2 - 4)$ _____

54. $5 + x(3 - 4x) - x^2(x + 4)$ _____

In Problems 55–64, simplify the square roots.

55. $\sqrt{40} = \sqrt{4 \cdot 10} = \sqrt{4}\sqrt{10} = \underline{2\sqrt{10}}$ 56. $\sqrt{25} =$ _____

57. $\sqrt{24} =$ _____   58. $\sqrt{45} =$ _____

59. $\dfrac{7}{\sqrt{2}} =$ _____   60. $\sqrt{\dfrac{2}{5}} =$ _____

61. $\sqrt{4x^3} =$ _____   62. $\sqrt{\dfrac{8}{2x}} =$ _____

63. $\dfrac{x\sqrt{y}}{\sqrt{xy}}$ _____   64. $\sqrt{32}\sqrt{2x^3} =$ _____

Chapter Test

Solve each of the following problems and match your answer with one of the responses.

1. Compute: $4 - \dfrac{-2}{3} =$

    (a) $\dfrac{10}{3}$ (b) $\dfrac{14}{3}$ (c) $\dfrac{-10}{3}$ (d) $\dfrac{-14}{3}$ (e) None of the these

2. Compute: $4\dfrac{2}{3} + 2\dfrac{4}{5} =$

    (a) 0   (b) $\dfrac{52}{15}$ (c) $\dfrac{112}{15}$ (d) $\dfrac{28}{15}$ (e) None of these

3. Compute: $\frac{1}{2} - \frac{3}{4} \times \frac{8}{9} =$
   (a) $\frac{1}{6}$ (b) $-\frac{1}{6}$ (c) $-\frac{1}{2}$ (d) $\frac{5}{36}$ (e) None of these

4. Consider the following statements P and Q:
   P: The base in 2^5 is 2.
   Q: The base in -3^7 is 3.
   Which of the above statements are true?
   (a) P only   (b) Q only   (c) P and Q   (d) None

5. Consider the following statements P and Q:
   P: The base in $3x^9$ is 3x.
   Q: $\frac{x^5}{x^{15}} = -x^{10}$.
   Which of the above statements are true?
   (a) P only   (b) Q only   (c) P and Q   (d) None

6. Compute: $(3a^3)^2 =$
   (a) $9a^6$ (b) $6a^9$ (c) $6a^6$ (d) $9a^9$ (e) None of the these

7. Simplify: $(3y^3)^2 (3y^3)^2 =$
   (a) $81y^{10}$ (b) $81y^{12}$ (c) $18y^{12}$ (d) $12y^{10}$ (e) None of these

8. Consider the following statements P and Q:
   P: $(-4x^4)^5 = -x^{20}$.
   Q: $(3x^2)^2 = 3x^4$.
   Which of the above statements are true?
   (a) P only   (b) Q only   (c) P and Q   (d) None

9. Consider the following statements P and Q:
   P: $x^2 x^0 = x^2$.
   Q: $x^{-2} = \frac{1}{x^2}$.
   Which of the above statements are true?
   (a) P only   (b) Q only   (c) P and Q   (d) None

10. Consider the following statements P and Q:
    P: $(x^{-3})^{-2} = x^{-6}$.
    Q: $-2x^{-2} = \frac{1}{2x^2}$.
    Which of the above statements are true?
    (a) P only   (b) Q only   (c) P and Q   (d) None

11. Use scientific notation to find the answer:
    $\frac{4.2 \times 10^2}{1.4 \times 10^{-2}}$
    (a) 3×10^4 (b) $.3 \times 10^0$ (c) $.3 \times 10^{-4}$ (d) 3×10^1 (e) 3×10^3

12. Consider the following statements P and Q:
    P: $\sqrt{0.081} = 0.9$
    Q: $\sqrt{0.25} = 0.5$
    Which of the above statements are true?
    (a) P only (b) Q only (c) P and Q (d) None

13. Simplify: $\sqrt{18} =$
    (a) $9\sqrt{2}$ (b) 9 (c) $2\sqrt{3}$ (d) $3\sqrt{2}$ (e) None of these

14. Simplify: $\sqrt{\frac{1}{75}}$
    (a) $\frac{41}{75}$ (b) $\frac{1}{25}$ (c) $\frac{1}{3}$ (d) $\frac{1}{5\sqrt{3}}$ (e) None of these

15. Simplify: $\left(\frac{-2x^2y}{x}\right)^2 =$
    (a) $-4x^2y^2$ (b) $4x^3y^2$ (c) $-2x^2y^2$ (d) $4x^2y^2$ (e) None of these

16. Simplify: $(3)^{-2} =$
    (a) 9 (b) $\frac{1}{9}$ (c) $-\frac{1}{9}$ (d) -9 (e) None of these

17. Simplify completely: $\frac{x^{-6}y^{-2}z^{-2}}{x^{-3}y^{-1}z^{-4}} =$
    (a) $\frac{x^3y^{-1}}{z^{-2}}$ (b) $\frac{x^3z^2}{y}$ (c) $\frac{1}{x^3yz^2}$ (d) $\frac{z^2}{x^3y}$ (e) None of these

18. Consider the following statements P and Q:
    P: The literal part of $2x^2$ is x^2.
    Q: The expression $4x^2 - 2x + 7$ has three terms.
    Which of the above statements are true?
    (a) P only (b) Q only (c) P and Q (d) None

19. Simplify: $(4x - 3) - 5(2x + 1) =$
    (a) $-6x - 2$ (b) $-6x - 8$ (c) $-6x + 8$ (d) $-6x + 2$ (e) None of these

20. Determine if the entry in column A is less than, greater than, or equal to the entry in column B.
    Column A   Column B
    x^3 [ ] x^2 ($x < -1$)
    (a) = (b) < (c) > (d) Can't be determined

Chapter 3

Linear Equations and Inequalities

The concept of the solution of equations and inequalities is one of the most widely used concepts for solving practical problems.

In this chapter we will discuss:

(a) types of equations and their solutions,
(b) solution of simple linear equations,
(c) solution of equations having unknowns on both sides of the equation,
(d) solution of equations involving grouping symbols and fractions,
(e) solution of linear inequalities.

3.1 Types of Equations

An equation is a statement of equality between two algebraic expressions. An equation may be classified into three categories.

1. *Conditional Equation.* It is a statement which is true only for some values of the unknown. For example, $x + 4 = 7$ is a conditional statement since it is true only for $x = 3$

   $x^2 = 9$ is a conditional statement since it is true only for $x = 3$ and $x = -3$.
   $(3^2) = 9, (-3)^2 = 9$

   $3x + 5 = 2x - 1$ is a conditional statement since it is true only for $x = -6$.
   $3(-6) + 5 = 2(-6) - 1$ or $-13 = -13$

2. *Identity.* It is a statement which is true for all values of the unknown. For example,

   $x + 2 = x + 2$ is an identity since it is true no matter what value you assign to the variable x.

   $2(x + 4) = 2x + 8$ is an identity since this statement is true for all values of x.

   $y(y + 2) - y^2 = 2y$ is an indentity since this statement is true for all values of y.

> In general, when you equate an expression to its simplified form you get an identity, i.e., an equality statement true for all permissible values of the unknown.

3. *False Statement or a Contradiction.* It is a statement which is not true for any value of the unknown. For example,

$0(x) = 4$ is a false statement since $0(x) = 0$, and not 4, for all values of x.
$7 - 3 = 3$ is a false statement.

Throughout this book an *equation* means a conditional equation—a statement true only for some values of the unknown. In an equation the value(s) of the unknown that makes the statement true is called the *solution* of the equation.

Example 1: *Solution of an equation*

(a) $x + 4 = 7$ is an equation, and $x = 3$ is the only value of x which makes this statement true. In the future we will say $x = 3$ is the only value of x which *satisfies* the equation. Thus, $x = 3$ is the solution of $x + 4 = 7$.

(b) $x^2 = 4$ is an equation. There are two values of x, 2 and -2, which satisfy this equation. Thus, $x = 2$ and $x = -2$ are the solutions of the equation $x^2 = 4$.

Notice that $x + 4 = 7$ has only one solution and $x^2 = 4$ has two solutions. In general, if in an equation there is only one unknown and the highest exponent of the unknown is one then the equation has only one solution. If the highest exponent of the unknown is 2 then it has two solutions, etc.

Example 2: *Number of solutions*

(a) $x + 9 = 2x + 7$ is an equation. It has only one unknown "x" and its highest exponent is 1. Hence, this equation has only one solution. Check by inspection that $x = 2$ is the only solution of this equation. $2 + 9 = 2 \cdot 2 + 7$ is true.

(b) $x^2 - 2x = x + 18$ is an equation. There is only one unknown in this equation and its highest exponent is 2. Thus, this equation has two solutions. $x = 6$ is a solution of this equation since

$(6)^2 - 2(6) = 6 + 18$
$36 - 12 = 24$

$x = -3$ is also a solution of this equation since

$(-3)^2 - 2(-3) = -3 + 18$
$9 + 6 = 15$

There is no other value of x which satisfies this equation.

An equation in which, after simplification, the highest exponent of the unknown is 1 is called a *linear equation*. Notice that a *linear equation has only one solution*. In this Chapter, we shall deal only with linear equations. In the next section, we will develop a procedure for solving linear equations.

Problem Set 3.1

In Problems 1–5, identify the type of statement; conditional equation, identity, or a false statement.

Statement	Type
1. $x(x + 2) = x^2 + 2x$	Identity
2. $5x - 4 = 9$	

3. $0(x - 4) = -4$ _____

4. $2x + x = 3x$ _____

5. $x^2 = 16$ _____

In Problems 6–13, indicate the number of solutions of the equation.

Equation	Number of Solutions	Equation	Number of Solutions
6. $x^2 + 4 = 20$	2	7. $x + 3 = 7$	_____
8. $2x - 3 = 5x - 4$	_____	9. $y^2 - y + 3 = 0$	_____
10. $2y^2 - 8y = 0$	_____	11. $x^2 - 4x = x^2 + 4$	_____
12. $x(x + 1) = 3x - 4$	_____	13. $x^2(x - 3) = x^2 - 4x$	_____

In Problems 14–30, find by inspection, the solutions of the equation.

Equation	Solutions	Equation	Solutions
14. $x - 5 = 13$ $18 - 5 = 13$ $13 = 13$	18	15. $2x = 8$	_____
16. $x + 8 = 10$	_____	17. $4x + 1 = 9$	_____
18. $5x - 12 = 13$	_____	19. $.5x = 2.5$	_____
20. $\frac{4}{5}x = 4$	_____	21. $\frac{3}{7}x = 9$	_____
22. $2.1x - 2 = 2.2$	_____	23. $\frac{3}{8}x - 1 = \frac{1}{2}$	_____
24. $x^2 = 25$	_____	25. $x^2 - 1 = 8$	_____
26. $(x - 1)^2 = 4$	_____	27. $(x + 3)^3 = -1$	_____
28. $x^2 - 4 = 12$	_____	29. $2x - 1 = 1 - 2x$	_____
30. $3x - 4 = 4 - x$	_____		

3.2 Solutions of Simple Linear Equations

As discussed in Section 3.1, an equation is a statement of equality between two expressions which is true only for some values of the unknown—only for one value in case it is a linear equation. The truth of such statements is not affected if we add the same number to both sides of the equation. Let us see what happens if we add the same number to both sides of an equation.

Equation	Add to both sides	Result
$9 = 9$	$9 + 4 = 9 + 4$	$13 = 13$
$15 = 15$	$15 + (-3) = 15 + (-3)$	$12 = 12$
$x + 2 = 3$	$x + 2 + (-2) = 3 + (-2)$	$x = 1$

The equation obtained by adding the same number to both sides of the given equation is called an *equivalent equation*. Equivalent equations have the same solutions. Consider for example;

$x - 3 = 4.$ Add 3 to both sides: $x - 3 + 3 = 4 + 3$ or $x = 7$

Thus, $x - 3 = 4$ and $x = 7$ are equivalent equations. But, since the solution of $x = 7$ is 7 ($7 = 7$ is true), the solution of $x - 3 = 4$ is also 7. Why did we pick the number 3 to be added to both sides of this equation? Because, by adding 3 we isolate x on one side of the equation and on the other side of the equation we have the number which is the solution of the given equation.

Again consider for example,

$x + 2 = 7$ Add (-2) to both sides: $x + 2 + (-2) = 7 + (-2)$ or $x = 5$

Thus $x + 2 = 7$ and $x = 5$ are equivalent equations. But since the solution of $x = 5$ is 5 ($5 = 5$ is true), the solution of $x + 2 = 7$ is 5.

Example 1: *The solution of equations of the form $x - a = b$ or $x + a = b$*

(a) $\begin{aligned} x - 5 &= 7 \\ +5 &= +5 \\ \hline x &= 12 \end{aligned}$ Add 5 to both sides

(b) $\begin{aligned} x + 4 &= 9 \\ -4 &= -4 \\ \hline x &= 5 \end{aligned}$ Add -4 to both sides

(c) $\begin{aligned} x - 4.5 &= 2.7 \\ +4.5 &= +4.5 \\ \hline x &= 7.2 \end{aligned}$ Add 4.5 to both sides

If we multiply or divide both sides of an equation by the same *non-zero* number then we get an equivalent equation. For example,

Equation	Multiply/Divide both sides	Result
$5 = 5$	$5 \times 3 = 5 \times 3$	$15 = 15$
$12 = 12$	$\dfrac{12}{4} = \dfrac{12}{4}$	$3 = 3$
$4x = 16$	$\dfrac{4x}{4} = \dfrac{16}{4}$	$x = 4$

Thus, $4x = 16$ and $x = 4$ are equivalent equations. But, since the solution of $x = 4$ is 4, the solution of $4x = 16$ is also 4. Why did we pick the number 4 for division on both sides of the equation? Because, by dividing both sides by 4 we isolate x on one side of the equation and on the other side of the equation we have the solution.

Example 2: *The solution of equations of the form ax = b*

(a) $5x = 10$
$\frac{5x}{5} = \frac{10}{5}$ Divide both sides by 5
$x = 2$ Check: $5(2) = 10$ is true.

(b) $-3x = 21$
$\frac{-3x}{-3} = \frac{21}{-3}$ Divide both sides by -3
$x = -7$ Check: $-3(-7) = 21$ is true

(c) $\frac{x}{2} = 3$
$2 \cdot \frac{x}{2} = 2 \cdot 3$ Multiply both sides by 2
$x = 6$ Check: $\frac{1}{2}(6) = 3$ is true

(d) $\frac{2x}{3} = 7$
$\frac{3}{2} \cdot \frac{2}{3} \cdot x = \frac{3}{2} \cdot 7$ Multiply both sides by $\frac{3}{2}$
$x = \frac{21}{2}$ Check: $\frac{2}{3} \cdot \frac{21}{2} = \frac{42}{6} = 7$ is true

Most of the problems involving percents can be reduced to the problem of solving an equation of the form ax = b. For illustrations, read the following examples carefully.

Example 3: What number is 20% of 420? Let that number be "x". Then 20% of 420 = x

.20 times 420 = x   ("of" always means "times")
$x = (.20)(420) = 84$

Thus, 20% of 420 is 84

Example 4: 35% of what number is 175? Let that number be "x". Then, 35% of x = 175

$\frac{35}{100}x = 175$

$\frac{100}{35} \cdot \frac{35}{100} \cdot x = \frac{100}{35} \cdot 175$ Multiply both sides by $\frac{100}{35}$.

or, $x = 500$

Check: 35% of 500 = $\frac{35}{100} \times 500 = 175$ is true.

Example 5: What % of 120 is 30? Let x denote the unknown.

x% of 120 = 30

$\frac{x}{100} \cdot 120 = 30$ or $\frac{120}{100}x = 30$

$\frac{100}{120} \cdot \frac{120}{100} \cdot x = \frac{100}{120} \cdot 30$ Multiply both sides by $\frac{100}{120}$

$$x = \frac{100}{120} \cdot 30 = 25$$

Check: 25% of 120 = $\frac{25}{100} \times 120 = 30$ is true.

Example 6: An increase from $50 to $60 is what percent increase? The actual increase is $60 − $50 = $10. Thus, there is a $10 increase in $50. The problem can be restated as: "What percent of $50 is $10?" Let x denote the unknown.

$$x\% \text{ of } 50 = 10$$
$$\frac{x}{100} \cdot 50 = 10 \quad \text{or} \quad \frac{x}{2} = 10$$
$$\text{Then, } 2 \cdot \frac{x}{2} = 2 \cdot 10 \quad \text{Multiply both sides by 2}$$
$$x = 20$$

Hence an increase from $50 to $60 is a 20% increase.

The solution of an equation of the form ax + b = c can be obtained by using the idea of obtaining an equivalent equation with the unknown isolated on one side of the equation.

$$\begin{aligned} ax + b &= c \\ -b &= -b \end{aligned} \quad \text{Add } -b \text{ to both sides}$$
$$ax = c - b$$
$$\frac{1}{a} \cdot ax = \frac{1}{a}(c - b) \quad \text{Multiply by } \frac{1}{a} \text{ or divide by a on both sides}$$
$$x = \frac{c - b}{a}.$$

Thus, the solution of $ax + b = c$ is $\frac{c - b}{a}$.

Example 7: *Solution of equations of the form ax + b = c*

(a)
$$\begin{aligned} 2x - 3 &= 5 \\ +3 &= +3 \end{aligned} \quad \text{Add 3 on both sides}$$
$$2x = 8$$
$$\frac{1}{2} \cdot 2x = \frac{1}{2} \cdot 8 \quad \text{Multiply by } \frac{1}{2}, \text{ or divide by 2 on both sides}$$
$$x = 4$$

Check: 2(4) − 3 = 5 or 8 − 3 = 5 is true

(b)
$$\begin{aligned} 3x + 5 &= 11 \\ -5 &= -5 \end{aligned} \quad \text{Add } -5 \text{ on both sides}$$
$$3x = 6$$
$$\frac{1}{3} \cdot 3x = \frac{1}{3} \cdot 6 \quad \text{Multiply by } \frac{1}{3} \text{ or divide by 3 on both sides}$$
$$x = 2$$

Check: 3(2) + 5 = 11 or 6 + 5 = 11 is true

(c) $2.5x + 3.4 = 11.9$
    $ -3.4 = -3.4$ Add -3.4
    $2.5x = 8.5$
    $\dfrac{2.5x}{2.5} = \dfrac{8.5}{2.5}$ Divide by 2.5
    $x = \dfrac{8.5}{2.5} = \dfrac{85}{25} = \dfrac{17}{5}$

Check: $2.5x + 3.4 = 2.5\left(\dfrac{17}{5}\right) + 3.4 = \dfrac{25}{10} \cdot \dfrac{17}{5} + 3.4$
$= \dfrac{17}{2} + 3.4 = 8.5 + 3.4 = 11.9$

Problem Set 3.2

In Problems 1–20, solve for the unknown.

	Equation	Solution		Equation	Solution
1.	$x + 5 = 7$ $x = 7 - 5$ $x = 2$	2	2.	$x + 9 = 13$	
3.	$x - 7 = 9$		4.	$x - \dfrac{1}{2} = \dfrac{3}{2}$	
5.	$-x + 4 = 5$		6.	$7 - x = 4$	
7.	$x + 2.5 = 7.5$		8.	$7x - 1.8 = 11.2$	
9.	$2y = 18$		10.	$3x = 1.2$	
11.	$\dfrac{a}{2} = 7$		12.	$\dfrac{2a}{3} = \dfrac{3}{4}$	
13.	$-\dfrac{x}{4} = \dfrac{7}{4}$		14.	$\dfrac{3}{4}p = -\dfrac{7}{2}$	
15.	$2x + 3 = 7$		16.	$3x - 1 = 5$	
17.	$2.5 = x + 1.5$		18.	$2.5 - 2x = 4.7$	
19.	$\dfrac{2}{3}x + \dfrac{1}{3} = \dfrac{2}{5}$		20.	$\dfrac{7}{2} = \dfrac{1}{3} - \dfrac{5x}{6}$	

In Problems 21–30, identify the unknown, set up an equation, and find the value of the unknown.

	Problem	Equation	Show work here	Solution
21.	*What number is* 30% of 400?	$x = .30 \times 400$	$x = 120$	120
22.	What number is 20% of 250?			
23.	15% of what number is 45?			
24.	35% of what number is 245?			
25.	What percent of 420 is 35?			
26.	What percent of 1250 is 250?			
27.	A change of price from $25 to $15 is what percent change?			
28.	An increase from 30 to 50 is what percent increase?			
29.	In a 30% off sale you pay $40 for an article. What is the marked price?			
30.	After a 4% increase in taxes you pay $312 in taxes. If your income remains the same, how much taxes were paid before the increase?			

3.3 Solution of Equations Having the Unknown on Both Sides of the Equation

Some linear equations may have the terms containing the unknowns on both sides of the equation. For example

$$5x + 3 = 2x - 7$$

In such equations, as a first step, you should isolate the terms containing the unknowns on one side of the equation and the constants on the other side.

Example 1: Solve $3x + 5 = x - 2$

$$\begin{aligned} 3x + 5 &= x - 2 \\ -x &= -x \\ \hline 2x + 5 &= -2 \\ -5 &= -5 \\ \hline 2x &= -7 \\ x &= \frac{-7}{2} \end{aligned}$$

Add $-x$ to both sides

Add -5 to both sides

Divide both sides by 2.

You should realize that this is not the only way to solve this equation. You could have made the following choices:

$$\begin{aligned} 3x + 5 &= x - 2 \\ -3x &= -3x \\ \hline 5 &= -2x - 2 \\ 2 &= + 2 \\ \hline 7 &= -2x \\ -\frac{7}{2} &= x \end{aligned}$$

Add $-3x$ to both sides

Add 2 to both sides

Divide both sides by -2

Example 2: Solve $4x - 5 = 2x + 8$

$$\begin{aligned} 4x - 5 &= 2x + 8 \\ -2x &= -2x \\ \hline 2x - 5 &= + 8 \\ +5 &= + 5 \\ \hline 2x &= 13 \\ x &= \frac{13}{2} \end{aligned}$$

Add $-2x$ to both sides

Add 5 to both sides

Divide both sides by 2

Check:

$$4x - 5 = 4\left(\frac{13}{2}\right) - 5$$
$$= 26 - 5 = 21$$

$$2x + 8 = 2\left(\frac{13}{2}\right) + 8$$
$$= 13 + 8 = 21$$

Example 3: Solve $3x - 4 = 7x + 8$

$$\begin{aligned} 3x - 4 &= 7x + 8 \\ -3x &= -3x \\ \hline -4 &= 4x + 8 \\ -8 &= - 8 \\ \hline -12 &= 4x \end{aligned}$$

or $4x = -12$

$$x = \frac{-12}{4} = -3$$

Divide both sides by 4

Check:

$$3x - 4 = 3(-3) - 4$$
$$= -9 - 4 = -13$$

$$7x + 8 = 7(-3) + 8$$
$$= -21 + 8 = -13$$

Sometimes equations may involve grouping symbols. In such equations we first use the distributive law to remove the grouping symbols.

Example 4: Solve $3(x + 2) = 2x - 9$

$$3(x + 2) = 3x + 6 \qquad \text{Distributive law}$$

Therefore, the given equation is

$$
\begin{array}{rl}
3x + 6 = & 2x - 9 \\
-2x = & -2x \\
\hline
x + 6 = & -9 \\
-6 = & -6 \\
\hline
x = & -15
\end{array}
$$

Add $-2x$ on both sides

Add -6 on both sides

Check:

$$3(x + 2) = 3(-15 + 2) \qquad 2x - 9 = 2(-15) - 9$$
$$= 3(-13) = -39 \qquad = -30 - 9 = -39$$

Example 5: Solve $4 - 2(x - 3) = 2x + 3(2x - 5)$

Since $-2(x - 3) = -2x + 6$ Distributive Law
and $+3(2x - 5) = 6x - 15$ Distributive Law

the given equation may be rewritten as

$$4 - 2x + 6 = 2x + 6x - 15$$

By combining like terms on both sides we get

$$
\begin{array}{rl}
-2x + 10 = & 8x - 15 \\
+2x = & +2x \\
\hline
10 = & 10x - 15 \\
+15 = & +15 \\
\hline
25 = & 10x
\end{array}
$$

Add $2x$ on both sides

Add 15 on both sides

or $10x = 25$

$$x = \frac{25}{10} = \frac{5}{2}$$

When the solution is a large number or a fraction, checking the solution may involve difficult computations. If you feel confident to handle the computation, fine; otherwise, you may use a calculator to check your solution. Below we give some illustrations of checking solutions with a calculator. In these illustrations and at all other places in this book we shall make use of a calculator that obeys Rule 1.1 (see page 6) for the order of operations. In order to be sure you have the right type of calculator, compute

$$5 \;\boxed{+}\; 3 \;\boxed{\times}\; 4 \;\boxed{=}\;.$$

If your calculator displays 32 instead of 17, then it does not follow the rule for the order of operations we have agreed to use. For checking solutions with a calculator, store the solution number in the memory of your calculator and press the RCL key whenever you want to replace the unknown in the equation by the solution. Compute the Left Hand Side (LHS) and the Right Hand Side (RHS) of the equation, respectively, and both times make note of the number on the display. These numbers should be the same if the solution is correct.

In Example 1, the solution of the equation $3x + 5 = x - 2$ was $-\frac{7}{2}$.

Calculator Check:

$$7 \;\boxed{+/-}\; 2 \;\boxed{=}\; \boxed{\text{STO}}$$

$3x + 5 =$
$3 \;\boxed{\times}\; \boxed{\text{RCL}}\; \boxed{+}\; 5 \;\boxed{=}\; -5.5$

$x - 2 =$
$\boxed{\text{RCL}}\; \boxed{-}\; 2 \;\boxed{=}\; -5.5$

In Example 2, the solution of the equation $4x - 5 = 2x + 8$ was $\frac{13}{2}$.

Calculator Check:

$$13 \;\boxed{\div}\; 2 \;\boxed{=}\; \boxed{\text{STO}}$$

$4x - 5 =$
$4 \;\boxed{\times}\; \boxed{\text{RCL}}\; \boxed{-}\; 5 \;\boxed{=}\; 21$

$2x + 8 =$
$2 \;\boxed{\times}\; \boxed{\text{RCL}}\; \boxed{+}\; 8 \;\boxed{=}\; 21$

In Example 3, the solution of the equation $3x - 4 = 7x + 8$ was -3.

Calculator Check:

$$3 \;\boxed{+/-}\; \boxed{=}\; \boxed{\text{STO}}$$

$3x - 4 =$
$3 \;\boxed{\times}\; \boxed{\text{RCL}}\; \boxed{-}\; 4 \;\boxed{=}\; -13$

$7x + 8 =$
$7 \;\boxed{\times}\; \boxed{\text{RCL}}\; \boxed{+}\; 8 \;\boxed{=}\; -13$

In Example 4, the solution of the equation $3(x + 2) = 2x - 9$ was -15.

Calculator Check:

$$15 \;\boxed{+/-}\; \boxed{=}\; \boxed{\text{STO}}$$

$(3x + 2) =$
$3 \;\boxed{\times}\; \boxed{(}\; \boxed{\text{RCL}}\; \boxed{+}\; 2 \;\boxed{)}$
$\boxed{=}\; -39$

$2x - 9 =$
$2 \;\boxed{\times}\; \boxed{\text{RCL}}\; \boxed{-}\; 9 \;\boxed{=}\; -39$

In Example 5, the solution of the equation $4 - 2(x - 3) = 2x + 3(2x - 5)$ was $\frac{3}{2}$.

Calculator Check:

$$3 \;\boxed{\div}\; 2 \;\boxed{=}\; \boxed{\text{STO}}$$

$4 - 2(x - 3) =$
$4 \;\boxed{-}\; 2 \;\boxed{\times}\; \boxed{(}\; \boxed{\text{RCL}}\; \boxed{-}\; 3 \;\boxed{)}$
$\boxed{=}\; 7$

$2x + 3(2x - 5) =$
$2 \;\boxed{\times}\; \boxed{\text{RCL}}\; \boxed{+}\; 3 \;\boxed{\times}$
$\boxed{(}\; 2 \;\boxed{\times}\; \boxed{\text{RCL}}\; \boxed{-}\; 5 \;\boxed{)}$
$\boxed{=}\; 7$

Problem Set 3.3

Solve the following equations and check the solutions.

	Equation	Solution	Check		
1.	$3x - 5 = 2x + 9$	14	$3x - 5 = 3(14) - 5 = 37$		
	$\underline{-2x = -2x}$		$2x + 9 = 2(14) + 9 = 37$		
	$x - 5 = 9$				
	$\underline{ 5 = 5}$				
	$x = 14$				
2.	$4x - 9 = 7x + 7$	_____	$4x - 9 =$ _____	$=$	_____
			$7x + 7 =$ _____	$=$	_____
3.	$2x - 5 = -3x + 4$	_____	$2x - 5 =$ _____	$=$	_____
			$-3x + 4 =$ _____	$=$	_____
4.	$-5x - 9 = -3x + 9$	_____	$-5x - 9 =$ _____	$=$	_____
			$-3x + 9 =$ _____	$=$	_____
5.	$4(3x + 2) + 7 = 3$	_____	$4(3x + 2) + 7 =$ _____	$=$	_____
6.	$2(3x - 2) = 2x + 4$	_____	$2(3x - 2) =$ _____	$=$	_____
			$2x + 4 =$ _____	$=$	_____
7.	$5(4 - x) = x - 2(x + 4)$	_____	$5(4 - x) =$ _____	$=$	_____
			$x - 2(x + 4) =$ _____	$=$	_____
8.	$2x - 4(x - 3) = 8x - 9$	_____	$2x - 4(x - 3) =$ _____	$=$	_____
			$8x - 9 =$ _____	$=$	_____
9.	$2 + 3(x - 4)$	_____	$2 + 3(x - 4) =$ _____	$=$	_____
	$= 7 - 4(x - 3)$		$7 - 4(x - 3) =$ _____	$=$	_____
10.	$3 - 3(x - 2)$	_____	$3 - 3(x - 2) =$ _____	$=$	_____
	$= 2 - 4(x - 5)$		$2 - 4(x - 5) =$ _____	$=$	_____
11.	$3x - 4 + 2x - 2(x - 3)$	_____	$3x - 4 + 2x - 2(x - 3) =$ _____	$=$	_____
	$= 7x - 13$		$7x - 13 =$ _____	$=$	_____
12.	$5x - 2(x - 4) = 2x - 1$	_____	$5x - 2(x - 4) =$ _____	$=$	_____
			$2x - 1 =$ _____	$=$	_____

13. $4x - 7 = 3x - 5(x - 4)$ _____ $4x - 7 =$ _____ = _____

   $3x - 5(x - 4) =$ _____ = _____

14. $2(x - 7) = x - 2(2x - 5)$ _____ $2(x - 7) =$ _____ = _____

   $x - 2(2x - 5) =$ _____ = _____

15. $4x - 3(x - 5)$ _____ $4x - 3(x - 5) =$ _____ = _____
    $= 2x - 3(x - 4)$

   $2x - 3(x - 4) =$ _____ = _____

16. $x - 3(2 - 5x) = 2x - 1$ _____ $x - 3(2 - 5x) =$ _____ = _____

   $2x - 1 =$ _____ = _____

17. $2x + 4(3 - 2x)$ _____ $2x + 4(3 - 2x) =$ _____ = _____
    $= 3(1 - 4x)$

   $3(1 - 4x) =$ _____ = _____

18. $2(3x - 4) - 3(2x - 4)$ _____ $2(3x - 4) - 3(2x - 4) =$ _____ = _____
    $= 2x - 1$

   $2x - 1 =$ _____ = _____

19. $5x - \frac{1}{2}(2x - 4)$ _____ $5x - \frac{1}{2}(2x - 4) =$ _____ = _____
    $= 7 - 8x$

   $7 - 8x =$ _____ = _____

20. $3.2x = 2.56$ _____ $3.2x =$ _____ = _____

21. $2.4x - 1.5 = 5.7$ _____ $2.4x - 1.5 =$ _____ = _____

22. $3.4x - 2.3 = 1.2 - 0.1x$ _____ $3.4x - 2.3 =$ _____ = _____

   $1.2 - 0.1x =$ _____ = _____

23. $1.2x - 7.5 = 0.2x + 1.5$ _____ $1.2x - 7.5 =$ _____ = _____

   $0.2x + 1.5 =$ _____ = _____

24. $1.2(5x - 2) = 15.6$ _____ $1.2(5x - 2) =$ _____ = _____

25. $2.3x + 2(1.3x - 4)$ _____ $2.3x + 2(1.3x - 4) =$ _____ = _____
    $= 4 - 0.1x$

   $4 - 0.1x =$ _____ = _____

3.4 Equations Involving Grouping Symbols and Fractions

Some linear equations may have grouping symbols, fractions or both. If an equation has grouping symbols and fractions, you can avoid possible errors by first simplifying the grouping symbols by use of the distributive law. Notice that *a fraction bar with two or more terms in the numerator is also a grouping symbol*. For example, consider the following *identities*.

$$\frac{2x-5}{4} = \frac{2x}{4} - \frac{5}{4} \quad \text{or} \quad \frac{2x-5}{4} = \frac{1}{4}(2x-5) = \frac{1}{4}(2x) - \frac{1}{4}(5) = \frac{2x}{4} - \frac{5}{4}.$$

Recall, in Section 3.2 we observed that if we multiplied both sides of an equation by any *non zero number* we got an equivalent equation. Thus, in a given equation we can make the equation free of fractions by multiplying both sides by a suitable number. As you can see the most suitable number to accomplish this task is the Least Common Denominator of all the fractions on both sides of the equation. You can avoid errors if you follow the steps in the following rule.

Rule 3.1 Solution of Linear Equation

Step 1. Simplify grouping symbols by use of the distributive law.
Step 2. Make the equation free of fractions by multiplying both sides by the LCD.
Step 3. Simplify each side of the equation by combining like terms.
Step 4. Isolate terms containing the unknown on one side of the equation and the constants on the other side as in Section 3.2.
Step 5. In Step 4 the given equation is reduced to an equivalent equation of the form $ax = b$. Divide both sides by a to solve for x.
Step 6. Check the solution.

Example 1: Solve for x. $3 + 2(x - 3) = 2(3x - 4)$

$3 + 2x - 6 = 6x - 8$		Step 1
$2x - 3 = 6x - 8$		Step 3
$-2x = -2x$		Add $-2x$ to both sides
$-3 = 4x - 8$		
$+8 = +8$		Add 8 to both sides
$5 = 4x$		Step 4
or $\quad 4x = 5$		Step 5
$x = \dfrac{5}{4}$		

Calculator Check: Step 6

5 [÷] 4 [=] [STO]

$3 + 2(x - 3) =$
3 [+] 2 [×] [(] [RCL] [−] 3
[)] [=] -0.5

$2(3x - 4) =$
2 [×] [(] 3 [×] [RCL] [−] 4
[)] [=] -0.5

Example 2: Solve for y. $\dfrac{5y-3}{5} - 4 = \dfrac{2y}{3}$

Since $\dfrac{5y-3}{5} = \dfrac{5y}{5} - \dfrac{3}{5} = y - \dfrac{3}{5}$ the given equation may be written as

$y - \dfrac{3}{5} - 4 = \dfrac{2y}{3}$ Step 1 (the fraction bar is also a grouping symbol)

The LCD is 15

$15(y - \dfrac{3}{5} - 4) = 15\left(\dfrac{2y}{3}\right)$ Step 2

$15y - 15 \cdot \dfrac{3}{5} - 15 \cdot 4 = 15 \cdot \dfrac{2y}{3}$

▶ Notice Step 2 amounts to multiplying each term on both sides by the LCD. We are using the distributive law.

$15y - 9 - 60 = 10y.$

$\begin{aligned} 15y - 69 &= 10y \\ -10y \phantom{{}-69} &= -10y \\ \hline 5y - 69 &= 0 \\ +69 &= +69 \\ \hline 5y &= 69 \end{aligned}$

Step 3 Add $-10y$ to both sides

Add 69 to both sides
Step 4

$y = \dfrac{69}{5}$ Step 5

Calculator Check: Step 6

69 [÷] 5 [=] [STO]

$\dfrac{5y-3}{5} - 4 =$ $\qquad\qquad\dfrac{2y}{3} =$

[(] 5 [×] [RCL] [−] 3 [)] [÷]   2 [×] [RCL] [÷] 3 [=] 9.2

5 [−] 4 [=] 9.2

Example 3: Solve for T. $T + \dfrac{1}{2}(3T - 5) = 4 - \dfrac{5 - 2T}{3}$

Since $\dfrac{1}{2}(3T - 5) = \dfrac{3T}{2} - \dfrac{5}{2}$

and $-\dfrac{5 - 2T}{3} = -\dfrac{5}{3} + \dfrac{2T}{3}$ (Why?)

the given equation may be written as

$T + \dfrac{3T}{2} - \dfrac{5}{2} = 4 - \dfrac{5}{3} + \dfrac{2T}{3}$ Step 1

The LCD is 6

$6T + 6 \cdot \dfrac{3T}{2} - 6 \cdot \dfrac{5}{2} = 6 \cdot 4 - 6 \cdot \dfrac{5}{3} + 6 \cdot \dfrac{2T}{3}$ Step 2

$$6T + 9T - 15 = 24 - 10 + 4T$$

$$15T - 15 = 14 + 4T$$
$$\underline{-4T \qquad = \qquad -4T}$$

Step 3
Add $-4T$ to both sides

$$11T - 15 = 14$$
$$\underline{+15 = +15}$$
$$11T = 29$$

Add 15 to both sides
Step 4

$$T = \frac{29}{11}$$

Step 5

Calculator Check:

Step 6

29 ÷ 11 = STO

$T + \frac{1}{2}(3T - 5) =$

RCL + 1 ÷ 2 × (3

× RCL − 5) =

4.0909091

$4 - \frac{5 - 2T}{3}$

4 − (5 − 2 × RCL

) ÷ 3 = 4.0909091

Problem Set 3.4

Solve the following equations and check your solutions with your calculator.

Equation   Solution   Check

1. $4x - \frac{1}{2}(3x - 5) = 4 + \frac{2x - 3}{5}$ $\frac{3}{7}$ LHS: 3.5714286

$4x - \frac{3x}{2} + \frac{5}{2} = 4 + \frac{2x}{5} - \frac{3}{5}$ Step 1   RHS: 3.5714286

LCD = 10

$40x - 15x + 25 = 40 + 4x - 6$ Step 2

$25x + 25 = 4x + 34$ Step 3

$21x = 9$ Step 4

$x = \frac{9}{21} = \frac{3}{7}$.

2. $2(3x - 7) = 4x - 7$ LHS:

       Step 1   RHS:

       Step 4
       Step 5

3. $\frac{2x}{3} - \frac{1}{2} = x - \frac{5}{6}$ LHS:

       Step 2   RHS:

       Step 4
       Step 5

4. $\dfrac{5x - 3}{2} = 2x - 7$ _____   LHS: _____

    Step 1   RHS: _____

    Step 2
    Step 4
    Step 5

5. $\dfrac{2}{3}\left(x + \dfrac{5}{2}\right) = \dfrac{7x}{3} - 4$ _____   LHS: _____

    Step 1   RHS: _____

    Step 2
    Step 3
    Step 4
    Step 5

6. $3x + \dfrac{5(x - 3)}{2} = 3 - \dfrac{x + 2}{3}$ _____   LHS: _____

    Step 1   RHS: _____

    Step 3
    Step 4
    Step 5

7. $x - \dfrac{2x}{3} + \dfrac{1}{2} = \dfrac{7}{2} - \dfrac{5x}{6}$ _____   LHS: _____

    Step 2   RHS: _____

    Step 3
    Step 4
    Step 5

8. $3x - \dfrac{3}{4}(4 - 5x) = 3 + \dfrac{2x - 3}{2}$ _____   _____

9. $\dfrac{3x - 1}{2} + \dfrac{x - 3}{4} = \dfrac{1}{2}$ _____   _____

10. $\dfrac{x + 1}{2} + \dfrac{x + 2}{3} = \dfrac{1}{3}$ _____   _____

11. $\dfrac{x - 3}{3} - \dfrac{4x - 1}{8} = \dfrac{1}{8}$ _____   _____

12. $\frac{5}{4}(x - 2) - \frac{7}{6}(x - 3) = \frac{2}{3}$ _____ _____

13. $\frac{2}{3}(x + 1) - \frac{3}{4}(x - 1) = 1$ _____ _____

14. $\frac{4}{5}(x + 2) - \frac{5}{6}(x + 1) = \frac{1}{2}$ _____ _____

15. $\frac{7}{6}(x - 3) - \frac{8}{9}(x - 4) = -\frac{1}{2}$ _____ _____

3.5 Linear Inequalities

Equations use the notion of equality between the two expressions. Another basic notion is that of ordering two expressions. For example,

$5 < 7$	$2x + 4 \leq x + 3$	$8 > 6$	$3x - 6 \geq 2x - 4$
↓	↓	↓	↓
less than	less than or equal	greater than	greater than or equal to

Such order relations between two expressions are called *inequalities*. We may call an *inequality a statement about the order between two expressions*. A solution of an inequality is a value of the unknown which makes the statement true. For example,

$x = -1$ is a solution of $2x + 4 < x + 5$, since $2(-1) + 4 < -1 + 5$ or $2 < 4$ is true.

Check by inspection that any number less than or equal to 1, e.g. $\frac{1}{2}$, 0, -2, -2.5, -3, ... is also a solution of this inequality.

In general an inequality has infinitely many solutions, *but not every number is a solution*. In the inequality $2x + 4 < x + 5$, no number greater than 1 is a solution. For example,

the number "2" is not a solution of $2x + 4 < x + 5$, since $2 \cdot 2 + 4 < 2 + 5$ or $8 < 7$ is not true.

In this section we discuss solutions of those inequalities where the expressions involved are *linear*. We call such inequalities, *linear inequalities*. A linear inequality is solved in a manner which is very similar to the methods we used to solve linear equations. We now make some observations that will help us in solving linear inequalities.

Observation 1: Notice that

$5 < 7$ is true,
$5 + 2 < 7 + 2$ is also true,
and $5 - 4 < 7 - 4$ is also true.

Thus, in general, if we add any number to or subtract any number from both sides of an inequality statement, the inequality is not affected. That is, we obtain an equivalent inequality.

Observation 2: Again, we know

$$10 > 6 \text{ is true,}$$
$$10 \cdot 2 > 6 \cdot 2 \text{ is also true,}$$
$$\text{and } \frac{10}{2} > \frac{6}{2} \text{ is also true.}$$

Thus, in general, if we multiply or divide both sides of an inequality by any positive number, the inequality is not affected.

Observation 3: The above italicized statements in 1 and 2 hold true even for equations. But

▶ *If we multiply or divide both sides of an inequality by a negative number the order between the two expressions thus obtained is reversed.*

For example, we know that

$$7 > 3 \text{ is true}$$
$$\text{but } 7(-2) > 3(-2)$$
$$\text{or } -14 > -6 \text{ is false.}$$

The true statement is

$$-14 < -6.$$

Thus, the result of multiplying both sides of an inequality by a negative number is that the inequality is reversed.

Example 1: *Use of Observations 1–3*

(a) $5 < 7$, therefore $9 < 11$ (add 4 to both sides)
(b) $9 > 3$, therefore $6 > 0$ (add -3 to both sides)
(c) $-4 > -5$, therefore $5 > 4$ (add 9 to both sides or multiply both sides by -1)
(d) $3 > 2$, therefore $12 > 8$ (multiply both sides by 4)
(e) $3 > 2$, therefore $-3 < -2$ (multiply both sides by -1)

Solutions of inequalities may be described graphically by points on the number line.

Example 2: *Graphical representation of solutions*
Graphically, all points to the left of -1 on the number line represent solutions of $x < -1$.

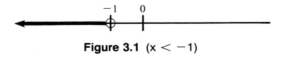

Figure 3.1 ($x < -1$)

In general, we solve inequalities the same way as we solve equations, keeping Observations 1–3 in mind.

Example 3: Solve for x: $3x + 4 < 10$

$$\begin{array}{r} 3x + 4 < 10 \\ -4 \quad -4 \end{array}$$ Add -4 to both sides

$$3x < 6$$
$$\frac{3x}{3} < \frac{6}{3}$$ Divide by 3
$$x < 2$$

Figure 3.2 (x < 2)

Thus, all real numbers less than 2 are solutions of the inequality 3x + 4 < 10. On a number line, all numbers corresponding to the points to the left of 2 are the solutions. Graphically the solutions are represented by the points on the thick portion of the line in Figure 3.2. A small open circle around 2 indicates that the number 2 itself is not a solution.

Example 4: Solve for y: $3y + 4 \geq 2(3 - y)$

$$
\begin{array}{rl}
3y + 4 \geq & 2(3 - y) \\
3y + 4 \geq & 6 - 2y \\
+2y & +2y \\
\hline
5y + 4 \geq & 6 \\
-4 & -4 \\
\hline
5y \geq & 2 \\
y \geq & \dfrac{2}{5}
\end{array}
$$

Add 2y to both sides

Add −4 to both sides

Divide by 5

Thus, all real numbers greater than or equal to $\dfrac{2}{5}$ or .40 are solutions of this inequality. Graphically, the solutions are represented by the point S and all other points to the right of S, as described in Figure 3.3 by the thick line.

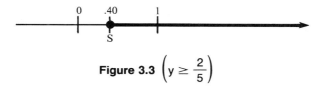

Figure 3.3 $\left(y \geq \dfrac{2}{5}\right)$

A small darkened circle at .4 is used to indicate that the number .4 is in the solution set.

Example 5: Solve for x: $7 - 2x \geq 3x + 4$

$$
\begin{array}{rl}
7 - 2x \geq & 3x + 4 \\
-3x & -3x \\
\hline
7 - 5x \geq & 4 \\
-7 & -7 \\
\hline
-5x \geq & -3 \\
\dfrac{-5x}{-5} \leq & \dfrac{-3}{-5}
\end{array}
$$

Add −3x to both sides

Add −7 to both sides

Divide by −5 and reverse the direction of the inequality

Notice the change in the order relation, from "≥" to "≤", when we divide by a negative number.

$x \leq \dfrac{3}{5}$ or .60

In graphic form, the solution is represented by the thick line in Figure 3.4.

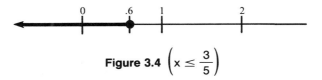

Figure 3.4 $\left(x \leq \dfrac{3}{5}\right)$

Problem Set 3.5

Complete the statements in Problems 1–8.

1. The symbol "$\geq$" stands for _____ .

2. The symbol "$>$" stands for _____ .

3. The symbol "$\leq$" stands for _____ .

4. The symbol "$<$" stands for _____ .

5. An inequality, in general, has _____ number of solutions.

6. It is _____ that if we add or subtract any number on both sides of an inequality the order
   (*true or false*)
   of inequality is not changed.

7. It is _____ that if we multiply or divide both sides of an inequality by any positive number,
   (*true or false*)
   the order of inequality is not changed.

8. It is _____ that if we multiply or divide both sides of an inequality by any negative number
   (*true or false*)
   the order of inequality is not changed.

In Problems 9–25, solve the inequality and also indicate your solution graphically by a thick directed (with arrow) line.

Inequality	Show work here	Answer $x\ (\geq, >, <, \leq)$ __

9. $2x + 3 \geq 4$ $2x \geq 4 - 3$ $\underline{\ x\ } \geq \underline{\ \tfrac{1}{2}\ }$
$2x \geq 1,\ x \geq \tfrac{1}{2}$

10. $2x \geq 6$ 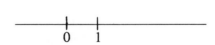 $\underline{\ x\ } \ \underline{\ \ \ \ \ \ \ \ }$

11. $3x + 1 \leq 7$ 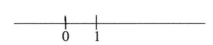 $\underline{\ x\ } \ \underline{\ \ \ \ \ \ \ \ }$

12. $-x \geq 2$ 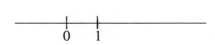 $\underline{\ x\ } \ \underline{\ \ \ \ \ \ \ \ }$

13. $-2x < 4$  $\underline{\ x\ } \ \underline{\ \ \ \ \ \ \ \ }$

14. $-2x + 4 \leq 9$

15. $2x + 3 \geq x - 5$

16. $5y - 4 < 3y + 5$

17. $3x + 4 - x < 2 + x$

18. $7x - 5 + 2x < 6 - 2x$

19. $4(3 - x) < 7 + 3(2 - x)$

20. $3(x - 1) < 4 - (1 - x)$

21. $\dfrac{x + 3}{3} \geq 2 + \dfrac{3 - x}{4}$

22. $2x + \dfrac{3}{2} < x - \dfrac{1}{2}$

23. $3\left(x - \dfrac{1}{2}\right) + x \geq -2x - \dfrac{5}{2}$

24. $\dfrac{3x + 1}{4} - \dfrac{7x + 2}{8} > \dfrac{3}{4}$

25. $\dfrac{3x + 2}{4} - \dfrac{x - 5}{6} \leq \dfrac{1}{6}$

Chapter Summary

1. **Conditional equation**   A statement of equality between two expressions that is true only for some values of the unknown.

2. **Identity**   A statement of equality that is true for all permissible values of the unknown. When we equate one expression to a simplified form of the expression, we get an identity.

3. **Contradiction**   A statement that is never true.

4. **An equation**   An equation means a conditional equation.

5. **Solutions of an equation**   The value or values of the unknown which satisfy the equation or for which the equation statement is true.

6. **Linear Equation**   An equation in which the highest exponent of the unknown is one. A linear equation has only one solution.

7. **Finding solution(s)**   See Rule 3.1.

8. **Linear inequality**   A statement of order between two linear expressions.

9. **Inequality symbols** $\leq$ stands for less than or equal to
   $<$ stands for less than
   $\geq$ stands for greater than or equal to
   $>$ stands for greater than.

10. **Multiplying or dividing by a negative number**   *Obs. 3:* The order of an inequality is changed if we multiply or divide both sides by a negative number.

11. **Solutions of an inequality**   A value of the unknown that makes the order statement true. An inequality, in general, has infinitely many solutions.

Review Problems

Complete the statements in Problems 1–7.

1. The statement $3(x + 2) = 3x + 6$ is an _____ .
   (equation or identity)

2. The statement $3x + 4 = 2x - 5$ is an _____ .
   (equation or identity)

3. The statement $-2x^2 = 8$ is an _____ .
   (equation or identity)

4. The equation $3x - 5 = 9x - 7$ has _____ solution(s).
   one or two

5. The equation $3x^2 + 4x - 9 = 0$ has _____ solution(s).
   one or two

6. $x = 2$ _____ a solution of $3x + 1 = 2x + 3$.
   is or is not

7. $x = -5$ _____ a solution of $x^2 = 25$.
    is or is not

In Problems 8–17, solve for the unknown.

	Equation	Solution		Equation	Solution
8.	$x + 2 = 5$	_____	9.	$x - 7 = 11$	_____
10.	$2x - 7 = 5$	_____	11.	$5x + 3 = -2$	_____
12.	$4 - x = 18$	_____	13.	$9 - 3x = 7$	_____
14.	$\frac{3x}{5} = 12$	_____	15.	$.05x = 2.5$	_____
16.	$\frac{3x}{5} - \frac{1}{2} = \frac{3}{2}$	_____	17.	$\frac{8x}{3} - \frac{5}{7} = \frac{2}{3}$	_____

In Problems 18–21, identify the unknown, set up an equation, and find the value of the unknown.

18. Fifteen percent of what number is 75?
    Equation: Ans. _____

19. What percent of 720 is 36?
    Equation: Ans. _____

20. An increase from 45 to 54 is what percent increase?
    Equation: Ans. _____

21. In a 40 percent off sale you pay $35 for an article. What is the original price?
    Equation: Ans. _____

In Problems 22–30, solve for the unknown and check your solution.

	Equation	Solution	Check		
22.	$3x + 4 = 2x + 5$	_____	LHS:	_____	= _____
			RHS:	_____	= _____
23.	$2x - 4 = 5x + 9$	_____	LHS:	_____	= _____
			RHS:	_____	= _____

24. $4 - 3x = 5 + 4x$ _____ LHS: _____ = _____

   RHS: _____ = _____

25. $2(x - 3) = 4x - 5$ _____ LHS: _____ = _____

   RHS: _____ = _____

26. $\dfrac{2x}{3} - \dfrac{1}{2} = 3x - \dfrac{4}{3}$ _____ LHS: _____ = _____

   RHS: _____ = _____

27. $2x + \dfrac{3}{2} = \dfrac{1}{2}(5 - 3x)$ _____ LHS: _____ = _____

   RHS: _____ = _____

28. $4x + 2\left(3x - \dfrac{5}{2}\right) = \dfrac{7}{2} - 2x$ _____ LHS: _____ = _____

   RHS: _____ = _____

29. $5 + \dfrac{2x - 5}{4} = \dfrac{1}{2}(3x - 4)$ _____ LHS: _____ = _____

   RHS: _____ = _____

30. $2x - \dfrac{3 - 4x}{5} = \dfrac{x}{2} - 4$ _____ LHS: _____ = _____

   RHS: _____ = _____

In Problems 31–38, solve the inequality and also indicate your solution graphically by a thick directed (with arrow) line.

   Inequality   Show work here   Answer
                                               x (≥, >, <, ≤) __

31. $2x \leq 4$ x _____

32. $x + 1 \leq 5$ x _____

33. $x - 5 \geq 9$ x _____

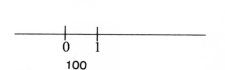

34. $2x - 5 > 9$ x _____

35. $-2x < 4$ 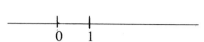 x _____

36. $4 - 2x \geq 3 + 5x$ 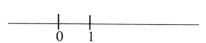 x _____

37. $5 - 2(x - 3) < 3x - 4$ 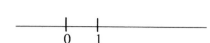 x _____

38. $2x - \frac{3}{2} \geq x - 2\left(5 - \frac{2x}{3}\right)$  x _____

Chapter Test

Solve each of the following problems and match your answer with the responses.

1. Multiply $\sqrt{8} \times \sqrt{3}$ and simplify.
   (a) $2\sqrt{3}$ (b) $4\sqrt{3}$ (c) $2\sqrt{6}$ (d) $3\sqrt{3}$ (e) $2\sqrt{8}$

2. Subtract $5\sqrt{5} - \sqrt{20}$
   (a) $3\sqrt{5}$ (b) $-3\sqrt{5}$ (c) $5 - \sqrt{15}$ (d) $-25\sqrt{20}$ (e) $10\sqrt{15}$

3. If $z = 3x^2 + 6xy + y^2$ and $x = 2$ and $y = 4$, then $z =$
   (a) 36   (b) 64   (c) 70   (d) 76   (e) 100

4. Consider the following statements P and Q:
   P: $(x^{-2})^{-3} = x^6$.
   Q: $2x^{-2} = \frac{2}{x^2}$.
   Which of the above statements are true?
   (a) P only   (b) Q only   (c) P and Q   (d) None

5. Compute: $(-3^2)^2 =$
   (a) -729 (b) 81   (c) -81 (d) 729   (e) None of these

6. What percent of 120 is 30?
   (a) 36   (b) 4   (c) 25   (d) 400   (e) None of these

7. Given: $2p = 14$
   Determine if the entry in column A is less than, greater than, or equal to the entry in column B.
   Column A   Column B
      $p + 7$ [ ]   7
   (a) >   (b) <   (c) =   (d) Cannot be determined

8. If $5PQ = 8n$, then $\frac{PQ}{4} =$

    (a) $2n$ (b) $\frac{8n}{5}$ (c) $3n$ (d) $\frac{2n}{5}$ (e) None of these

9. Solve for y: $-6y - 2 = 1 + 3y$

    (a) 1 (b) $-\frac{1}{3}$ (c) $\frac{1}{3}$ (d) -3 (e) None of these

10. Solve for x: $3(x - 2) - 2x = 4 - 3x$.

    (a) $-\frac{1}{2}$ (b) -1 (c) $-\frac{5}{2}$ (d) $\frac{5}{2}$ (e) None of these

11. Solve: $5x - 2 = 3x + 6$

    (a) 4 (b) 2 (c) $\frac{1}{2}$ (d) 1 (e) $\frac{1}{4}$

12. If $2x + 3\left(2x + \frac{2}{3}\right) = 18$ then $x =$

    (a) 2 (b) -2 (c) 3 (d) -3 (e) None of these

13. If $2x - \frac{(3 - 2x)}{3} = 2$ then $x =$

    (a) $-\frac{3}{4}$ (b) $\frac{3}{4}$ (c) $-\frac{9}{4}$ (d) $\frac{9}{4}$ (e) None of these

14. Solve: $\frac{x + 1}{4} - \frac{2x - 9}{10} = \frac{3}{2}$

    (a) 43 (b) 7 (c) 53 (d) 17 (e) None of these

15. If $\frac{x + 2}{5} = 9$, then $\frac{x - 3}{5} =$

    (a) 4 (b) 6 (c) $8\frac{2}{5}$ (d) 8 (e) $7\frac{2}{5}$

16. Consider the following statements P and Q:

    P: If $8x + 3 = -2x + 8$ then $x = \frac{1}{2}$.

    Q: If $\frac{2}{10}x = \frac{1}{5}$ then $x = 1$.

    Which of the above statements are true?

    (a) Q only   (b) P and Q   (c) P only   (d) None of these

17. If $3 + x = 7$ and $7 - y = 1$, determine if the entry in column A is less than, greater than, or equal to the entry in column B.

    Column A   Column B
        x   [ ]   y

    (a) >   (b) <   (c) =   (d) Cannot be determined

18. Solve: $3x - 1 < 5x + 5$

    (a) $x > -3$ (b) $x < -2$ (c) $x < -3$ (d) $x > -\frac{1}{2}$

19. The solution of $\frac{4}{5}(m-2) < \frac{5}{6}(2m+3)$ is

    (a) $m > \frac{27}{26}$ (b) $m < \frac{-27}{26}$ (c) $m < \frac{-123}{26}$ (d) $m > \frac{-123}{26}$

20. Determine if the entry in column A is less than, greater than, or equal to the entry in column B.

Column A		Column B
The greatest integer x such that $2x + 1 < 5$	[ ]	4

    (a) > (b) < (c) = (d) Cannot be determined

Chapter 4

Introduction to Word Problems

The concepts we developed in the last chapter are very powerful tools for solving problems which we encounter in real life situations. While it is true that the simple problems can often be solved mentally without the use of algebra, it is also true that the use of algebra makes it easy to solve difficult problems.

The problems we use here for illustrative purposes are generally very simple, but build the foundation for the solution of more difficult problems. In this Chapter we will discuss

- (a) translating English statements into algebraic statements,
- (b) a general approach to the solution of word problems with examples on numbers, percentages, ages, and coins,
- (c) problems on ratio and proportion,
- (d) mixture problems,
- (e) distance problems,
- (f) problems involving perimeters of geometric figures, and
- (g) general applications of algebra.

4.1 From English to Symbols

The first step involved in solving any word problem is to understand the problem. You must know the meaning of all the technical terms used in the problem.

➡ Do not try to reduce the problem to symbolic form until you are clear about what is given and what is being asked for in the problem.

A problem which we can solve by the use of algebra will have one or more unknown quantities. When we have identified these unknowns we usually assign them a letter. It is a good idea to assign letters to unknowns so that the assignment conveys meaning. For example, s is a good name for an unknown speed, d is a good name for an unknown distance, and I is a good name for an unknown interest. It is also common to use x to represent an unknown quantity.

> Remember these letters do not represent names of persons or objects. They represent unknown numbers.

When a problem has more than one unknown we often assign a letter, say x, to one of the unknowns and try to express all the other unknowns in terms of the one unknown, x. We illustrate this in the following example.

Example 1: *The use of letters to represent unknowns*

(a) Let two numbers be given, one of which is 5 more than the other. If the smaller one is x, then the larger number is $x + 5$.

(b) One number is twice another number. The numbers may be represented by x and 2x.

(c) The age of a son is x years. The age of the father is 10 years more than five times the age of the son. Then, the age of the father $= 5x + 10$.

(d) If you deposit $x in a savings account that gives you 9% annual interest, then after one year you will have

$x +$ the interest
$= x + 9\%$ of x
$= 1 \cdot x + .09x = 1.09x.$

(e) If my age at the present is x years, then my age after 15 years will be $(x + 15)$ and my age 10 years ago was $(x - 10)$.

Problem Set 4.1

In Problems 1–25 write the numbers in terms of x.

		Answer
1.	3 more than two times x	$2x + 3$
2.	Seventeen less than x	
3.	Twelve less than x	
4.	Seven more than x	
5.	Five times x	
6.	Two more than three times x	
7.	Twice the number x is increased by 9	
8.	One-seventh of x	
9.	Seven more than two-fifths of x	
10.	One-third of two more than x	
11.	Three-eighth of five less than two times x	

12. A son is x years old. What is his father's age if he is 3 years more than two times the age of the son. _____

13. Refer to Problem 12. What will be the ages of the son and the father after 5 years? son _____

    father _____

14. In one year, x dollars in a savings account at 6% interest will increase to _____

15. The marked price of an article is x dollars. How much will you pay for this article in a 40% off sale? _____

16. The marked price of an article is x dollars. How much will you pay if the sales tax charged is 4%? _____

17. The value of x twenty-two cent stamps equals how many cents? _____

18. The value of x quarters equals how many cents? _____

19. What is the perimeter of a square which has a side equal to x feet? _____

20. What is the area of a rectangle with adjacent sides of x feet and (x + 5) feet? _____

21. At the rate of x miles per hour, how far can a car travel in 5 hours? _____

22. Linda can ride her bicycle at an average speed of 15 miles per hour. How long will it take her to travel x miles? _____

23. If Johnson drives at 50 miles per hour, how far will he go in t hours? _____

24. How much silver is there in x pounds of 95% silver? _____

25. How much alcohol is there in (x − 4) gallons of a 65% alcohol solution? _____

4.2 Translating Word Problems into Equations

The importance of linear equations and their solution will be apparent to you in this Section. Many problems that you encounter may be translated into linear algebraic equations. Once this is accomplished, the rest of the job is easy. Just use the tools developed in Chapter 3 to solve the algebraic equation.

There is no standard procedure for translating word problems into algebraic equations but the following guidelines and illustrative examples should help you to accomplish the task.

1. Read the problem carefully, several times if necessary, and *be sure*
    (a) *you understand the meaning of all the technical terms in the problem,*
    (b) *you know what is being asked for in the problem and what information is given.*
2. Identify one of the unknowns by some letter, then express the remaining unknowns in the problem in terms of this letter. *Most of the times, it is convenient*
    (a) *to use a letter for the unknown whose value you want to find,*
    (b) *to construct tables to express other unknowns in terms of the assigned letter.*

3. In the above two steps you analyzed the problem and expressed the unknowns involved in terms of a letter. Translate the relation between the unknowns into an algebraic equation.
4. Solve the equation. Identify the values of the unknowns by use of the solution of the equation. Check whether these values make sense.

Example 1: *The sum of two consecutive integers is 17. Find the integers.*

Step 1. Meaning of the phrase *consecutive integers*. Two integers which differ by one are called consecutive integers.

Step 2. Let one integer be x, then the other integer is $x + 1$.

Step 3. The sum of the two integers $= x + (x + 1)$
$$= 2x + 1$$
This is given to be equal to 17.
Thus, $2x + 1 = 17$.

Step 4.
$$2x + 1 = 17$$
$$-1 = -1$$
$$2x = 16$$
$$x = 8$$
Thus, the integers are 8, and $8 + 1$. That is, 8 and 9.
Check: Their sum $= 8 + 9 = 17$.

Example 2: *The price of an Apple II computer in April 1983 was $1620. This was 40% less than its price in December 1981. What was its price in December 1981?*

Step 1. To find the price in December 1981.
Between December 1981 and April 1983, the price decreased by 40% of its price in 1981. The price in April 1983 is given to be $1620.

Step 2. Let the price in December 1981 be $= x$
Then the price in April 1983
$=$ (price in December 1981) $-$ 40% of (price in December 1981)
$= x - 40\%$ of x

Step 3. Since the price in April 1983 is given to be $1620 we get $x - 40\%$ of $x = 1620$

Step 4.
$$1 \cdot x - .40x = 1620$$
$$(1 - .4)x = 1620$$
$$.6x = 1620$$
$$x = \frac{1620}{.6} = \frac{16200}{6}$$
$$= 2700$$
Thus, the price in December 1981 was $2700. The answer should be larger than $1620 and it is, so we have some confidence that our answer is correct.

Example 3: *Smith is three years older than Linda. The sum of their ages is 85. What are their ages?*

Step 1. It is clear that Smith is older than Linda.

Step 2. Let the age of Linda be x. Then the age of Smith is $x + 3$.

▶ In a problem if you are to find two unknown quantities then it is generally convenient to represent the smaller of the two quantities by a letter.

Step 3. The sum of their ages is 85.
The age of Linda + the age of Smith = 85
   x   +   x + 3   = 85

Step 4. 2x + 3 = 85
         − 3 = −3
         2x = 82
         x = 41

Thus, the age of Linda = 41 and the age of Smith = 41 + 3 = 44
Check: The sum is 41 + 44 = 85.

Example 4: *A waitress has 16 coins in her pocket. She has three more dimes than nickels and two fewer quarters than nickels. How many coins of each kind does she have?*

Step 1. Number of dimes = number of nickels + 3
Number of quarters = number of nickels − 2

Step 2. Let the number of nickels = x,
then the number of dimes = x + 3,
and the number of quarters = x − 2.

Step 3. nickels + dimes + quarters = 16
x + (x + 3) + (x − 2) = 16
x + x + 3 + x − 2 = 16
3x + 1 = 16

Step 4. − 1 = −1
3x = 15
x = 5

Thus, the number of nickels = 5
The number of dimes = 5 + 3 = 8
The number of quarters = 5 − 2 = 3
Check: The total number of coins = 5 + 8 + 3 = 16.

Example 5: *George can do a job in 6 hours and John can do the same job in 4 hours. How long will it take to finish the job, if both of them work together?*

Let both of them take x hours to finish the job.

George does $\frac{1}{6}$ of the job in one hour, therefore in x hours he will do $\frac{x}{6}$ of the job.

John does $\frac{1}{4}$ of the job in one hour, therefore in x hours he will do $\frac{x}{4}$ of the job.

$$\frac{x}{6} + \frac{x}{4} = \text{one complete job}$$
$$\frac{2x}{12} + \frac{3x}{12} = 1$$
$$\frac{5x}{12} = 1$$
$$x = \frac{12}{5}$$

Thus, both together can finish the job in $2\frac{2}{5}$ hours or 2 hours, 24 minutes.

Problem Set 4.2

In Problems 1–6, designate one number as x and express the other numbers in terms of x. Set up and solve an equation to find the numbers.

1. One number is 12 larger than another number. If their sum is 30, find the two numbers.

    1st number: x   2nd number: x + 12
    Equation: x + x + 12 = 30
                2x + 12 = 30
                2x = 18
                x = 9
    The numbers are 9 and 21

2. One number is 5 less than another number. If their sum is 19, find the numbers.

    1st number: _____ 2nd number: _____
    Equation:

    The numbers are _____ and _____

3. One number is four times another and their sum is 35. Find the numbers.

    1st number: _____ 2nd number: _____
    Equation:

    The numbers are _____ and _____

4. One number is one-half another and their sum is 51. Find the numbers.

    1st number: _____ 2nd number: _____
    Equation:

    The numbers are _____ and _____

5. Find two consecutive whole numbers whose sum is 61.

    Answer _____ , _____

6. Find two consecutive odd whole numbers whose sum is 44.

    Answer _____ , _____

In Problems 7–10, the assignment of a letter for one of the unknowns is already made for you. Find the other unknowns in terms of this letter. Set up an equation, solve it and find the numbers. (Let us denote the numbers by N_1, N_2, N_3)

7. One number is three more than another number. A third number is two less than the smaller of the first two numbers. If the sum of the three numbers is 31, what are the numbers?

   | N_1 | N_2 | N_3 | $N_1 + N_2 + N_3 = 31$
   | :---: | :---: | :---: |
   | x + 3 | x | x − 2 |

   Equation: $x + 3 + x + x - 2 = 31$
   $$3x + 1 = 31$$
   $$3x = 30$$
   $$x = 10$$

   The numbers are: <u>13</u>, <u>10</u>, <u>8</u>
   Check: $13 + 10 + 8 = 31$

8. One number is one less than twice another. A third number is seven more than the smaller of the first two numbers. What are the three numbers if their sum is 26?

   | N_1 | N_2 | N_3 | $N_1 + N_2 + N_3 =$
   | :---: | :---: | :---: |
   | _____ | x | _____ |

   Equation:

   The numbers are: ____ , ____ , ____
   Check:

9. The sum of three numbers is 45. The first number is twice the second number and the third number is three times the first number. Find the numbers.

   | N_1 | N_2 | N_3 | $N_1 + N_2 + N_3 =$
   | :---: | :---: | :---: |
   | _____ | x | _____ |

   Equation:

   The numbers are: ____ , ____ , ____

10. The sum of three numbers is 16. The second number is twice the first and the third number is four less than the first. Find the numbers.

In Problems 12–14, set up the problems as in Problem 11 and find the ages.

11. Dan is four years older than Mike. In five years the sum of their ages will be 30. What are their ages at present?

	Dan	Mike
Now	x + 4	x
After 5 yrs	x + 9	x + 5

    $(x + 9) + (x + 5) = 30$
    $$2x + 14 = 30$$
    $$2x = 16$$
    $$x = 8$$

    Ages: *Dan* is $8 + 4$, 12 years old.   *Mike* is 8 years old.

12. Rick is four years older than Bob. Seven years ago the sum of their ages was 14. How old is each of them now?

13. A sister is four years older than her brother. Two years ago the sum of their ages was 20. What are their ages now?

    Sister: _____

    Brother: _____

14. Linda is three years more than twice the age of Jane. In five years the sum of their ages will be 28. How old are they now?

    Linda: _____

    Jane: _____

In Problems 15–18, solve the problem. (Fill in the blanks)

15. The price of gasoline increased by 200 percent between December 1977 and January 1980. If the price per gallon in January 1980 was $1.23 per gallon, what was the price per gallon in December 1977?

    | p_{77} | p_{80} | $x + 200\%$ of $x = 123$ cents |
    | x | x + 200% of x | $x + 2.00x = 123$ |
    | | | $3x = 123$ |
    | | | $x = 41$ |

    p_{77} = price in 1977 = 41¢ per gallon

16. In a 30% off sale, you pay $21.70 for an article. What is its actual price?

    Let the actual price =   Discount =

    What you pay =   = 21.70

    Solve the equation:

    Actual price = _____

111

17. You pay $8.32 for an article. What is the marked price if the rate of sales tax is 4 percent?

    Let the marked price =   Sales tax =

    Solve the equation: = 8.32

                                  Marked price = _____

18. Johnson deposited some money in his savings account at a 6 percent annual rate of interest. What was his deposit if he has $2120 in his savings account after one year?

    Let the deposit =   one year's interest =

    His account after one year =   = $2120

    Solve the equation:

In Problems 19–22, find how many coins of each kind a waitress has. Fill in the blanks in the tables, set up an equation and solve.

19. A waitress has two more dimes than nickels, five fewer quarters than nickels and the total number of coins is 27. (N denotes the number of nickels, D denotes the number of dimes, and Q denotes the number of quarters.)

    N   D   Q   N + D + Q = 27
    x   x + 2   x − 5
    Equation: $x + (x + 2) + (x - 5) = 27$
                        $3x - 3 = 27$
                            $3x = 30$
                              $x = 10$

                      | N = 10, D = 12, Q = 5 |

    Check: Total = 10 + 12 + 5 = 27

20. A waitress has three times as many dimes as nickels, one more quarter than nickels, and a total of 21 coins.

    N   D   Q   N + D + Q =
    x   _____   _____
    Equation:

        Solve: N = ____, D = ____, Q = ____

        Check:

21. A waitress has two more dimes than nickels, three less quarters than dimes, and 31 coins in total.

    N   D   Q   N + D + Q =
    x   _____   _____
    Equation:

        Solve: N = ____, D = ____, Q = ____

        Check:

22. A cashier has 35 more pennies than nickels, 21 less dimes than nickels, 15 more quarters than dimes and 104 coins in total.

       P   N   D   Q   P + N + D + Q =
     _____ x _____ _____

     Equation:

     Solve: P = ____, N = ____, D = ____, Q = ____

     Check:

In Problems 23–24, set up an equation and solve.

23. Len can complete a job in 3 hours and Dan can complete the same job in 2 hours. How long will it take both of them to complete the job?

                                                                            Answer _____

24. David can complete a job in 8 hours, Jane can complete the same job in 6 hours and Mary can do the same in 10 hours. How long will it take for all the three together to complete the job?

                                                                             Answer _____

4.3 Ratio and Proportion Problems

A ratio is a fraction which describes a comparison between two quantities. Consider the following examples.

(a) If 25 out of 30 students in a class passed a test, then the ratio of students who passed, to the total number of students is

$$\frac{25}{30} = \frac{5}{6} \quad \text{(reducing the fraction)}.$$

(b) In the following triangle,

     5 cm
              3 cm
     4 cm

The ratio of the shortest side to the longest side is $\frac{3}{5}$.

(c) If a car travels 72 miles on 5 gallons of gas, then the ratio of miles to gallons is

$$\frac{72}{5}.$$

(d) If a car travels 129 miles in three hours then the ratio of miles to hours is

$$\frac{129}{3}.$$

This ratio also describes the rate at which the car travels and is 43 miles per hour.

(e) If five pounds of sugar costs $1.30, then the ratio of cost to weight is

$$\frac{1.30}{5}.$$

This also describes the cost per pound as $.26.

If two pairs of numbers have the same ratio, they are said to be proportional. An equation that describes such a relationship is called a *PROPORTION*. For example;

(a) The pairs (3,6) and (5,10) are proportional. The proportion is

$$\frac{3}{6} = \frac{5}{10}$$

(b) The pairs (10 pounds, $4) and (20 pounds, $8) are proportional. The proportion is

$$\frac{10}{4} = \frac{20}{8}$$

(c) The pairs (150 miles, 3 hours) and (200 miles, 4 hours) are proportional. The proportion is

$$\frac{150}{3} = \frac{200}{4}$$

In general, if (a,b) and (c,d) are proportional then the proportion equation is

$$\frac{a}{b} = \frac{c}{d}$$

Example 1: *If 10 pounds of potatoes cost $4.50, how many pounds of potatoes can you buy for $1.80?*

The ratio of $4.50 to 10 pounds is the price per pound of the potatoes. Since the price of potatoes per pound remains the same, any other ratio of the price to weight will be proportional. If p denotes the number of pounds you can buy for $1.80, then

$$\frac{4.50}{10} = \frac{1.80}{p}$$

Multiply both sides by 10p,

$$10p \cdot \frac{4.50}{10} = 10p \cdot \frac{1.80}{p}$$
$$4.50p = 18.0$$
$$p = \frac{18.0}{4.50} = 4$$

Example 2: *Suppose a map is scaled so that $\frac{1}{2}$ inch represent 20 miles. How many inches will represent 125 miles?*

The ratio of inches to miles = $\frac{\frac{1}{2}}{20} = \frac{1}{40}$

The proportional ratio for x inches and 125 miles = $\frac{x}{125}$. Thus, $\frac{x}{125} = \frac{1}{40}$ (I)

We solve this equation for x. Multiply both sides of this equation by 40 · 125.

$$40 \cdot 125 \cdot \frac{x}{125} = 40 \cdot 125 \cdot \frac{1}{40}$$
$$40 \cdot x = 125 \cdot 1 \quad \text{(II)}$$
$$x = \frac{125}{40} = \frac{25}{8} = 3\frac{1}{8} \text{ inches.}$$

▶ Notice that we could obtain result II by cross multiplication in Equation I. That is,

$$\frac{x}{125} = \frac{1}{40} \quad \text{or} \quad 40x = 125 \cdot 1$$

The next example involves the knowledge of similar triangles.

Two triangles are similar if their angles are identical. If two triangles are similar then their corresponding sides are proportional. For example, the two triangles ABC and DEF are similar.

Figure 4.1

We can find the length of the side DE by using proportions.

$$\frac{4}{8} = \frac{x}{10} \quad \text{or} \quad 40 = 8x, \; x = 5.$$

Example 3: *Suppose a 3 foot pole casts a shadow of length 4 feet. How long is the shadow for a 6 foot pole?*

From the following figure you can see this is a case of similar triangles.

Thus, $\dfrac{\ell}{6} = \dfrac{4}{3} \quad \text{or} \quad 3\ell = 24$
$\ell = 8.$

Example 4: *A motor bike travels 220 miles on two gallons of gasoline. At the same rate of consumption of the gasoline, how far can the bike travel on 3.6 gallons?*

Let the bike travel d miles on 3.6 gallons. Then, since the rate of consumption is the same, the two ratios

$$\frac{220}{2} \quad \text{and} \quad \frac{d}{3.6}$$

are proportional. Thus,

$$\frac{220}{2} = \frac{d}{3.6} \quad \text{or} \quad 220(3.6) = 2d$$
$$d = 110(3.6) = 396 \text{ miles.}$$

Example 5: *Percent problems as proportion problem*

(a) What is 25% of 300?
The ratio of 25 to 100 = ratio of a number, call it x, to 300

$$\frac{25}{100} = \frac{x}{300} \quad \text{or} \quad 100x = 25 \cdot 300$$
$$x = 75$$

(b) 35 is 28% of what number?
The ratio of 28 to 100 = ratio of 35 to a number x

$$\frac{28}{100} = \frac{35}{x} \quad \text{or} \quad 28x = 3500$$
$$x = \frac{3500}{28} = 125$$

(c) 12 is what percent of 60?
The ratio of a number to 100 = ratio of 12 to 60

$$\frac{x}{100} = \frac{12}{60} \quad \text{or} \quad 60x = 1200$$
$$x = \frac{1200}{60} = 20$$

Problem Set 4.3

In Problems 2–5, find the ratios.

                                                        Answer

1. A car travels 175 miles in $3\frac{1}{2}$ hours. Find the speed in miles per hour.   50 miles per hour

$$\frac{175}{3\frac{1}{2}} = \frac{175}{\frac{7}{2}} = 175 \cdot \frac{2}{7}$$

2. If 5 pounds of sugar costs $1.25, find the cost per pound of sugar.   _____

3. If 5 pounds of almonds costs $2.25, find the ratio of cost to weight. (Note this ratio is the cost per pound)   _____

4. A work study student earns $15 for 4 hours of work. Find the ratio of his earnings to the number of hours he works.   _____

5. Out of 200 students who took a test, 140 passed. Find the ratio of passing students to those who took the test.   _____

In Problems 6–10, solve the proportions to find the value of the unknown.

	Proportion	Show work here	Answer
6.	$\frac{x}{2} = \frac{11}{4}$	$4x = 22, \; x = \frac{22}{4}$	$\frac{11}{2}$
7.	$\frac{5}{2} = \frac{a}{4}$		_____
8.	$\frac{4}{x} = \frac{10}{25}$		_____

9. $\dfrac{14}{3} = \dfrac{x}{4.5}$ _____

10. $\dfrac{22}{9} = \dfrac{55}{x}$ _____

In Problems 11–24, set up proportions and solve for the unknowns.

11. Fifteen gallons of gasoline cost $16.50. What is the cost of seven gallons of gasoline?

    $\dfrac{16.5}{15} = \dfrac{x}{7}$
    $15x = (16.50)7$
    $x = 7.7$ Ans. _____$7.70_____

12. A car traveled 364 miles in seven hours. If it keeps moving at the same speed, how much time will it take to travel another 468 miles?   Ans. _____

13. Suppose a map is scaled so that 1 inch represents 30 miles. How many miles do 2.5 inches represent?   Ans. _____

14. Suppose a map is scaled so that $\dfrac{1}{2}$ inch represent 25 miles. How many inches of the map represents $87\dfrac{1}{2}$ miles?   Ans. _____

15. Smith drove 160 miles in 2 hours. At this rate, how many hours will it take her to travel 440 miles?   Ans. _____

16. Aceto earned $34 in 8 hours. How much should he expect to earn in 13 hours? (round the answer to cents)   Ans. _____

17. If 25 pounds of grass seed covers 3000 square feet of lawn, how many pounds of grass seed are needed to cover 4000 square feet of lawn?   Ans. _____

18. $1\dfrac{1}{2}$ ounces of candy costs 40 cents. What is the cost of 16 ounces of candy? (round the answer to cents)   Ans. _____

19. A twenty-five foot tall pole casts a 15 foot long shadow. How long will be the shadow for a 15 foot tall pole?   Ans. _____

20. What number is 15% of 48?   Ans. _____

21. 72 is 20% of what number? Ans. _____

22. 15 is what percent of 25? Ans. _____

23. Thirty percent of what number is 75? Ans. _____

24. You pay $1.02 as sales tax. What is the sales tax rate if the marked price is $25.50? Ans. _____

4.4 Mixture Problems

In some situations you are given a mixture of different items each of which has different values. You are required to find one or more unknowns involved in these problems. Most such types of mixture problems can be solved conveniently by the use of tables containing the list of items, their weights and values. We illustrate the procedure for solving different types of mixture problems in the following examples.

Example 1: *There are two types of noodles in a mixture, a thin type and a thick type. The thin type costs 15¢ per pound and the thick type cost 25¢ per pound. If 12 pounds of this mixture cost $2.50, find how many pounds of each kind were in the mixture?*

The sum of the weights of the two mixtures is 12 pounds; therefore, if there are x pounds of thin noodles, there must be 12 − x pounds of thick noodles.

The information about the problem can be summarized in the following table.

	Quantity (lbs)	Cost $ per lb	Total cost ($)
Thin	x	$.15	$.15x
Thick	12 − x	$.25	$.25 (12 − x)

Total cost of the thin noodles + Total cost of the thick noodles = Total cost of the mixture

Equation: $.15x + .25(12 − x) = 2.50$

Solve:
$$.15x + 3 − .25x = 2.5$$
$$(.15 − .25)x = 2.5 − 3$$
$$−.10x = −.50$$
$$\text{or } x = \frac{.50}{.10} = 5$$

Thin noodles = 5 pounds
Thick noodles = 12 − x = 12 − 5 = 7 pounds

Example 2: *A waitress has three more nickels than dimes and six more quarters than dimes. If the total value of the coins is $3.65, find the number of coins of each kind.*

	# of coins	Value (¢)	Total value of each kind
N	x + 3	5	5(x + 3)
D	x	10	10x
Q	x + 6	25	25(x + 6)

Total value of N + Total value of D + Total value of Q = Total value of the mixture of coins

$$\text{Equation: } 5(x + 3) + 10x + 25(x + 6) = 365$$
$$5x + 10x + 25x + 15 + 150 = 365$$
$$40x + 165 = 365$$
$$40x = 200$$
$$x = 5$$

D = 5, N = 8, Q = 11
Check: $5 \times 10 + 8 \times 5 + 11 \times 25 = 365$

Example 3: *John has $40,000, part of which he invests at 8% interest and the rest at 7%. His total annual income from these investments is $3100. How much did he invest at each rate?*

	Amount $	Rate of interest	Interest
Invest. 1	x	8%	.08x
Invest. 2	40,000 − x	7%	.07(40,000 − x)

Interest from Investment 1 + Interest from Investment 2 = Total annual income

$$\text{Equation:} \quad .08x + .07(40,000 - x) = 3100$$
$$\text{Solve:} \quad .08x + .07 \times 40,000 - .07x = 3100$$
$$.01x + 2800 = 3100$$
$$.01x = 300$$
$$x = \frac{300}{.01} = 30,000$$

Investment 1 (8% interest) = $30,000
Investment 2 (7% interest) = $10,000

Example 4: *Mixture problems*

(a) Find the amount of salt in 20 grams of a 40 percent salt solution. The amount of salt in the solution = 40% of 20 grams = $.40 \times 20 = 8$ grams

(b) You are given 20 grams of a 40% of salt solution. How many grams of water must be added to this solution to make it 30% solution? Let the amount of water to be added to 20 grams of the 40% salt solution for making it a 30% salt solution be x grams. The information about this problem is condensed in the following table.

Stage	Salt concentration (%)	Weight (grams)	Amount of salt (grams)
Initial	40%	20 grams	8 grams
Water Added	0%	x grams	0
Final	30%	20 + x grams	.30(20 + x)

Since we only add water which contains no salt, the amount of salt before and after adding water is the same. Therefore, we have: $.30(20 + x) = 8$ grams, or

$$\text{Solve:} \quad 6 + .30x = 8$$
$$-6 \qquad\qquad -6$$
$$.30x = 2$$
$$x = \frac{2}{.3} = \frac{20}{3} = 6\frac{2}{3} \text{ grams.}$$

(c) You are given 20 grams of a 40% salt solution. How many grams of salt must be added to this solution to increase its concentration to 50%? Let the amount of salt to be added to 20 grams of the 40% solution for making it a 50% solution be x grams.

Stage	Concentration (%)	Weight (grams)	Amount of salt (grams)
Initial	40%	20 grams	8 grams
Salt	100%	x grams	x grams
Final	50%	(20 + x)	.50(20 + x)

Initially, there are 8 grams of salt in the solution. Then, x grams of pure salt is added. The final solution should have $8 + x$ grams of salt.

Equation: $8 + x = .5(20 + x)$

Solve:
$$8 + x = 10 + .5x$$
$$-.5x \quad\quad -.5x$$
$$8 + .5x = 10$$
$$-8 \quad\quad -8$$
$$.5x = 2,$$
$$x = \frac{2}{.5} = \frac{20}{5} = 4$$

Problem Set 4.4

In Problems 1–12, construct a table, with an equation and solve for the unknown.

1. A 4-pound mixture of beef and pork is worth $2.10 per pound. If beef costs $2.25 per pound and pork costs $1.65 per pound, then how many pounds of each kind is in the mixture?

	Weight	Value	Total price
Beef	x	$2.25	2.25x
Pork	4 − x	$1.65	1.65(4 − x)

   Equation: $2.25x + 1.65(4 - x) = 4 \times 2.10$
   $$.60x = 1.80$$
   $$x = 3$$

   Beef = 3 pounds
   Pork = 1 pound

2. A 15 gallon mixture of white and pink paint costs $12 per gallon. If white paint costs $10 per gallon and pink paint costs $16 per gallon, then how many gallons of white and pink paint are in the mixture?

	Weight	Value	Total price
Pink			
White			

   Equation:

   Solve:

   Pink: _____

   White: _____

3. Orange juice costs 12¢ per ounce, and pineapple juice costs 25¢ per ounce. A thirty-pound mixture of these two juices costs 14¢ per ounce. How many ounces of each kind is used in the mixture? (1 pound = 16 ounces)

Orange: _____

Pineapple: _____

4. Cashew nuts cost $4.00 per pound and raisins cost $4.50 per pound. A four pound mixture of these two costs $4.25 per pound. How many pounds of each type is used in the mixture?

Cashew nuts: _____

Raisins: _____

5. A cashier has six more nickels than dimes, two fewer quarters than nickels. If the total value of these coins is $4.50, find the number of each kind.

N = _____

D = _____

Q = _____

6. A cashier has twice as many nickels as pennies, three more dimes than nickels and four fewer quarters than pennies. If the total value of these coins is $7.70, find the number of each kind.

P = _____ N = _____

D = _____ Q = _____

7. Erica invested $30,000, part of it at 10% interest and the rest at 8%. How much did she put in each of these investments if her total annual income was $2600?

At 10%: _____

At 8%: _____

8. Dan made two investments totalling $15,000. On one investment he made a 20% profit and on the other he suffered a 4% loss. How much did he put in each investment if his total profit was $360?

At 20%: _____

At −4%: _____

9. Barb invested $12,000 at 8% and now wants to invest an additional amount at 13% to make 9% on her total investment. How much does she need to invest at 13%?

Ans. _____

10. How much water should be added to 20 grams of a 15% salt solution to make it a 10% salt solution?

Ans. _____

11. How much salt should be added to 25 grams of a 12% solution to make it a 20% solution?

Ans. _____

12. How much salt should be added to 40 grams of a 20% solution to make it a 25% solution?

Ans. _____

4.5 Distance Problems

In this Section, we will discuss problems which deal with distances covered in a certain amount of time when the objects move at a certain rate.

The rate at which the object moves, also called the *speed,* is often expressed as *"feet per second"* or *"miles per hour".*

The time of travel is often expressed in terms of "seconds" or "hours".

And, the distance covered is often expressed in terms of "feet" or "miles."

The relation between these three quantities is given by the following equation:

$$\boxed{\begin{array}{c} \text{Distance} = \text{Rate} \cdot \text{Time} \\ d = r \cdot t \end{array}}$$

Some of the simple distance problems can be solved directly by use of this equation. This equation contains three variables d, r, and t. If any two of them are given, then the value of the third can be obtained by solving the equation.

Example 1: *The use of* $d = r \cdot t$

(a) Sam rode a horse for $3\frac{1}{2}$ hours moving at a constant speed of 16 miles per hour. How much distance did he cover?

$d = r \cdot t$ $\quad\quad\quad r = 16$ miles per hour $\quad\quad t = 3\frac{1}{2}$ hours

$d = 16 \cdot \frac{7}{2} = 56$

$= 56$ miles

➡ Notice if the rate is in miles per hour then time must be in hours and the distance will be in miles.

(b) If you cover 102 miles in two hours, what is your average speed?

$d = r \cdot t$ $\quad\quad d = 102$ miles $\quad\quad t = 2$ hours

$102 = r \cdot 2$

$r = \frac{102}{2} \frac{\text{miles}}{\text{hours}}$

$= 51$ miles per hour

(c) How much time will you take in covering a distance of 180 miles, if you travel at the rate of 45 miles per hour?

$d = r \cdot t$ $\quad\quad\quad r = 45$ miles per hour $\quad\quad d = 180$ miles

$180 = 45 \cdot t$

$t = \frac{180}{45} = 4$ hours

The problems in Example 1 above are simple and you may have solved them in your head. But, if the problem is not that simple and there is a lot of information in the problem which you have to use for its solution, then, as in mixture problems, constructing a table will be quite useful.

Example 2: *Ali took 10 hours to drive to her farm and 12 hours to drive back. How far is the farm if the average speed while going is five miles per hour faster than the average speed driving back.*

Notice that the distance covered to the farm and back again is the same. This will lead us to an equation.

	Speed	Time	Distance d = rt
Going	x + 5	10	10(x + 5)
Returning	x	12	12x

Thus, $10(x + 5) = 12x$
$10x + 50 = 12x$
$2x = 50$
$x = 25$

The distance = 12x   or   = 10(x + 5)
= 10(25)   = 10(30)
= 300 miles   = 300 miles

Example 3: *Paul jogged at 5 mph towards Cynthia. She jogged towards him at 4 mph, but started 1 hour earlier than he did. How long did each of them jog before they met if they started 13 miles apart?*

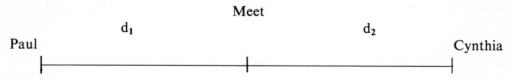

Notice that the total distance covered by both of them before they meet is the same as the distance between them in the start. That is

$d_1 + d_2 = 13$ miles

	Speed	Time	Distance
Paul	5 mph	x hours	5x
Cynthia	4 mph	x + 1 hours	4(x + 1)

Thus, $5x + 4(x + 1) = 13$ miles
$5x + 4x + 4 = 13$
$9x = 9$
$x = 1$

Paul jogged for 1 hour, and Cynthia jogged for 2 hours

Problem Set 4.5

In Problems 1–4, solve for the unknown using the result $d = r \cdot t$.

1. Find the distance covered by a car in $4\frac{1}{2}$ hours if it had been moving at an average speed of 46 miles per hour.

    $r = 46$ mph $\qquad t = 4\frac{1}{2} = \frac{9}{2}$ hours

    Thus, $d = r \cdot t = 46 \cdot \frac{9}{2} = 207$ miles

2. If a car covers 420 miles in 8 hours, find the average speed of the car.

    Speed: _____

3. How much time will it take to travel 540 miles at an average speed of 45 mph?

    Time: _____

4. How much time will it take to travel 330 miles at the speed of 55 mph?

    Time: _____

In Problems 5–10, construct a table, design an equation and solve.

5. It took me 12 hours to drive to New York and 14 hours to drive back home. What is the distance from home to New York if my speed going was 6 mph faster than my speed while coming back?

	Speed	Time	Distance
New York	x + 6	12	12(x + 6)
Back	x	14	14x

    $12(x + 6) = 14x$, $2x = 72$, $x = 36$ mph
    distance $= 14 \times 36 = 504$ miles

6. It takes me $1\frac{1}{2}$ hour to go to the college and 2 hours to come back home from the college. If my average speed when going to the college is 2 mph faster than my speed when coming back, find my speed for going to the college.

	Speed	Time	Distance
To college			
Back			

Answer: _____

7. Sheila drove at 40 mph towards Mary's house. Mary drove at 30 mph towards Sheila's house and started two hours earlier. If their homes are 270 miles apart, how long will it take them to meet each other after Sheila starts towards Mary's house?

	Speed	Time	Distance
Sheila			
Mary			

Shelia: _____

Mary: _____

8. Cindy drove at 50 mph towards John. John drove at 55 mph towards Cindy and started one hour later than Cindy. If John drives for three hours before he meets Cindy, find the distance between their starting points. (You do not need an equation in this question.)

	Speed	Time
Cindy		
John		

Answer: _____

9. John and Mona started riding a bike from the same point at the same time. How long will it take John to be 4 miles ahead if John rides at 12 mph and Mona rides at 9 mph?

	Speed	Time	Distance
John			
Mona			

Answer: _____

10. Smith and Linda started jogging together. Both jogged together for about one hour at the same speed of 5 mph. After that Smith jogged faster by 1 mph and Linda jogged slower by 1 mph. How long will it take them to be 1 mile apart, after they started jogging at different speeds? *(See if you can solve it without setting up an equation. Of course an equation can be used to solve the problem.)*

	Speed	Time	Distance
Smith			
Linda			

Answer: _____

4.6 Problems on Geometric Figures

The prior knowledge you need to solve problems involving geometric figures is provided in the following definitions.

1. **Perimeter of a Rectangle:** The length around the rectangle is called its perimeter. If ℓ is the length and w is the width of a rectangle, as shown in the figure below, then the length around it equals

    $= \ell + w + \ell + w = \ell + \ell + w + w$
    $= 2\ell + 2w$ (combining like terms)

    Thus, P(perimeter) = 2ℓ(length) + 2w(width)

    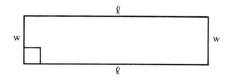

2. **Perimeter of a square:** A square is just a rectangle whose length and width are the same (ℓ = w). Then

    P = 2ℓ + 2w = 2ℓ + 2ℓ = 4ℓ

    Thus, P(perimeter) = 4 · (length of any side) = 4ℓ

Example 1: *Find the dimensions of a rectangle whose perimeter is 22 feet and whose length is 3 feet longer than its width.*

Let the width = x feet
length = (x + 3) feet
Therefore, Perimeter = 2x + 2(x + 3)
That is, 2x + 2(x + 3) = 22
2x + 2x + 6 = 22
4x + 6 = 22
− 6   −6
4x = 16
x = 4 feet

Thus, width = 4 feet, length = 7 feet.

Example 2: *A rectangular field has a 2 foot wide pavement around the outside of the field. If the length of the field is three times its width and the outer perimeter is 96 feet, find the perimeter of the field.*

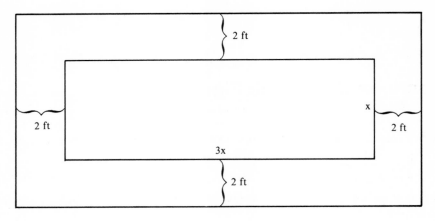

Let the dimensions of the field be x and 3x. Then, the dimensions of the outer boundary are (x + 4) and (3x + 4). Since the outer perimeter is 96 feet, we get

$$2(x + 4) + 2(3x + 4) = 96$$
$$2x + 8 + 6x + 8 = 96$$
$$8x + 16 = 96$$
$$8x = 80$$
$$x = 10 \text{ feet}$$

Therefore, the dimensions of the field are 10 feet and 30 feet. Hence, the inner perimeter = 2 · 10 + 2 · 30 = 20 + 60 = 80 feet

Problem Set 4.6

In Problems 1–10, set up equations and solve.

1. The length of a rectangle is twice its width. Find its dimensions if the perimeter is 84 feet.

    Length: _____

    Width: _____

2. The length of a rectangle is 7 feet more than its width. Find its dimensions if the perimeter is 94 feet.

    Length: _____

    Width: _____

3. The width of a rectangle is 5 feet less than its length. Find its dimensions if its perimeter is 54 feet.

    Length: _____

    Width: _____

4. The length of a rectangle is 4 meters more than its width. A square has one of its sides equal to the perimeter of this rectangle. Find the dimensions of the rectangle if the perimeter of the square is 128 meters.

Length: _____

Width: _____

5. The width of a rectangle is 2 feet less than its length. Two-thirds of the length and one-half of the width together make 6 feet. Find the dimensions of the rectangle.

Length: _____

Width: _____

6. The length of a rectangle is two feet less than three times its width. What are the dimensions of the rectangle if its perimeter is 36 feet?

Length: _____

Width: _____

7. The width of a rectangle is three feet more than one-half its length. What are the dimensions of the rectangle if its perimeter is 66 feet?

Length: _____

Width: _____

8. A rectangular field has a 3 foot wide sidewalk around the outside of the field. If the length of the field is twice its width and the outer perimeter is 84 feet, find the perimeter of the field.

Perimeter: _____

9. A rectangular field has a 3 foot wide sidewalk along its long sides and a 2 foot wide sidewalk along its short sides. If the long side is 18 feet more than the short side and the outer perimeter (including the walks) is 100 feet, find the area of the field.

Area: _____

10. The length of a rectangle is twice its width. The sum of one-third the length and one-half the width is 49 feet. Find the dimensions of the rectangle.

Length: _____

Width: _____

4.7 General Applications of Algebra

In this Section, we discuss some interesting results which require skill in simplifying algebraic expressions, factoring and the solution of quadratic equations. We will prove the results completely if they involve concepts already discussed. Otherwise, we will formulate the proofs in symbols and indicate the concepts to be used to complete the proofs.

The format of discussion in this Section is different from the other sections. There will be no examples backed by problem sets of similar exercises. Our objective, in this Section, is to apprise you of the applicability of algebraic concepts to other than linear equations. The following illustrations are designed keeping this in mind.

Illustration 1: *Number patterns*

Consider any number of two or more digits. For example

    (1) 54   (2) 239   (3) 5427

Step 1. Reverse the digits of these numbers

    (1) 45   (2) 932   (3) 7245

Step 2. For each pair subtract the smaller number from the larger number.

$$
\begin{array}{lll}
(1)\begin{array}{r}54\\-45\\\hline 9\end{array} & (2)\begin{array}{r}932\\-239\\\hline 693\end{array} & (3)\begin{array}{r}7245\\-5427\\\hline 1818\end{array}
\end{array}
$$

We observe that all the numbers in Step 2 are multiples of 9. *We claim this is always true no matter what number we chose in the beginning.*

Let ABC be any number of three digits, where A, B, and C are the digits of the number and not the factors. Since A, B, and C can be any digit, the number ABC is the most general form of a number of three digits.

    The place value of A is 100
    The place value of B is 10
    The place value of C is 1

    Therefore, $ABC = A \cdot 100 + B \cdot 10 + C$ (I)

Step 1. Reverse the digits of the number ABC

    $CBA = C \cdot 100 + B \cdot 10 + A$ (II)

Step 2. Subtract (II) from (I)

$$
\begin{aligned}
ABC - CBA &= A \cdot 100 + B \cdot 10 + C \\
&\; -(C \cdot 100 + B \cdot 10 + A) \\
&\; \overline{100A - 100C + C - A}
\end{aligned}
$$

$$
\begin{aligned}
ABC - CBA &= 99A - 99C \qquad \textit{Combining like terms}\\
&= 99(A - C)\\
&= 9[11(A - C)]
\end{aligned}
$$

Thus, the number we obtain in Step 2 is a multiple of 9 for any value of A, B, and C.

Problems You May Like to Try

Prove that the following results are always true. (Hint: use letters for numbers that can take any value.)

1. Prove the result of Illustration 1 for a four digit number.
2. If the sum of the digits of a number is a multiple of 9 then that number is also a multiple of 9. For example, observe that
$$369 = 9 \cdot 41$$
Also, $3 + 6 + 9 = 18 = 9 \cdot 2$

3. Take any number of three digits such that the difference between the digits with place value 1 and 100 is at least two. (Say 239)
   Then Step 1. Reverse the digits of the number chosen (932)
   Step 2. Of the two numbers subtract the smaller from the larger ($932 - 239 = 693$)
   Step 3. Reverse the digits of the number in Step 2 (396)
   Step 4. Add the numbers of Step 2 and Step 3 ($693 + 396$) =
   The number in Step 4 is always 1089.

4. Is the result of Problem 3 true if you take a number of four digits with the same restriction?

Illustration 2: *Squaring multiples of 5*

Consider the following products
$$15^2 = 15 \times 15 = 225$$
$$25^2 = 25 \times 25 = 625$$
$$35^2 = 35 \times 35 = 1225$$

Can you identify the pattern? One pattern you can observe is

$$a5^2 = b25, \text{ where } b = a(a + 1) \qquad \text{(I)}$$
↓
a is a digit in the tens place

Check: $\underline{25} = 6\underline{25} \qquad 6 = 2(2 + 1)$
$\underline{35}^2 = 1\underline{225} \qquad 12 = 3(3 + 1)$

We claim that result (I) is true for all a's. To prove this claim we have to prove that

$$a5^2 = \underline{b}25 = 100b + 25$$
or $(10a + 5)^2 = 100a(a + 1) + 25$ for all a.

You have to wait until Chapter 6 (Rule 6.1) to prove this result.

Compute the following without long multiplication using the pattern of Illustration 2.

1. $(95)^2 =$ 2. $(105)^2 =$
3. $(4.5)^2 =$ 4. $(1.05)^2 =$

Illustration 3: *Discount and sales tax problem*

Suppose you buy a pair of jeans at 10 percent discount. The marked price is $25. The sales tax is 4 percent. Would you like to pay the tax on the discounted price or get the discount on the total price (marked price + sales tax).

Let us calculate your payment both ways.

I. (Marked price − Discount) + S.Tax on the discounted price
= (25 − 10% of 25) + 4% of the discounted price
= (25 − 2.50) + .04(25 − 2.50)
= 22.50 + .04(22.50)
= 22.50 + .90 = $23.40

II. (Marked price + Tax) − Discount on the total price
= (25 + 4% of 25) − 10% of the total price
= 26 − .10(26)
= 26 − 2.6
= $23.40

Thus, you pay the same amount whether you pay the tax before the discount or after the discount. This may be true only for the specific numbers used in this problem for the marked price discount and the sales tax. What do you think? *We claim this is always true irrespective of the price, discount or the sales tax.*

Let the marked price be $p, the discount be d percent and the sales tax be r percent.

I. (Marked price − Discount) + S.Tax on the discounted price
$= (p - d\% \text{ of } p) + r\% \text{ of the discounted price}$
$= \left(p - \frac{dp}{100}\right) + \frac{r}{100}\left(p - \frac{dp}{100}\right)$
$= p - \frac{dp}{100} + \frac{rp}{100} - \frac{rdp}{10000}$ . . . *Distributive Law*

II. (Marked price + Sales Tax) − Discount on the Total Price
$= (p + r\% \text{ of } p) - d\% \text{ of the total price}$
$= \left(p + \frac{rp}{100}\right) - \frac{d}{100}\left(p + \frac{rp}{100}\right)$
$= p + \frac{rp}{100} - \frac{dp}{100} - \frac{rdp}{10000}$ . . . *Distributive Law*

You see the results in I and II are identical algebraic expressions and therefore have the same value for all p, d and r.

Can you determine which way it is profitable for the storekeeper to calculate your payments?

Review Problems

In Problems 1–15, set up equations and solve.

1. The width of a rectangular lot is 8 feet less than three-fifths of the length. The perimeter of the lot is 400 feet. Find the length and width of the lot.

Length: _____

Width: _____

2. Find a number such that five-sixths of the number is 4 more than two-thirds of the number.

Answer: _____

3. The width of a rectangle is 2 centimeters less than its length. One half of the length plus two-thirds of the width equals 8 centimeters. Find the length and width of the rectangle.

Length: _____

Width: _____

4. Find three consecutive integers such that the sum of the first plus one-third of the second plus three-eighths of the third is 25.

Integers: ___ ___ ___

5. Carol bought a dress at a 35% discount sale for $32.50. What was the original price of the dress?

Answer: _____

6. Jim bought a pair of slacks at a 30% discount sale for $28. What was the original price of the slacks?

Answer: _____

7. The owner of a pizza parlor wants to make a profit of 30% of the cost for each pizza sold. If it costs $3 to make a pizza, at what price should it be sold?

Answer: _____

8. John can finish a job in 14 hours and Mann can finish the same job in 10 hours. How much time will it take to complete the job if both of them work together?

Answer: _____

9. A total of $4000 was invested, part of it at 8% interest and the remainder at 9%. If the total yearly interest amounted to $350, how much was invested at each rate?

At 8%: _____

At 9%: _____

10. Sally is five years older than Margie. In four years the sum of their ages will be 69. How old are they now?

Sally: _____

Margie: _____

11. I have one more dime than nickels and four fewer quarters than nickels. The total value of the coins is $1.90. How many coins of each denomination do I have?

N = ____ , D = ____ , Q = ____

12. I have the same number of nickels as dimes, five fewer quarters than nickels, and their total value is $3.15. How many coins of each denomination do I have?

N = ____ , D = ____ , Q = ____

13. Beef costs $2.10 per pound and pork costs $1.60 per pound. How many pounds of each went into a 10 pound mixture worth $2.00 per pound?

Pork: _____

Beef: _____

14. Ben drove toward Becky at 40 mph. Becky started four hours after Ben did, and drove toward him at 25 mph. If they started out 355 miles apart, how long was Ben driving before they met?

Time: _____

15. It took me 12 hours to drive to my cousin's farm, and 14 hours to drive home. How far away is the farm if my average speed going there was 5 mph more than my average speed driving home?

Distance: _____

Chapter Test

Solve each of the following problems and match your answer with the response.

1. Simplify: $\dfrac{1}{(-4)^{-3}} =$
   (a) 64   (b) 32   (c) -32 (d) -64

2. Simplify: $(-3x^{-3} y^2)^{-3} =$
   (a) $-\dfrac{x^9}{27y^6}$ (b) $\dfrac{x^9}{9y^6}$ (c) $\dfrac{9y^9}{x^9}$ (d) $\dfrac{27x^9}{y^6}$ (e) $\dfrac{x^9}{27y^6}$

3. Consider the following statements P and Q:
   P: If $8 - 2x = 4x - 4$ then $x = 2$.
   Q: If $\frac{7x}{2} = 7$ then $x = 2$.
   Which of the above statements are true?
   (a) Q only   (b) P and Q   (c) P only   (d) None of these

4. Determine if the entry in column A is less than, greater than, or equal to the entry in column B.

Column A		Column B
the least integer x such that $5x + 2 > 38$	[ ]	7

   (a) >   (b) <   (c) =   (d) Cannot be determined

5. If $15 \leq x \leq 25$ and $y - x = 5$, what is the greatest possible value of $x + y$?
   (a) 55   (b) 60   (c) 53   (d) 47   (e) 28

6. Determine which of the following equations or inequalities has (-2) as a solution.
   1. $x - 2 = 0$ 2. $(x + 1)(x + 3) < -1$ 3. $6x < 3x - 4$
   (a) 1 only   (b) 1 and 2 only   (c) 2 and 3 only   (d) 1 and 3 only   (e) None

7. A son is x years old. What is his father's age if he is 3 years more than two times the age of the son?
   (a) $3 + 2x$ (b) $(3 + 2)x$ (c) $3 - 2x$ (d) $3(2 + x)$ (e) None of these

8. Change the following statement to symbols. Sixty pounds more than Linda's weight (h) is 210 pounds.
   (a) $\frac{h + 60}{2} = 210$ (b) $h + 60 = 210$ (c) $h - 60 = 210$ (d) None of these

9. Consider the following statements P and Q.
   P: If $\frac{3x}{a} = b$ then $x = ab + 3$.
   Q: The difference of z and the sum of 3 and y is $3 + y - z$.
   Which of the above statements are true?
   (a) Q only   (b) P and Q   (c) P only   (d) None of these

10. The sum of three numbers is 45. The first number is twice the second number and the third number is three times the first number. Find the numbers.
    (a) 10, 5, 30   (b) 15, 20, 10   (c) 24, 12, 9   (d) None of these

11. Linda is three more than twice the age of Jane. In five years the sum of their ages will be 28. How old are they now?
    (a) 15, 3   (b) 10, 8   (c) 9, 9   (d) 13, 5   (e) None of these

12. Johnson deposited some money in his savings account at a 6 percent annual rate of interest. What was his deposit if he has 2120 dollars in his savings account after one year?
    (a) 2120   (b) 2200   (c) 2100   (d) 2000   (e) None of these

13. A car travels 175 miles in $3\frac{1}{2}$ hours. Find the speed in miles per hour.
    (a) 50   (b) $\frac{1}{50}$ (c) 612.5   (d) 0.2

14. Cindy drove at 50 mph towards John. John drove at 55 mph towards Cindy and started one hour later than Cindy. If John drives for three hours before he meets Cindy, find the distance between their starting points.
    (a) 360 miles   (b) 270 miles   (c) 345 miles   (d) None of these

15. The width of a rectangle is 5 feet less than its length. Find its dimension if its perimeter is 54 feet.
    (a) Length = 16 ft, width = 11 ft   (b) Length = 11 ft, width = 16 ft   (c) Length = 34.5 ft, width = 29.5 ft   (d) Length = 11 ft, width = 6 ft   (e) None of these

Chapter 5

Basic Operations on Algebraic Expressions

In Section 2.6, some elementary concepts about algebraic expressions were introduced. In this Chapter we briefly review Section 2.6 and also discuss,

(a) Multiplication of multinomials,
(b) Division of expressions, and
(c) Long division with polynomials.

5.1 Review

Expressions such as $3x - 5$, $4x^2 - 2x$ are examples of polynomial expressions. The expression $5x^2 - 7x + 2$ has three terms, $5x^2$, $-7x$, and 2. The term $-7x$ has two parts, -7 called the *coefficient* and x called the *literal* part. Some polynomials have special names.

Monomial: A polynomial having only one term, such as $5x$, $3y^2$, $4xyz$, and $-\frac{3y}{2}$.

Multinomial: A polynomial having two or more terms, such as $2x + 3y$ and $5x^2y - 7xy - 9x$.

Example 1: *On definitions*

(a) The expression $9x^3 - 3x^2 + 4x - 5$ is a multinomial and has four terms.
(b) The coefficient of $-3x^2$ is -3, and its literal part is x^2.

The terms in a polynomial expression having the same literal part are called *like terms*. If an expression has two or more like terms then it can be simplified by combining the like terms into a single term. The like terms can be combined by adding their coefficients and affixing the common literal part to the sum.

Example 2: *On simplifying polynomial expressions*

(a) $4x^2 - 7x - x^2 - 9x + 4 = 4x^2 - x^2 - 7x - 9x + 4$
$$= (4-1)x^2 + (-7-9)x + 4$$
$$= 3x^2 - 16x + 4$$

(b) $3x^3 - 7x^2 - x^3 - x + 4x = 3x^3 - x^3 - 7x^2 + 4x - x$
$$= (3-1)x^3 - 7x^2 + (4-1)x$$
$$= 2x^3 - 7x^2 + 3x$$

The addition of two or more polynomial expressions, as discussed in Section 2.6 can be performed easily—just add the terms of the two expressions and simplify by combining like terms. We describe here another method of adding expressions, which we will use in Section 5.2 when multiplication of two multinomials is discussed.

Write the two expressions, one over the other, arranging the terms so that like terms are in line with each other. Then add by combining like terms.

Example 3: *On adding expressions*

(a) Add $2x^2 - 3x + 4$ and $3x^2 - 5x - 7$

$$\begin{array}{r} 2x^2 - 3x + 4 \\ \oplus\ 3x^2 - 5x - 7 \\ \hline 5x^2 - 8x - 3 \end{array}$$

(b) Add $5x^3 - 3x^2 - 7x + 5$ and $3x - 4x^3 - 9$

$$\begin{array}{r} 5x^3 - 3x^2 - 7x + 5 \\ \oplus\ -4x^3 - 0x^2 + 3x - 9 \\ \hline x^3 - 3x^2 - 4x - 4 \end{array}$$

In Section 2.7, we discussed how to subtract one expression from another by use of the Distributive Law. We discuss here another method of subtraction. This method will be used in Section 5.4 when we discuss long division of polynomials.

Write the expression to be subtracted below the other expression so that like terms are in line with each other. Then change the signs of all the terms of the expression to be subtracted, and add as in Example 3. Remember, changing the sign of an expression and adding is equivalent to adding the negative of the expression. And, this is what subtraction means.

Example 4: *On subtracting one expression from the other*

(a) Subtract $4x - 3$ from $3x - 4$

$$\begin{array}{r} 3x - 4 \\ \ominus\ 4x - 3 \\ -\quad + \\ \hline -x - 1 \end{array}$$

(b) Subtract $2x^2 - 3x + 4$ from $7 - x - 2x^2$

$$\begin{array}{r} -2x^2 - x + 7 \\ \ominus\ \ 2x^2 - 3x + 4 \\ -\quad +\quad - \\ \hline -4x^2 + 2x + 3 \end{array}$$

Example 5: *Combined operations*

$$(3x^2 - 5xy + y^2) + (xy - 6y^2) - (2x^2 + 3y^2)$$

$$\begin{array}{r} 3x^2 - 5xy + y^2 \\ \oplus \underline{ xy - 6y^2} \\ 3x^2 - 4xy - 5y^2 \\ \oplus \underline{ 2x^2 + 3y^2} \\ \ominus \ominus \\ \hline x^2 - 4xy - 8y^2 \end{array}$$

In Section 2.7 we discussed how a monomial can be multiplied by a binomial or by a multinomial by use of the Distributive Law. The distributive law states that if a, b and c are any three numbers then

$$a \cdot (b + c) = a \cdot b + a \cdot c$$

Example 6: *Use of the Distributive Law*

(a) $2(x + 4) = 2 \cdot x + 2 \cdot 4 = 2x + 8$

➡ $\quad\quad 2(x + 4) \neq 2x + 4$

(b) $x(2x + 3) = x(2x) + x \cdot 3 = 2x^2 + 3x$
(c) $3x^2(x^2 + 4x + 5) = 3x^2(x^2) + 3x^2(4x) + 3x^2 \cdot 5 = 3x^4 + 12x^3 + 15x^2$

Problem Set 5.1

Complete the statements in Problems 1–5.

1. The expression $3x^2 - 5x - 7$ has _____ terms.

2. A binomial has exactly _____ terms.

3. A monomial has only _____ term.

4. The coefficient of x^2 in $x^3 - 7x^2 + 5x - 8$ is _____ .

5. Like terms in an expression have the same _____ part.

In Problems 6–11, perform the indicated operations.

6. $\begin{array}{r} x^2 - 2x + 7 \\ \oplus \underline{2x^2 - 3x - 5} \\ 3x^2 - 5x + 2 \end{array}$

7. $\begin{array}{r} x^2 + 4x - 9 \\ \ominus \underline{2x^2 - x + 7} \\ - + - \\ \hline -x^2 + 5x - 16 \end{array}$

8. $\begin{array}{r} 2x^2 - 9x - 7 \\ \oplus \underline{x^2 - x + 5} \end{array}$

9. $\begin{array}{r} x^3 - 7x + 5 \\ \ominus \underline{2x^3 + 2x - 11} \end{array}$

10. $\begin{array}{r} 2x^2y - 3x^3 + 2x^2 \\ \oplus \underline{-x^2y - 7x^2} \end{array}$

11. $\begin{array}{r} 4x^2 - 7xy - 5 \\ \ominus \underline{2x^2 - 7 + 4x} \end{array}$

12. Subtract $3x - 5$ from 4

   Ans. _____

13. Subtract $2x^2 - 7x + 4$ from $5x - 3$

   Ans. _____

In Problems 14–20, use the distributive law to perform the multiplication.

14. $3(x - 2) =$ $\underline{\quad 3x - 6 \quad}$ 15. $x(2x - 4) =$ _____

16. $3x(2x - 5) =$ _____   17. $x^2(3x - 7) =$ _____

18. $2(x^2 - 7x + 5) =$ _____   19. $x^2(2x^2 - xy + 3y^2) =$ _____

20. $3x^4(5x^3 - 7x^2 - 8x - 4) =$ _____

In Problems 21–25, simplify the expressions.

Expression	Show work here	Simplified form
21. $x^2 - 2x - 2(x - 3) =$	$x^2 - 2x - 2x + 6$	$\underline{x^2 - 4x + 6}$
22. $x(2x - 3) - 3(2x - 3) =$		
23. $3x^2 - 5 - x(3x - 5) =$		
24. $x(2x - 5) - 4(3x + 7) =$		
25. $x^2(x - 7) - 3x + 2x(x^2 - 7) =$		

5.2 Multiplication of Multinomials

In Section 5.1, we reviewed how to find the product of a monomial and a multinomial by use of the Distributive Law. We now use the same concept for finding the product of two multinomials.

Illustration 1: Find the product $(2x + 3)(3x + 5)$.

$(2x + 3)(3x + 5) = (2x + 3)(AE)$ where AE stands for $3x + 5$.
$\quad (2x + 3)(AE) = 2x(AE) + 3(AE)$ Distributive law
$\quad\quad\quad\quad\quad\quad = 2x(3x + 5) + 3(3x + 5)$
$\quad\quad\quad\quad\quad\quad = 6x^2 + 10x + 9x + 15$ Distributive law
$\quad\quad\quad\quad\quad\quad = 6x^2 + 19x + 15$ Combine like terms

We could proceed as follows:

$$(2x + 3)(3x + 5) = (AE)(3x + 5) \text{ where AE now stands for } 2x + 3.$$
$$\begin{aligned}(AE)(3x + 5) &= (AE)3x + (AE)5 \\ &= (2x + 3)3x + (2x + 3)5 \\ &= 6x^2 + 9x + 10x + 15 \\ &= 6x^2 + 19x + 15\end{aligned}$$

Either way, we get the same answer.

▶ Note: If you learned to compute such products by use of the *foil* method, you may certainly use that method. However, we do not use it here because it is limited to use for those products in which both factors have only two terms. Furthermore, while the *foil* method is an easy rule to remember, it masks the importance of understanding and being able to use the distributive law.

Illustration 2: Find the product $(x + 4)(x^2 + 8x - 7)$

$$(x + 4)(x^2 + 8x - 7) = (x + 4)(AE)$$

where AE stands for the Algebraic Expression $x^2 + 8x - 7$.

$$\begin{aligned}(x + 4)(AE) &= x(AE) + 4(AE) &&\text{Distributive Law} \\ &= x(x^2 + 8x - 7) + 4(x^2 + 8x - 7) \\ &= x^3 + 8x^2 - 7x + 4x^2 + 32x - 28 &&\text{Distributive Law} \\ &= x^3 + (8 + 4)x^2 + (-7 + 32)x - 28 &&\text{Combining like terms} \\ &= x^3 + 12x^2 + 25x - 38 &&\text{(I)}\end{aligned}$$

We could also write

$$(x + 4)(x^2 + 8x - 7) = AE(x^2 + 8x - 7)$$

where AE stands for $x + 4$

$$\begin{aligned}AE(x^2 + 8x - 7) &= (AE)x^2 + (AE)8x - (AE)7 \\ &= (x + 4)x^2 + (x + 4)8x - (x + 4)7 \\ &= x^3 + 4x^2 + 8x^2 + 32x - 7x - 28 \\ &= x^3 + 12x^2 + 25x - 28 &&\text{(II)}\end{aligned}$$

Either way we get the same result.

Notice that we obtained result (I) by multiplying $x^2 + 8x - 7$ first with x and then with 4 and added the two products. In obtaining (I) and (II) above, we displayed our calculations in a horizontal format. The same calculations can be performed in a vertical format and it has the advantage of being simpler to do. The secret is to keep like terms in line.

$$\begin{array}{r} x^2 + 8x - 7 \\ x + 4 \\ \hline \text{times } x = \quad x^3 + 8x^2 - 7x \\ \oplus \text{ times } 4 = \quad 4x^2 + 32x - 28 \\ \hline x^3 + 12x^2 + 25x - 28 \end{array}$$

The vertical method is definitely more convenient if both the expressions contain more than two terms. For example,

$$(2x^2 + 3x - 4)(3x^2 - 2x + 4) =$$

$$\begin{array}{r} 2x^2 + 3x - 4 \\ 3x^2 - 2x + 4 \\ \hline \end{array}$$

times $3x^2 = \quad 6x^4 + 9x^3 - 12x^2$
$\oplus$ times $(-2x) = \quad -4x^3 - 6x^2 + 8x$
$\oplus$ times $4 = \quad\quad\quad\quad\quad 8x^2 + 12x - 16$
$\quad\quad\quad\quad\quad\quad 6x^4 + 5x^3 - 10x^2 + 20x - 16$

Example 1: *Find the product by the vertical method*

(a) $(3x - 4)(4x - 5) =$

$$\begin{array}{r} 3x - 4 \\ 4x - 5 \\ \hline \end{array}$$

times $4x = \quad 12x^2 - 16x$
$\oplus$ times $(-5) = \quad\quad -15x + 20$
$\quad\quad\quad\quad\quad 12x^2 - 31x + 20$

Also by using distributive law we have

$(3x - 4)(4x - 5) = 3x(4x - 5) - 4(4x - 5)$
$\quad\quad\quad\quad\quad\quad\quad = 12x^2 - 15x - 16x + 20$
$\quad\quad\quad\quad\quad\quad\quad = 12x^2 - 31x + 20$

(b) $(3x - y)(x^2 - 2xy + 4) =$

$$\begin{array}{r} x^2 - 2xy + 4 \\ 3x - y \\ \hline \end{array}$$

times $3x = \quad 3x^3 - 6x^2y + 12x$
$\oplus$ times $-y = \quad\quad -x^2y \quad\quad + 2xy^2 - 4y$
$\quad\quad\quad\quad\quad 3x^3 - 7x^2y + 12x + 2xy^2 - 4y$

(c) $(2x^2 + 4x - 7)(-x^2 + 9x - 5) =$

$$\begin{array}{r} 2x^2 + 4x - 7 \\ -x^2 + 9x - 5 \\ \hline \end{array}$$

times $-x^2 = \quad -2x^4 - 4x^3 + 7x^2$
$\oplus$ times $9x = \quad\quad\quad 18x^3 + 36x^2 - 63x$
$\oplus$ times $-5 = \quad\quad\quad\quad\quad -10x^2 - 20x + 35$
$\quad\quad\quad\quad\quad -2x^4 + 14x^3 + 33x^2 - 83x + 35$

Example 2: *The product of more than two factors by use of the distributive Law*

(a) $x(2x - 1)(3x + 4) = [x(2x - 1)](3x + 4)$
$\quad\quad\quad\quad\quad\quad\quad = (2x^2 - x)(3x + 4)$
$\quad\quad\quad\quad\quad\quad\quad = 2x^2(3x + 4) - x(3x + 4)$
$\quad\quad\quad\quad\quad\quad\quad = 6x^3 + 8x^2 - 3x^2 - 4x$
$\quad\quad\quad\quad\quad\quad\quad = 6x^3 + 5x^2 - 4x.$

(b) $2x - 3(x - 2)(4 - x^2) = 2x - 3[x(4 - x^2) - 2(4 - x^2)]$
$\quad\quad\quad\quad\quad\quad\quad\quad = 2x - 3[4x - x^3 - 8 + 2x^2]$
$\quad\quad\quad\quad\quad\quad\quad\quad = 2x - 3[-x^3 + 2x^2 + 4x - 8]$
$\quad\quad\quad\quad\quad\quad\quad\quad = 2x + 3x^3 - 6x^2 - 12x + 24$
$\quad\quad\quad\quad\quad\quad\quad\quad = 3x^3 - 6x^2 - 10x + 24$

Problem Set 5.2

In Problems 1–25, find the products.

1. $(3x^2)(4x) =$ ____12x³____
2. $x^3(4x^4y^3) =$ _____
3. $(-2x)(3x^4y^3) =$ _____
4. $(-2y)(-3x^2) =$ _____
5. $x(2x - 3) =$ _____
6. $2x(4 - 2x) =$ _____
7. $2 - x(x - 3) =$ _____
8. $3x - 2(4x - 5) =$ _____
9. $(2x - 3)(x - 4) =$ 9. _____
10. $(x - 5)(3 + x) =$ 10. _____
11. $(x - y)(2x - y) =$ 11. _____
12. $(2a - b)(b - 5a) =$ 12. _____
13. $(2x + 3y)(4x - 5y) =$ 13. _____
14. $(3x - 5y)(2x - 3y) =$ 14. _____
15. $(y - 4x)(2x - 7y) =$ 15. _____
16. $(5x - 7y)(2x + y) =$ 16. _____
17. $(x^2 - 2x - 5)(2x - 3) =$ 17. _____
18. $(2x^2 - 3xy + 2y^2)(4x - 5y) =$ 18. _____
19. $(a^2 - 3a + b)(2b - 5) =$ 19. _____
20. $(x^2 + x)(x^2 - 2x + 3) =$ 20. _____
21. $(y^2 - y - 4)(2y^2 - 3y + 5) =$ 21. _____
22. $(x^2 + x - 1)(2x^2 - 5) =$ 22. _____
23. $(x^2 + 4)(3x^2 - 4x - 5) =$ 23. _____
24. $(x^2 - 2x + 4)(2x^2 + 5x - 4) =$ 24. _____
25. $(3x^2 - 5xy + 4y^2)(x^2 - 2xy - y^2) =$ 25. _____

5.3 Division by Monomials

When we divide two monomials the problem is simply one of reducing fractions by use of the laws of exponents and by use of our knowledge of reducing numerical fractions. For example,

$$\frac{4x^2y}{2xy} = \frac{(2xy)(2x)}{2xy}$$

Look for the factors common to the numerator and the denominator. Cancel these common factors.

$$\frac{4x^2y}{2xy} = \frac{2xy}{2xy} \cdot (2x) = 1 \cdot 2x = 2x.$$

In problems on division we assume that the unknown never takes on a value which makes the denominator zero.

Example 1: *Divide a monomial by a monomial*

(a) $\dfrac{4x^2}{2x} = \dfrac{2 \cdot 2 \cdot x \cdot x}{2x} = \dfrac{2x}{2x} 2x = 1 \cdot 2x = 2x$

(b) $\dfrac{15x^2y^3}{5xy^2} = \dfrac{5x \cdot y^2 \cdot 3xy}{5x \cdot y^2} = 3xy$

(c) $\dfrac{4x^2y}{12x^3} = \dfrac{x^2y}{3x^3} = \dfrac{x^2y}{x^2 \cdot 3x} = \dfrac{y}{3x}$

To divide a multinomial by a monomial, *we divide each term of the dividend (numerator) by the divisor (denominator).*

That is, $\dfrac{3x^2 + 5xy}{15x} = \dfrac{3x^2}{15x} + \dfrac{5xy}{15x}$. The reason for this is that division by 15x is the same thing as multiplication by its reciprocal $\dfrac{1}{15x}$,

$$\frac{3x^2 + 5xy}{15x} = \frac{1}{15x}(3x^2 + 5xy)$$

$$= \frac{1}{15x}(3x^2) + \frac{1}{15x}(5xy) \qquad \text{Distributive Law}$$

$$= \frac{3x^2}{15x} + \frac{5xy}{15x}$$

$$= \frac{x}{5} + \frac{y}{3}. \qquad \text{Reducing fractions}$$

▶ In general, $\dfrac{a + b + c}{d} = \dfrac{a}{d} + \dfrac{b}{d} + \dfrac{c}{d}$, where $d \neq 0$.

Example 2: *Divide a multinomial by a monomial*

(a) $\dfrac{2x + 4}{2} = \dfrac{2x}{2} + \dfrac{4}{2} = x + 2$

(b) $\dfrac{3x^2 + 6x}{3x} = \dfrac{3x^2}{3x} + \dfrac{6x}{3x} = x + 2$

(c) $\dfrac{4x^3 + 3xy - 2y^2}{6xy} = \dfrac{4x^3}{6xy} + \dfrac{3xy}{6xy} - \dfrac{2y^2}{6xy} = \dfrac{2x^2}{3y} + \dfrac{1}{2} - \dfrac{y}{3x}$

Problem Set 5.3

In Problems 1–12, simplify.

1. $\dfrac{x^2}{x} =$ _____

2. $\dfrac{2x^3}{2x} =$ _____

3. $\dfrac{y^2}{y^5} =$ _____

4. $\dfrac{-x^2}{x} =$ _____

5. $\dfrac{x}{-x^2} =$ _____

6. $\dfrac{-2x}{x^4} =$ _____

7. $\dfrac{12xy^2}{4x} =$ _____

8. $\dfrac{-2x^3}{x^4} =$ _____

9. $\dfrac{-25a^2b^2}{(-5a)^2} =$ _____

10. $\dfrac{-21x^3y^4}{(-xy)^4} =$ _____

11. $\dfrac{(x-2)^3}{(x-2)^5} =$ _____

12. $\dfrac{x^2(x-4)^2}{4x(x-4)^2} =$ _____

In Problems 13–14, mark the correct response.

13. $\dfrac{x-2}{x} =$
    (a) -1 (b) -2 (c) $1 - \dfrac{2}{x}$ (d) none of these

14. $\dfrac{2x-4}{2} =$
    (a) $x-4$ (b) $2x-2$ (c) $x-2$ (d) none of these

In Problems 15–25, divide.

15. $\dfrac{2x+4}{2} = \dfrac{2x}{2} + \dfrac{4}{2} =$ 15. __x + 2__

16. $\dfrac{12-x}{3} =$ 16. _____

17. $\dfrac{x-y}{x} =$ 17. _____

18. $\dfrac{2x - 4x^2}{2x} =$ 18. _____

19. $\dfrac{15x^3 - 10x^2}{5x} =$ 19. _____

20. $\dfrac{4x^2 - 8x}{12x} =$ 20. _____

21. $\dfrac{21ab - 35a^2b^3}{14ab} =$ 21. _____

22. $\dfrac{4 - 2x + 6x^2}{2} =$ 22. _____

23. $\dfrac{5x^3 - 2x^2 + 10}{5x} =$ 23. _____

24. $\dfrac{4x^2y - 6xy^2 + 8x^2y^2}{14xy} =$ 24. _____

25. $\dfrac{24r^2t - 8rt^4 + 4t^2}{-4rt} =$ 25. _____

5.4 Long Division

In order to divide a multinomial by a multinomial we use the method of long division. This method is, in fact, the same one we use for dividing two large numbers. Before we discuss the method of long division, let us do some ground work, define new words and have a fresh look at some of the basic facts of long division.

(1) *Definition of a Polynomial.* A multinomial of the type

$$a_n x^n + a_{n-1} x^{n-1} + a_{n-2} x^{n-2} + \ldots + a_1 x + a_0.$$

where the a_i's are all real numbers and n is a positive whole number, is called a *polynomial of degree n.* For example;

$3x^2 + 2x - 1$ is a polynomial of degree 2.
$2x - 3$ is a polynomial of degree 1.
$4x^3 - 3x^2 - 5x + 6$ is a polynomial of degree 3.

(2) The terms of the polynomials can be arranged in any order. That is,

$$3x^2 - 2x + 4 = 4 - 2x + 3x^2$$
$$= -2x + 3x^2 + 4$$
$$= -2x + 4 + 3x^2$$

are all the same polynomial of degree 2. But, it is usually desirable to write a polynomial so that the powers of the unknown appear in descending order. It is necessary to do this in division problems with polynomials. That is, $-2x + x^2 - 3$ should be expressed as $x^2 - 2x - 3$.

(3) If we divide 25 by 4, we get 6 as the quotient and 1 as the remainder and we can write

$$\dfrac{25}{4} = 6 + \dfrac{1 \rightarrow \text{remainder}}{4 \rightarrow \text{divisor}}$$

$\quad\quad\;\;\downarrow$
$\quad\;\;\text{quotient}$

(4) Any numeral can be expressed as a polynomial written in powers of 10. For example,

$537 = 5 \text{ hundreds} + 3 \text{ tens} + 7$
$\quad\;\; = 5 \cdot 10^2 + 3 \cdot 10 + 7$
$2378 = 2 \text{ thousands} + 3 \text{ hundreds} + 7 \text{ tens} + 8$
$\quad\quad = 2 \cdot 10^3 + 3 \cdot 10^2 + 7 \cdot 10 + 8$

(5) Consider the following two equivalent ways of performing division.

$$\dfrac{693}{33} = \dfrac{600 + 90 + 3}{30 + 3} = \dfrac{6 \cdot 10^2 + 9 \cdot 10 + 3}{3 \cdot 10 + 3}$$

```
         21
    33)693
      -66
       33
      -33
        0
```

```
              20 +  1
   30 + 3 )600 + 90 + 3
            600 + 60
            ─────────
                  30 + 3
                  30 + 3
                  ──────
                       0
```

Thus, we have $\frac{639}{33} = 21$ and also $\frac{600 + 90 + 3}{30 + 3} = 20 + 1 = 21$

If we replace 10 by x in $\frac{6 \cdot 10^2 + 9 \cdot 10 + 3}{3 \cdot 10 + 3}$ we get $\frac{6x^2 + 9x + 3}{3x + 3}$. Let us now compare the division indicated by these two expressions.

```
              20 +  1
   30 + 3 )600 + 90 + 3
            600 + 60
            ─────────
                  30 + 3
                  30 + 3
                  ──────
                       0
```

```
               2x + 1
   3x + 3 )6x² + 9x + 3
           6x² + 6x
           ─────────
                 3x + 3
                 3x + 3
                 ──────
                      0
```

Thus, we see that long division with polynomials is performed in the same way as we do it with numbers. While comparing the two divisions in observation 5, above, you may have noticed how important it is to write the polynomials in descending order of the powers before we perform the division. We further demonstrate the procedure of long division with polynomials in the following examples.

Example 1: *Divide $6x^2 + x - 2$ by $2x - 1$*

```
              3x + 2
    2x - 1 )6x² +  x - 2
            6x² - 3x
            ─────────
                  4x - 2
                  4x - 2
                  ──────
                       0
```

$\frac{6x^2}{2x} = 3x \rightarrow$ goes in the quotient

new dividend, repeat the process

$\frac{4x}{2} = 2 \rightarrow$ goes in the quotient

Thus, $\frac{6x^2 + x - 2}{2x - 1} = 3x + 2$

Example 2: *Divide $27x^3 - 27x^2 + 9x - 1$ by $3x - 1$*

$$\begin{array}{r} 9x^2 - 6x + 1 \\ 3x - 1 \overline{)27x^3 - 27x^2 + 9x - 1} \\ 27x^3 - 9x^2 \\ \hline -18x^2 + 9x \\ -18x^2 + 6x \\ \hline 3x - 1 \\ 3x - 1 \\ \hline 0 \end{array}$$

⊖ times $9x^2 =$
⊖ times $-6x =$
⊖ times $1 =$

$\dfrac{27x^3}{3x} = 9x^2 \rightarrow$ goes in the quotient

$\dfrac{-18x^2}{3x} = -6x \rightarrow$ goes in the quotient

$\dfrac{3x}{3x} = 1 \rightarrow$ goes in the quotient

Thus, $\dfrac{27x^3 - 27x^2 + 9x - 1}{3x - 1} = 9x^2 - 6x + 1$

Example 3: *Divide $3x^2 - 2x + 5$ by $x + 4$*

$$\begin{array}{r} 3x - 14 \\ x + 4 \overline{)3x^2 - 2x + 5} \\ 3x^2 + 12x \\ \hline -14x + 5 \\ -14x - 56 \\ \hline 61 \end{array}$$

⊖ times $3x =$
⊖ times $-14 =$

$\dfrac{3x^2}{x} = 3x$

$\dfrac{-14x}{x} = -14$

The quotient is $3x - 14$ and the remainder is 61. Thus, $\dfrac{3x^2 - 2x + 5}{x + 4} = 3x - 14 + \dfrac{61}{x + 4}$

Example 4: *Divide $x^3 - 8$ by $x - 2$*

The dividend $x^3 - 8$ is a polynomial of degree 3 but there are no terms with x^2 and x as the literal parts. In such situations it is helpful to take $x^3 + 0x^2 + 0x - 8$ as the dividend, instead of $x^3 - 8$. We do this to provide the missing terms with zero coefficient since addition of such terms do not alter the expression. You can easily see why it is convenient. It helps you keep like terms in order. Therefore, $(x^3 - 8) \div (x - 2) =$

$$\begin{array}{r} x^2 + 2x + 4 \\ x - 2 \overline{)x^3 + 0x^2 + 0x - 8} \\ x^3 - 2x^2 \\ \hline 2x^2 + 0x \\ 2x^2 - 4x \\ \hline 4x - 8 \\ 4x - 8 \\ \hline 0 \end{array}$$

⊖ times $x^2 =$
⊖ times $2x =$
times $4 =$

$\dfrac{x^3}{x} = x^2$

$\dfrac{2x^2}{x} = 2x$

$\dfrac{4x}{x} = 4$

Thus, the quotient is $x^2 + 2x + 4$ and the remainder is zero.

Example 5: *Divide $2x^3$ by $2x + 1$*

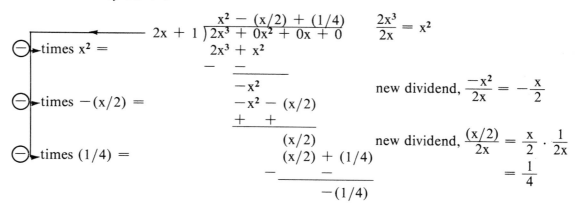

The quotient is $x^2 - \frac{x}{2} + \frac{1}{4}$ and the remainder is $-\frac{1}{4}$.

Thus, $\frac{2x^3}{2x + 1} = x^2 - \frac{x}{2} + \frac{1}{4} + \frac{-(1/4)}{2x + 1}$

Problem Set 5.4

In Problems 1–10, perform the indicated division by the long division method and find the quotient and the remainder.

Division	Show work here	Quotient	Remainder
1. $(x^3 - 1) \div (x - 1)$	$\begin{array}{r} x^2 + x + 1 \\ x - 1 \overline{)x^3 + 0x^2 + 0x - 1} \\ \underline{x^3 - x^2} \\ - + \\ x^2 + 0x \\ \underline{x^2 - x} \\ - + \\ x - 1 \\ \underline{x - 1} \\ - + \\ 0 \end{array}$	$x^2 + x + 1$	0
2. $(x^2 + 1) \div (x - 1)$		_____	_____
3. $(x^2 + 3x - 1) \div (x + 1)$		_____	_____
4. $(x^2 + x - 2) \div (x + 2)$		_____	_____

5. $(2x^2 - 3x + 5) \div (x - 3)$ _____ _____

6. $(2x - x^2 + 3) \div (x - 4)$ _____ _____

7. $(12x^3 + 17x^2 - 13x - 7) \div (2 + 3x)$ _____ _____

8. $(6x^3 + 2x^2 - 3x - 1) \div (3x + 1)$ _____ _____

9. $(x^5 - 1) \div (x - 1)$ _____ _____

10. $(2x^3 + 3x - 1) \div (x + 4)$ _____ _____

Chapter Summary

1. **Monomial** A polynomial expression having only one term.
2. **Multinomial** A polynomial expression having two or more terms.
3. **To add expressions** Just add and simplify by combining like terms.
4. **To subtract one expression from the other** Write the expression to be subtracted below the other expression, lining up the like terms, change the signs of the lower terms and add.
5. **To multiply two expressions** Multiple each term of one expression by the other expression by use of the distributive law, add and simplify.
6. **To divide a monomial by a monomial** Simply reduce the fraction as we do in numerical fractions.
7. **To divide a multinomial by a monomial** Divide each term of the multinomial by the monomial.
   $$\frac{a + b + c}{d} = \frac{a}{d} + \frac{b}{d} + \frac{c}{d}$$
8. **A polynomial** Is an expression of the type
   $$a_n x^n + a_{n-1} x^{n-1} + \ldots + a_o$$
   where all a_i's are real numbers and n is a positive integer, e.g.,
   $$2x + 1 \text{ and } 4x^2 - 3x + 1.$$
9. **To divide one polynomial by another** Arrange the terms of both the polynomials in descending order of the powers of the unknown and use long division as is done with numbers.

Review Problems

In Problems 1–6, perform the indicated operations.

1. $+\ \dfrac{\begin{array}{r}x^2 - 2x + 9\\ 2x^2 - 3x - 8\end{array}}{}$

2. $+\ \dfrac{\begin{array}{r}3x^3 - 14x^2 - 4x - 7\\ 3x^2 - 7x + 9\end{array}}{}$

3. $-\ \dfrac{\begin{array}{r}4x^2 + 3x - 2\\ -x^2 + 5x - 8\end{array}}{}$

4. $-\ \dfrac{\begin{array}{r}x^3 - 3x^2 + 4x - 5\\ x^2 - 2x + 7\end{array}}{}$

5. Subtract 7 from -5 Answer _____

6. Subtract $3x^2 + 4x - 7$ from $2x - 7$ Answer _____

In Problems 7–11, use the distributive law to perform the multiplication.

7. $2(2x - 7) =$ _____

8. $-2(7 - x) =$ _____

9. $8x(2x - 3) =$ _____

10. $-2x(3 - x) =$ _____

11. $2x^2(3x^2 - 4x + 5) =$ _____

Simplify the expressions in Problems 12–14.

Expression	Show work here	Simplified form
12. $2x - 3 + 2(x - 4) =$		_____
13. $2x(x - 4) - 3(x - 4) =$		_____
14. $3x^2(x^2 - 7x + 4) - x(x^2 - 4) =$		_____

In Problems 15–22, multiply by use of the distributive law.

Expression	Show work here	Simplified form
15. $(x + 1)(2x + 1) =$		_____
16. $(2x - 1)(5x - 3) =$		_____
17. $(2 - 3x)(4x - 5) =$		_____

18. $(2x + 4)(x^2 - 5x + 5) =$ _____

19. $(3x - 2)(2x^2 + 4x - 5) =$ _____

20. $(4x - 7)(3x^3 + 2x^2 - 5) =$ _____

21. $(3y^2 + 7)(5y^2 - 8y + 8) =$ _____

22. $(2a^2 - 7a + 9)(5a^2 + 7) =$ _____

In Problems 23–25, multiply by use of the vertical format.

23. $\otimes \dfrac{\begin{array}{r}2x^2 + 3x - 2\\ 2x - 4\end{array}}{}$ Answer _____

24. $\otimes \dfrac{\begin{array}{r}4x^2 - 3x + 5\\ x^2 - 2x + 1\end{array}}{}$ Answer _____

25. $\otimes \dfrac{\begin{array}{r}5x^3 - 3x^2 + 4x - 2\\ -x^2 + 2x - 3\end{array}}{}$ Answer _____

In Problems 26–32, perform the division.

26. $\dfrac{4x^2}{-2x} =$ _____

27. $\dfrac{-15x^3y}{5x} =$ _____

28. $\dfrac{40x^4y^5z^7}{-8x^2y^3} =$ _____

29. $\dfrac{2x + 4}{2} =$ _____

30. $\dfrac{2x^2 - 4x}{2x} =$ _____

31. $\dfrac{x^3 + 4x^2 - 12}{6x} =$ _____

32. $\dfrac{4x^2y - 6x^3y^2 + 12xy^4}{4x^2y^2} =$ _____

In Problems 33–35, perform the division by the long division method. Find the quotient and remainder.

                                                                     Quotient   Remainder

33. $2x + 1 \,\overline{\smash{\big)}\, 2x^2 - 5x - 3}$

34. $x - 2 \,\overline{\smash{\big)}\, 4x^3 - 2x^2 + 5x - 3}$

35. $2x - 1 \,\overline{\smash{\big)}\, 16x^4 - 32x^3 + 24x^2 - 8x + 1}$

Chapter Test

Solve each of the following problems and match your answer with the responses.

1. Simplify: $(3y^2)^2 (2y^2)^4 =$
   (a) $144y^{12}$ (b) $72y^{12}$ (c) $144y^{10}$ (d) $72y^{10}$ (e) None of these

2. Consider the following statements P and Q:
   P: $\sqrt{81} = 9$
   Q: $\sqrt{0.25} = 0.05$
   Which of the above statements are true?
   (a) P only (b) Q only (c) P and Q (d) None

3. Simplify: $\sqrt{18} =$
   (a) $9\sqrt{2}$ (b) $3\sqrt{2}$ (c) $-3\sqrt{2}$ (d) $-9\sqrt{2}$ (e) None of these

4. If $x - \dfrac{(1-x)}{2} = \dfrac{1}{4}$ then $x =$
   (a) $-\dfrac{1}{2}$ (b) $\dfrac{1}{3}$ (c) $\dfrac{1}{2}$ (d) $\dfrac{3}{2}$ (e) None of these

5. The solution of $\dfrac{2}{3}x - \dfrac{2(x+3)}{9} < x + \dfrac{5}{6}$ is
   (a) $x < \dfrac{3}{7}$ (b) $x < \dfrac{27}{7}$ (c) $x < \dfrac{-27}{10}$ (d) $x > \dfrac{-27}{10}$

6. The present ages of 3 people are consecutive whole numbers of years. Eleven years from now their average (arithmetic mean) age will be twice their average now. Determine if the entry in column A is less than, greater than, or equal to the entry in column B.

Column A		Column B
The present age of the youngest person	[ ]	5 years

    (a) >   (b) <   (c) =   (d) Can't be determined

7. Office equipment is purchased for 3,000 dollars and depreciates 10% per year. What is its value after two years of use.
   (a) 2400 dollars   (b) 2700 dollars   (c) 2430 dollars   (d) 2370 dollars

8. Subtract $6(1 + 2x + 3x^2)$ from $x^2 + 1$.
   (a) $-17x^2 - 12x - 5$ (b) $-17x^2 - 12x + 5$ (c) $-17x^2 + 12 - 5$ (d) $-17x^2 + 12x + 5$ (e) None of these

9. Consider the following statements P and Q:
   P: $(2a - b)^2 = 4a^2 - b^2$.
   Q: $2x(x + 4) = 2x^2 + 4$.
   Which of the above statements are true?
   (a) P only   (b) Q only   (c) P and Q   (d) None

10. Like terms in expression have the same _____ part
    (a) Binomial   (b) Literal   (c) Combined   (d) None of these

11. Subtract $3x - 6$ from 8
    (a) $14 - 3x$ (b) $2 - 3x$ (c) $3x - 14$ (d) None of these

12. Combine like terms: $(4x - 3) - (2x - 3) - (5x - 6) =$
    (a) $-3x$ (b) $-3x + 12$ (c) $-3x - 9$ (d) $-3x - 6$ (e) None of these

13. Simplify: $(2x - 3)(x - 6)$
    (a) $2x^2 - 3x - 18$ (b) $2x^2 - 15x + 18$ (c) $2x^2 - 15x - 18$ (d) None of these

14. Find the product: $(y + 2z)(y^2 - 2yz - z^2) =$
    (a) $y^3 + 8z^3$ (b) $y^3 - 3yz^2 + 2z^2$ (c) $y^3 + 3y^2z - 3yz^2 + 2x^2$ (d) None of these

15. $\frac{1}{3}\left(\frac{4}{5}a - \frac{2}{3}b\right) =$
    (a) $\frac{1}{2}a - \frac{1}{3}$ (b) $\frac{4}{15}a - \frac{2}{9}b$ (c) $\frac{1}{2}a + \frac{1}{3}b$ (d) $\frac{4}{15}a + \frac{2}{9}b$ (e) None of these

16. Multiply: $(x + 1)(x^2 - x + 1) =$
    (a) $x^3 - x^2 + 1$ (b) $x^3 + 1$ (c) $x^3 - 1$ (d) $x^3 - x^2 + 1$ (e) None of these

17. Divide: $\frac{49a^5b^2c^3}{7abc^2} =$
    (a) $7a^6b^3c^5$ (b) $\frac{7c}{a^5b}$ (c) $7a^4bc$ (d) $7a^6bc^5$ (e) None of these

18. Simplify: $\dfrac{-42a^2bc - 63ab^2c - 70abc^2}{-7abc}$
    (a) $6a + 9b + 10c$ (b) $-6a^2 - 9bc - 10c$ (c) $7a - 63b - 70c$ (d) None of these

19. Perform the long division $(x^3 + 1) \div (x + 1)$
    The quotient Q and remainder R are:
    (a) $Q = x^2 + 2x - 1, R = 1$ (b) $Q = x^2 + 2x + 1, R = 1$
    (c) $Q = x^2 - x + 1, R = 0$ (d) $Q = x^2 - x + 1, R = 2$ (e) None of these

20. Simplify $\dfrac{-3a^4 - 12a^3}{a + 4}$
    (a) $-3a^4 - 3a^3$ (b) $-3a^3$ (c) 0 (d) $3a^2$

Chapter 6

Special Products and Factoring

These two topics, special products and factoring, are so important that your success in this course will depend on how well these skills are mastered. The products of some expressions are called *special* because they occur frequently in the study of algebra. Factoring is a method we use to write an expression as a product of simple factors. We often encounter special products and the factored form of these expressions.

In this chapter we will discuss

(a) Five special products,
1. $(x + y)(x + y) = (x + y)^2$
2. $(x - y)(x - y) = (x - y)^2$
3. $(x - y)(x + y) = x^2 - y^2$
4. $(x + y)(x^2 - xy + y^2) = x^3 + y^3$
5. $(x - y)(x^2 + xy + y^2) = x^3 - y^3$

(b) Simple factors and the difference of two squares
(c) Critical numbers used for factoring trinomials
(d) Factoring trinomials of the type $x^2 + bx + c$
(e) Factoring trinomials of the type $ax^2 + bx + c$
(f) More on factors
(g) Sum and difference of two cubes
(h) Factoring by completing the square

6.1 Review

We have already discussed what we mean by the phrases; *polynomial expression, terms of a polynomial expression, coefficient of terms, monomial, binomial, trinomial,* and *multinomial.* Since we shall be using these terms quite frequently in this chapter, let us review these once again in the following examples.

Examples: *Definition of terms*

(a) The expression $5x^2 - 6x - 7$ is a trinomial having three terms, namely, $5x^2$, $-6x$, and -7. The coefficient of the term $-6x$ is -6 and its literal part is x. The term -7 has no literal part.

(b) The expression $6x^2 - 8xy$ is a binomial and has two terms, $6x^2$ and $-8xy$. The coefficient of $-8xy$ is -8 and its literal part is xy.

(c) The expression $6x^3 + 4x^2 - 3x + 8$ is a multinomial and has four terms, $6x^3$, $4x^2$, $-3x$ and 8.

(d) The expression $5x^3y$ is a monomial having one term $5x^3y$. Its literal part is x^3y and its coefficient is 5.

Problem Set 6.1

Complete the statements in the following problems.

1. The expression $6x^2 - 5xy + 9y^2$ is called a _____ .

2. The expression $5x^2 - 7xy$ has _____ terms and the literal part of $-7xy$ is _____ .

3. The expression $12xy - 7y^3 - 7y + 2$ has _____ terms and the coefficient of $-7y$ is _____ .

4. The coefficient of xy in $x^2 - 5xy + 6y^2$ is _____ and the literal part of the term $6y^2$ is _____ .

5. The literal part of the term $-2x^2$ in $12x^2y - 2x^2 + 4x$ is _____ .

6.2 Special Products

The special products that we consider in this section are the products of two particular types of binomials. In arithmetic there are many special products that we have to remember. For example, all multiplication tables are tables of special products. But, in algebra there are five types of products which occur frequently. In this section we shall discuss three of these five special products.

1. The product of two *Identical Binomials,* where the binomials are a sum of monomials, i.e.,

    $(x + y)(x + y)$ or $(x + y)^2$

2. The product of two *Identical Binomials,* where the binomials are the difference of monomials, i.e.,

    $(x - y)(x - y)$ or $(x - y)^2$

3. The product of the sum and the difference of two terms, i.e.,

    $(x + y)(x - y)$.

In the following illustrations we develop rules for finding these products. The other two special products are discussed in Section 6.9.

Illustrations:

1. $(x + y)^2 = (x + y)(x + y)$
    $= x(x + y) + y(x + y)$ Distributive Law
    $= x^2 + xy + yx + y^2$ Distributive Law
    $= x^2 + 2xy + y^2$ Combining like terms since $xy = yx$

    (1st term + 2nd term)^2 = (1st term)^2 + 2(product of two terms) + (2nd term)^2

> Note that $(x + y)^2 \neq x^2 + y^2$
> Check: Let $x = 2$ and $y = 3$
> $(2 + 3)^2 = 25$
> $2^2 + 3^2 = 13$ $(2 + 3)^2 \neq 2^2 + 3^2$

The formula $(x + y)^2 = x^2 + 2xy + y^2$ can be given a simple geometric interpretation. Consider the following figure:

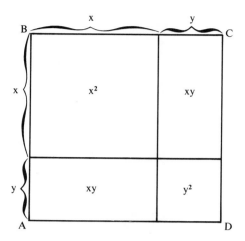

Figure 6.1 Geometric meaning of $(x + y)^2$.

The area of the square ABCD is given by

$$\text{Area of ABCD} = AB \cdot AD$$
$$= (x + y)(x + y)$$
$$= (x + y)^2.$$

The square has also been divided into four subareas and therefore

$$\text{Area of ABCD} = x^2 + xy + xy + y^2$$
$$= x^2 + 2xy + y^2.$$

2. $(x - y)^2 = (x - y)(x - y)$
   $= x(x - y) - y(x - y)$ Distributive Law
   $= x^2 - xy - yx + y^2$ Distributive Law
   $= x^2 - 2xy + y^2$ Combining like terms since $xy = yx$

   Thus, $(\text{1st term} - \text{2nd term})^2 = (\text{1st term})^2 - 2(\text{product of terms}) + (\text{2nd term})^2$.

3. $(x + y)(x - y) = x(x - y) + y(x - y)$
   $= x^2 - xy + yx - y^2 = x^2 - y^2$

   Thus, the product of the sum and the difference of two terms is equal to the difference of the squares of those two terms.

From the illustrations we can deduce the following rule

Rule 6.1

$(a + b)^2 = a^2 + 2ab + b^2$
$(a - b)^2 = a^2 - 2ab + b^2$
$(a - b)(a + b) = a^2 - b^2$

Example 1: *The applications of Rule 6.1*

(a) $(3 + 7)^2 = 3^2 + 2(3 \cdot 7) + 7^2$
$= 9 + 42 + 49 = 100$ Rule 6.1
Check: $10^2 = 100$

(b) $(x + 2)^2 = (x + 2)(x + 2)$
$= x(x + 2) + 2(x + 2)$ Distributive Law
$= x^2 + 2x + 2x + 4 = x^2 + 4x + 4$
Again, by use of Rule 6.1
$(x + 2)^2 = x^2 + 2(2 \cdot x) + 2^2 = x^2 + 4x + 4$

(c) $(2x^2 + 3y^3)^2 = (2x^2 + 3y^3)(2x^2 + 3y^3)$
$= 2x^2(2x^2 + 3y^3) + 3y^3(2x^2 + 3y^3)$ Distributive Law
$= 4x^4 + 6x^2y^3 + 6y^3x^2 + 9y^6$
$= 4x^4 + 12x^2y^3 + 9y^6$
Again, by use of Rule 6.1
$(2x^2 + 3y^3)^2 = (2x^2)^2 + 2(2x^2)(3y^3) + (3y^3)^2$
$= 4x^4 + 12x^2y^3 + 9y^6$

(d) $(x - 2)^2 = (x - 2)(x - 2)$
$= x(x - 2) - 2(x - 2)$ Distributive Law
$= x^2 - 2x - 2x + 4 = x^2 - 4x + 4$
Again, by use of Rule 6.1
$(x - 2)^2 = x^2 - 2(x \cdot 2) + 2^2$
$= x^2 - 4x + 4$

(e) $(5 + 3)(5 - 3) = 5^2 - 3^2 = 25 - 9 = 16$
Check: $8 \cdot 2 = 16$

(f) $(x + 2)(x - 2) = x^2 - 2^2 = x^2 - 4$

(g) $(2x + 3y)(2x - 3y) = (2x)^2 - (3y)^2$
$= 4x^2 - 9y^2$

(h) $(x - 1)(x + 1)(x^2 + 1) = (x^2 - 1)(x^2 + 1)$
$= (x^2)^2 - 1^2$
$= x^4 - 1.$

Problem Set 6.2

Find the products in Problems 1–12.

Factors	Show work here	Product
1. $(x + 2y)^2 =$	$x^2 + 2(x)(2y) + (2y)^2 =$	$x^2 + 4xy + 4y^2$
2. $(2x + y)(2x + y) =$		
3. $(3 + 2x)^2 =$		
4. $(x^2 + 1)^2 =$		
5. $(x - 1)^2 =$		
6. $(2x - 1)(2x - 1) =$		
7. $(2x^2 - 4)^2 =$		

8. $(2x - 3y)^2 =$ _____

9. $(x - 3)(x + 3) =$ _____

10. $(4x - y)(4x + y)$ _____

11. $(2x - 3)(2x + 3) =$ _____

12. $(3x^2 + y^2)(3x^2 - y^2) =$ _____

6.3 Simple Factors

In Chapter 2 and Sections 6.2, we obtained expressions representing the product of two given factors. For example,

$$x(x + 4) = x^2 + 4x$$

Of course, since this is an identity we may also write it as

$$x^2 + 4x = x(x + 4)$$

Here we are writing an expression $x^2 + 4x$ in the factored form "$x(x + 4)$" where "x" and "x + 4" are the factors of $x^2 + 4x$. In this section we shall learn how to rewrite a given expression in factored form.

Illustrations:
1. Factor $3x^2 - 6x$
   Step 1. Factor each term
   $3x^2 = 3 \cdot x \cdot x$
   $-6x = (-2) \cdot 3 \cdot x$
   Step 2. Identify the factors that are common in all the terms
   $3x^2 - 6x = \underline{3 \cdot x} \cdot x - 2 \cdot \underline{3 \cdot x}$
   Step 3. Pull the common factor out and enclose whatever is left in parentheses
   $3x^2 - 6x = 3x(x - 2)$
   Check: $3x(x - 2) = 3x(x) + 3x(-2) = 3x^2 - 6x$

2. Factor $5x^3y^2 - 10x^2y + 15xy^2$
   This expression has three terms
   Step 1. Factor each term
   $5x^3y^2 = 5 \cdot x \cdot x \cdot x \cdot y \cdot y$
   $-10x^2y = (-2) \cdot 5 \cdot x \cdot x \cdot y$
   $15xy^2 = 3 \cdot 5 \cdot x \cdot y \cdot y$
   Step 2. Identify the common factors
   $5x^3y^2 - 10x^2y + 15xy^2 = \underline{5xy}(x^2y) + \underline{5xy}(-2x) + \underline{5xy}(3y)$
   Step 3. Pull the common factor out and enclose whatever is left in parentheses
   $5x^3y^2 - 10x^2y + 15xy^2 = 5xy(x^2y - 2x + 3y)$

3. Factor $x(x + a) + b(x + a)$
   There are two terms in this expression. Each of these two terms has one common factor, that is $(x + a)$. Thus, we pull this common factor out and enclose the rest in parentheses.
   $x\underline{(x + a)} + b\underline{(x + a)} = (x + a)(x + b)$

   Notice that a factor common to all the terms of the given expression can be other than a monomial.

Example 1: *Factor*

(a) $5x + 10x^2 = 1 \cdot \underline{5 \cdot x} + 2 \cdot \underline{5 \cdot x} \cdot x = 5x(1 + 2x)$

(b) $4x^2 - 24x + 20 = \underline{2 \cdot 2} \cdot x \cdot x - \underline{2 \cdot 2} \cdot 2 \cdot 3 \cdot x + \underline{2 \cdot 2} \cdot 5 = 4(x^2 - 6x + 5)$

(c) $12x^3y^7 + 30x^4y^5 = \underline{6x^3y^5} \cdot 2y^2 + \underline{6x^3y^5} \cdot 5x = 6x^3y^5(2y^2 + 5x)$

Example 2: Factor $2x(x + 4) - 3(x + 4)$

This expression has two terms and $x + 4$ is a factor common in both the terms.
Thus, $2x(\underline{x + 4}) - 3(\underline{x + 4}) = (x + 4)(2x - 3)$.

Problem Set 6.3

In Problems 1–15, write the expressions in factored form.

	Expression	Factoring each term	Factors
1.	$6x^2y - 9x^3y^2 =$	$2 \cdot \underline{3 \cdot x \cdot x \cdot y} - 3 \cdot \underline{3 \cdot x \cdot x} \cdot x \cdot \underline{y} \cdot y$	$\underline{3x^2y(2 - 3xy)}$
2.	$6x - 24 =$		
3.	$2y^2 + 4 =$		
4.	$2xy - 4y^2 =$		
5.	$x^2y + y^2x =$		
6.	$xy^3 + x^3y =$		
7.	$21y^2 - 14xy =$		
8.	$4x^2y + 12xy^2 =$		
9.	$3x^2 - 6x^2y =$		
10.	$3x^3 - 12xy^2 - 9x^2 =$		
11.	$14x^4 + 21x^3 =$		
12.	$2x^2y^2 + y^3x =$		
13.	$15a^9b^5 + 25a^{20}b^4 =$		
14.	$12x^5 - 36x^7 =$		
15.	$14x^5 - 21x^4 + 35x^3 =$		

In Problems 16–27, write the expressions in factored form.

	Expression	Factors
16.	$x(x + 1) - 2(x + 1) =$	$\underline{(x + 1)(x - 2)}$
17.	$2x(4x - 5) - 3(4x - 5) =$	

18. $x(2x - y) - 7(2x - y) =$

19. $3x(4 - 2x) - 5(4 - 2x) =$

20. $(x + y)^3 + 3(x + y)^2 =$

21. $21(x - 4)^2 + 8(x - 4) =$

22. $2x^2(3x - 7) + y^2(3x - 7) =$

23. $5xy(3y^2 + 4) + 15x(3y^2 + 4) =$

24. $3x(x - 3y) - 5y(x - 3y) =$

25. $2x^2(2x - 7y) - y(7y - 2x) =$

26. $x(2x - y) - 3y(y - 2x) =$

27. $4x(2 - y) + 3(y - 2) =$

6.4 Factors of a Difference of Two Squares

We showed in Section 6.2 that
$$(x + y)(x - y) = x^2 - y^2$$
Therefore, it must be true that
$$x^2 - y^2 = (x + y)(x - y)$$
That is: A difference of the squares of two quantities equals the product of the sum and difference of these two quantities.

▶ $\quad x^2 - y^2 \neq (x - y)^2$

Example 1: $x^2 - y^2 = (x + y)(x - y)$

(a) $x^2 - 4 = (x)^2 - (2)^2 = (x + 2)(x - 2)$

(b) $4x^2 - 9y^2 = (2x)^2 - (3y)^2$
$\qquad = (2x + 3y)(2x - 3y)$

(c) $2x^2y - 8y^3 = 2y(x^2 - 4y^2)$
$\qquad = 2y(x - 2y)(x + 2y)$

Problem Set 6.4

Factor each of the following expressions.

	Expression	Show work here	Factored Form
1.	$2x^2 - 18 =$	$2(x^2 - 9) = 2(x^2 - 3^2) =$	$2(x + 3)(x - 3)$
2.	$y^2 - 16 =$		
3.	$3a^2 - 12 =$		
4.	$9x^2 - 4 =$		
5.	$64 - x^2 =$		
6.	$x^3 - x =$		
7.	$x^4 - x^2 =$		
8.	$4x^2 - 9y^2 =$		
9.	$25 - 16x^2 =$		
10.	$4 - x^2y^2 =$		
11.	$169 - y^2 =$		
12.	$4x^2 - 36y^2 =$		
13.	$ab^2 - a^3 =$		
14.	$30xy^3 - 120yx^3 =$		
15.	$x^4 - a^4 =$		
16.	$y^4 - 16x^4 =$		
17.	$3a^4 - 48 =$		
18.	$5x - 80x^5 =$		
19.	$243x^6 - 3x^2 =$		
20.	$25x^2 - (y - 2)^2 =$		
21.	$2(x - y)^2 - 8 =$		
22.	$4(2x - y) - (2x - y)^3 =$		
23.	$(3x - y)^3 + 9(y - 3x) =$		
24.	$9x^2 - 25(y - 2)^2 =$		

6.5 Critical Integers

Whenever we multiply two linear expressions with terms having the same literal part, we obtain a trinomial. For example $(2x + 3)(3x - 5) = 6x^2 - x - 15$. Of course, we may also write $6x^2 - x - 15 = (2x + 3)(3x - 5)$. The linear expressions "$2x + 3$", and "$3x - 5$" are called the factors of the trinomial $6x^2 - x - 15$. In this section and the next section we will develop methods for factoring an expression of the type $ax^2 + bx + c$, where a, b and c are integers. The method we use for factoring such a trinomial requires us to be able to find two numbers whose sum and product are given. Once these two numbers are obtained, factoring $ax^2 + bx + c$ becomes extremely simple. We shall call these numbers **critical integers**. In this section we discuss how to find critical integers.

Illustrations
1. Find two integers whose product is −6 and whose sum is −1.
    Step 1. *Determine the signs of the critical integers.*
        Since the product (p) is negative, the integers are of opposite signs—one Critical Integer (CI) is positive and the other is negative.
        Since the sum (s) is negative, the integer with larger absolute value is negative.
    Step 2. *Find all possible integer pairs of factors of −6*
        (−1,6), (1,−6), (2,−3), (−2,3).
    Step 3. *The pair whose sum is −1 is (−3,2)*
        Thus, the critical integers are −3, and 2.

2. Find the integers whose product is −12 and whose sum is 4.
    Step 1. Determine the signs of the critical integers.
        Since the product is negative, the two integers are of opposite signs.
        Since the sum is positive, the integer with larger absolute value is positive.
    Step 2. Find all possible integer pairs of factors of −12
        (−1,12), (1,−12), (2,−6), (−2,6), (3,−4), (−3,4).
    Step 3. The pair whose sum is 4 is (−2,6)
        Thus, the critical numbers are −2 and 6.

3. Find the integers whose product is 20 and whose sum is −9.
    Step 1. The product is positive, therefore the numbers are of like signs.
        Since the sum is negative, both the numbers are negative.
    Step 2. All possible integer factors of 20 which have negative factors and (−1,−20), (−2,−10), (−4,−5).
    Step 3. The pair whose sum is −9 is (−4,−5)
        Thus, the critical integers are −4 and −5.

Problem Set 6.5

In the following problems, find the critical integers for the given product and sum.

	Product-Sum	Critical	Integers		Product-Sum	Critical	Integers
1.	p = 3 s = 4	1	3	2.	p = 5 s = 6		
3.	p = 5 s = −6			4.	p = 6 s = −5		

5. p = −6
   s = −1 _____ _____

6. p = −6
   s = 1 _____ _____

7. p = 12
   s = 8 _____ _____

8. p = 12
   s = −8 _____ _____

9. p = −12
   s = −1 _____ _____

10. p = −12
    s = −11 _____ _____

11. p = 24
    s = −11 _____ _____

12. p = 24
    s = −14 _____ _____

13. p = 15
    s = −8 _____ _____

14. p = 48
    s = −19 _____ _____

15. p = −48
    s = −22 _____ _____

6.6 Factoring Trinomials of the Type $x^2 + bx + c$

In the expression $ax^2 + bx + c$, the coefficient of x^2 is called the *leading coefficient*. If the leading coefficient in this expression is one then the expression is $x^2 + bx + c$. Most of the expressions of this type can be factored easily by use of Critical Integers, as defined in Section 6.5, where the product is "c" and the sum is "b". *In this section we shall consider only expressions for which the CIs exist.*

In the following example, the procedure for factoring $x^2 + bx + c$ is illustrated.

Consider the expression $x^2 + 5x + 6$. The factors of this expression must be of the type $x + q$ and $x + r$. Why? That is, $x^2 + 5x + 6 = (x + q)(x + r)$. But,

$$(x + q)(x + r) = x(x + r) + q(x + r)$$
$$= x^2 + rx + qx + qr$$
$$= x^2 + (r + q)x + qr$$

Thus, $x^2 + 5x + 6 = x^2 + (r + q)x + qr$

In order for this equation to be true for all x we must have

$$r + q = 5$$
and $$rq = 6$$

Thus, if we know two numbers q and r such that their product is 6 and sum is 5 then we know the factors of $x^2 + 5x + 6$ are $x + q$ and $x + r$. As discussed earlier q and r are the critical integers. In this case the critical integers (q and r) are 3, and 2. We conclude

$$x^2 + 5x + 6 = (x + 3)(x + 2).$$

The observations made above for factoring trinomials with leading coefficient 1 can be summarized in the following rule.

Rule 6.2

To factor $x^2 + sx + p$:

Step 1. Find two numbers (the critical integers) whose product is p and whose sum is s.

Step 2. If the numbers found in Step 1 are a and b, then the factored form will be
$x^2 + sx + p = (x + a)(x + b)$

Example 1: Factoring trinomials of the form $x^2 + sx + p$.

(a) Factor $x^2 + 3x + 2$.
Step 1: Two numbers whose product is 2 and whose sum is 3 are 1 and 2.
Check: $1 \cdot 2 = 2, 1 + 2 = 3$
Step 2: $x^2 + 3x + 2 = (x + ?)(x + ?)$
$= (x + 1)(x + 2)$.

(b) Factor $x^2 - x - 6$.
Step 1: Two numbers whose product is -6 and whose sum is -1 are 2 and -3.
Check: $(2)(-3) = -6, 2 + (-3) = -1$.
Step 2: $x^2 - x - 6 = (x + ?)(x + ?)$
$= (x + 2)(x - 3)$.

(c) Factor $x^2 + 4x - 12$.
Step 1: Two numbers whose product is -12 and whose sum is 4 are 6 and -2.
Check: $(6)(-2) = -12, 6 + (-2) = 4$.
Step 2: $x^2 + 4x - 12 = (x + 6)(x - 2)$.

(d) Factor $x^2 - 9x + 20$.
Step 1: The critical integers are -5 and -4.
Check: $(-5)(-4) = 20, (-5) + (-4) = -9$.
Step 2: $x^2 - 9x + 20 = (x - 5)(x - 4)$.

Example 2: Expressions which cannot be factored in the rationals.

(a) Factor $x^2 + 2x + 2$.
We cannot find critical integers for the expression. That is, there are not two integers whose product is 2 and whose sum is 2. Therefore, this expression cannot be factored in the rationals.

(b) Factor $x^2 + 4$.
We cannot find critical integers for this expression. That is, there are not two integers whose product is 4 and whose sum is 0. Therefore, this expression cannot be factored in the rationals.

➡ $\quad x^2 + 4 \neq (x + 2)^2$

Example 3: Factoring trinomials of the form $x^2 + sxy + py^2$.

(a) Factor $x^2 - 4xy + 3y^2$.
We proceed exactly as we have done in Example 1. That is, we find two integers whose product is 3 and whose sum is -4.
Step 1: The critical integers are -1 and -3.
Check: $(-1)(-3) = 3, (-1) + (-3) = -4$.
Step 2: $x^2 - 4xy + 3y^2 = (x - ?y)(x - ?y)$
$= (x - y)(x - 3y)$.

(b) Factor $x^2 - 3xy - 18y^2$.
Step 1: The critical integers are 3 and -6.
Check: $(3)(-6) = -18, 3 + (-6) = -3$.
Step 2: $x^2 - 3xy - 18y^2 = (x + ?y)(x + ?y)$
$= (x + 3y)(x - 6y)$.

Problem Set 6.6

Factor the expressions in Problems 1–22.

    Expression   Factors   Expression   Factors

1. $x^2 + 3x + 2 =$ $(x + 2)(x + 1)$
    product (p) = 2   Critical integers (CI) are 2 and 1.
    sum (s) = 3

2. $x^2 + 4x + 3 =$ ( )( )
    p =   CI =
    s =

3. $x^2 + 5x + 6 =$ ( )( )
    p =   CI =
    s =

4. $x^2 - x - 6 =$ ( )( )
    p =   CI =
    s =

5. $x^2 + x - 6 =$ ( )( )
    p =   CI =
    s =

6. $x^2 - 2x - 8 =$ ( )( )
    p =   CI =
    s =

7. $x^2 - 2x - 48 =$ ( )( )
    p =   CI =
    s =

8. $x^2 - 2x - 15 =$ ( )( )
    p =   CI =
    s =

9. $x^2 - 8x + 15 =$ ( )( )
    p =   CI =
    s =

10. $x^2 - 10x + 24 =$ ( )( )
     p =   CI =
     s =

11. $x^2 - 3x - 18 =$ ( )( )
     p =   CI =
     s =

12. $h^2 - 2h - 24 =$ ( )( )
     p =   CI =
     s =

13. $m^2 + 13m + 36 =$ ( )( )
     p =   CI =
     s =

14. $x^2 - 13x + 36 =$ ( )( )
     p =   CI =
     s =

15. $x^2 + 9xy + 14y^2 =$ ( )( )
     p =   CI =
     s =

16. $x^2 - 9xy + 14y^2 =$ ( )( )
     p =   CI =
     s =

17. $c^2 + cd - 20d^2 =$ ( )( )
     p =   CI =
     s =

18. $x^2 - 10xy + 16y^2 =$ ( )( )
     p =   CI =
     s =

19. $x^2 - 4 =$ ( )( )
     p =   CI =
     s =

20. $x^2 - 9 =$ ( )( )
     p =   CI =
     s =

21. $2x^2 - 8 =$ ( )( )
     p =   CI =
     s =

22. $ax^2 - ax =$ ( )( )
     p =   CI =
     s =

In Problems 23–28, determine, using critical integers, those expressions which cannot be factored.

	Expression	Critical Integers	Conclusion
23.	$x^2 + 9$ $\quad$ $p = 9$ $\quad$ $s = 0$	no critical integers	Cannot be factored in rationals
24.	$x^2 + x + 1$ $\quad$ $p =$ $\quad$ $s =$		
25.	$x^2 + 4x + 5$ $\quad$ $p =$ $\quad$ $s =$		
26.	$x^2 + 3x + 3$ $\quad$ $p =$ $\quad$ $s =$		
27.	$x^2 - 5x + 6$ $\quad$ $p =$ $\quad$ $s =$		
28.	$x^2 - x + 1$ $\quad$ $p =$ $\quad$ $s =$		

6.7 Factoring a Trinomial of the Type $ax^2 + bx + c$

Even for factoring a trinomial $ax^2 + bx + c$ where the leading coefficient is not 1 ($a \neq 1$), the main step is to find the critical integers. But "p" in this case is $a \cdot c$ (the product of the leading coefficient and the constant term). The method explained in the following illustrations not only makes factoring of $ax^2 + bx + c$ easy and straightforward, but also assures that you have the right factors, because of a built in check point. In this Section we will consider the expressions for which the CIs exist. The expressions for which the CIs do not exist will be discussed in Section 6.10.

Illustrations:
1. Factor $2x^2 + 7x + 3$

   Step 1. Define "p" and "s" as follows
   $p = (2)(3) = 6 \quad \rightarrow$ the product of the leading coefficient and the constant term
   $s = 7 \quad \rightarrow$ the coefficient of the first degree term

   Our problem is to find the critical integers. That is, we want two integers whose sum is 7 and whose product is 6. In this case the critical integers are 6 and 1.

   Step 2. Use the critical integers to break the first degree term into two parts.

   $2x^2 + 7\underline{x} + 3 = 2x^2 + \underline{6x + 1 \cdot x} + 3$

   Step 3. Factor, separately, the first two and the last two terms.

   $\underline{2x^2 + 6x} + \underline{1 \cdot x + 3} = 2x(x + 3) + 1(x + 3)$

Step 4. **Check point and final factors**

Check Point

In Step 3, after factoring, there are two terms. Both of these terms have one factor which is a binomial. If this factor in both the terms is identical then you are doing all right. Or else, go back to Step 1 and find your error.

If the two terms in Step 3 have a common factor then, as in Section 6.3, pull this factor out and enclose whatever is left in parentheses.

$$2x(x + 3) + 1(x + 3) = (x + 3)(2x + 1)$$

Example 1: Factor $3x^2 + 8x - 3$

Step 1. $p = 3(-3) = -9, s = 8$
The critical integers are -1 and 9

All possible integer pairs whose product is -9 are $(1,-9), (-1,9), (3,-3),$ and $(-3,3)$. The pair whose sum is 8 is $(-1,9)$.

Step 2. $3x^2 + 8x - 3 = 3x^2 + 9x - 1 \cdot x - 3$
Step 3. $\qquad\qquad\qquad = 3x(x + 3) - 1(x + 3)$
Step 4. $\qquad\qquad\qquad = (x + 3)(3x - 1)$
Thus, $3x^2 + 8x - 3 = (x + 3)(3x - 1)$

Since the two terms have a common binomial factor we are doing alright.

Example 2: Factor $24x^2 + 14x - 3$

Step 1. $p = 24(-3) = -72, s = 14$
The critical integers are 18 and -4
Step 2. $24x^2 + 14x - 3 = 24x^2 + 18x - 4x - 3$
Step 3. $\qquad\qquad\qquad\;\; = 6x(4x + 3) - 1(4x + 3)$
Step 4. $\qquad\qquad\qquad\;\; = (4x + 3)(6x - 1)$
Thus, $24x^2 + 14x - 3 = (4x + 3)(6x - 1)$

Since the two terms have a common binomial factor, we are doing alright.

Problem Set 6.7

Factor, if possible, each of the following expressions.

	Expression	Steps 2–3	Factors
1.	$6x^2 + 7x + 2 =$ $p = 12, s = 7$ CIs are $\underline{\;4\;}\;\underline{\;3\;}$	$6x^2 + 4x + 3x + 2$ $= 2x(3x + 2) + 1(3x + 2) =$	$(3x + 2)(2x + 1)$
2.	$5y^2 - 14y + 8 =$ $p = \quad, s =$ CIs are $\underline{\quad}\;\underline{\quad}$		$(\quad)(\quad)$
3.	$2x^2 + 7x + 3 =$ $p = \quad, s =$ CIs are $\underline{\quad}\;\underline{\quad}$		$(\quad)(\quad)$
4.	$4x^2 - 23x + 15 =$ $p = \quad, s =$ CIs are $\underline{\quad}\;\underline{\quad}$		$(\quad)(\quad)$
5.	$10a^2 + 7a - 12 =$ $p = \quad, s =$ CIs are $\underline{\quad}\;\underline{\quad}$		$(\quad)(\quad)$

6. $6p^2 + 11p - 10 =$
   p = , s =
   CIs are ___ ___   ( )( )

7. $6a^2 - 13a + 6 =$
   p = , s =
   CIs are ___ ___   ( )( )

8. $12x^2 + 4x - 5 =$
   p = , s =
   CIs are ___ ___   ( )( )

9. $12x^2 - 7x - 12 =$
   p = , s =
   CIs are ___ ___   ( )( )

10. $6x^2 + 5x - 4 =$
    p = , s =
    CIs are ___ ___   ( )( )

11. $6a^2 + a - 5$ 12. $2x^2 + 7x - 15$

13. $2y^2 + 2y + 3$ 14. $9x^2 + 11x + 2$

15. $12x^2 - 11x + 2$ 16. $3x^2 - x - 10$

17. $4 - 12h + 9h^2$ 18. $15r^2 + 20r + 8$

19. $a - 7a^2 - 18$ 20. $2x^2 + 7x + 3$

21. $4x^2 + 9x + 2$ 22. $2x^2 + 5x - 12$

23. $4x^2 + 11x - 3$ 24. $4x^2 - 9x - 9$

25. $10x^2 - 7x - 6$ 26. $9x^2 - 28xy + 3y^2$

27. $6x^3 - 19x^2 + 3x$ 28. $18x^2 + 6x - 4$

29. $3 + 16x - 12x^2$ 30. $6 - 5x - 6x^2$

31. $3(x + y)^2 + 2(x + y) - 5$ 32. $3x^4 - 14x^3 - 5x^2$

33. $8x^6 - 7x^3 - 1$

6.8 More on Factors

The method of factoring $ax^2 + bx + c$ by the use of critical integers can be used for a wide variety of trinomials, e.g.;

$x^2 - 4$
$x^2 - 5x + 6$
$12x^2 + 7x - 12$

There are several other expressions which look different but can be factored in a similar way, by use of critical integers. In the following examples we factor some of these expressions.

Example 1: Factor completely $5x^2 - 25x + 30$

This expression is similar to the expressions factored earlier. The only difference in this case is that each term in the given expression has a common factor. In such a case we should pull out this common factor and then factor, if possible, the expression in parenthesis.

$$5x^2 - 25x + 30 = 5(x^2 - 5x + 6) = 5(x - 3)(x - 2)$$

▶ Whenever we can factor out a constant from a given expression, we should do that as a first step.

Example 2: Factor completely $3x^4 - 48$

$$\begin{aligned}3x^4 - 48 &= 3(x^4 - 16) \\ &= 3((x^2)^2 - 4^2) \\ &= 3(x^2 - 4)(x^2 + 4) \\ &= 3(x - 2)(x + 2)(x^2 + 4)\end{aligned}$$

Since $x^2 + 4$ cannot be factored any further, we get $3x^4 - 48 = 3(x - 2)(x + 2)(x^2 + 4)$

Example 3: Factor completely $x^4 - 5x^2 + 4$

Since only even powers of x are present it is often helpful to let $x^2 = y$.

$$\begin{aligned}x^4 - 5x^2 + 4 &= (x^2)^2 - 5(x^2) + 4 \\ &= y^2 - 5y + 4 = (y - 1)(y - 4)\end{aligned}$$

Thus, $x^4 - 5x^2 + 4 = (x^2 - 1)(x^2 - 4)$

We must check whether each expression in parentheses can be factored further or not. In this case both the factors can be factored—each being a difference of squares.

$$\begin{aligned}x^4 - 5x^2 + 4 &= (x^2 - 1)(x^2 - 4) \\ &= (x - 1)(x + 1)(x - 2)(x + 2).\end{aligned}$$

This same technique could have been used in Example 2. Try it.

Example 4: Factor completely $10x^2 - 11xy - 6y^2$

This type of expression can be factored in exactly the same way as

$ax^2 + bx + c$
$p = (-6)(10) = -60; s = -11$
The critical numbers are $-15, 4$ Step 1
$\begin{aligned}10x^2 - 11xy - 6y^2 &= 10x^2 - 15xy + 4xy - 6y^2 \\ &= 5x(2x - 3y) + 2y(2x - 3y) \\ &= (2x - 3y)(5x + 2y)\end{aligned}$ Step 2, Step 3, Step 4

Problem Set 6.8

Factor each of the following expressions.

	Expression	Factors		Expression	Factors
1.	$3x^2 + 18x + 15$ $3(x^2 + 6x + 5)$ CIs are 1, 5	$3(x + 1)(x + 5)$	2.	$2x^2 + 10x + 12$	_____
3.	$5x^2 - 25x + 30$	_____	4.	$2x^2 + 2x - 12$	_____
5.	$ax^2 - 2ax - 8a$	_____	6.	$9x^2 + 18x - 72$	_____
7.	$4x^2 - 36$	_____	8.	$x^2y - 9y$	_____
9.	$x^4 - 16$	_____	10.	$y^4 - 81$	_____
11.	$2x^4 - 32$	_____	12.	$4x^4 - 4y^4$	_____
13.	$x^4 - 2x^2 - 8$	_____	14.	$2y^4 + 2y^2 - 12$	_____
15.	$6x^2 + 21x + 9$	_____	16.	$12x^2 + 10x - 8$	_____
17.	$x^2 - 2xy - 48y^2$	_____	18.	$x^2 - 3xy - 18y^2$	_____
19.	$3x^4 - 26x^2 - 9$	_____	20.	$9x^4 + 35x^2 - 4$	_____

6.9 Factoring the Sum or Difference of Two Cubes

We discussed three special products in Section 6.2. Two other special products which occur quite often are

1. $(x - y)(x^2 + xy + y^2)$
2. $(x + y)(x^2 - xy + y^2)$

These two products when simplified are unexpectedly simple.

1. $(x - y)(x^2 + xy + y^2)$
    $= x(x^2 + xy + y^2) - y(x^2 + xy + y^2)$
    $= x^3 + \underline{x^2y} + \underline{\underline{xy^2}} - \underline{yx^2} - \underline{\underline{xy^2}} - y^3$
    $= x^3 - y^3$ Combining like terms

2. $(x + y)(x^2 - xy + y^2)$
    $= x(x^2 - xy + y^2) + y(x^2 - xy + y^2)$
    $= x^3 - \underline{x^2y} + \underline{\underline{xy^2}} + \underline{yx^2} - \underline{\underline{xy^2}} + y^3$
    $= x^3 + y^3$ Combining like terms

From these two examples we can state the following rule for factoring the sum or difference of two cubes.

Rule 6.3

Factors of the Sum or Difference of Two Cubes

$x^3 - y^3 = (x - y)(x^2 + xy + y^2)$
$x^3 + y^3 = (x + y)(x^2 - xy + y^2)$

Example 1: *Factor expressions of the form $x^3 - y^3$ or $x^3 + y^3$*

(a) $x^3 - 8 = x^3 - 2^3$
    $= (x - 2)(x^2 + 2x + 4)$

(b) $2x^3 - 54 = 2(x^3 - 27)$
    $= 2(x^3 - 3^3)$
    $= 2(x - 3)(x^2 + 3x + 9)$

(c) $4y^4 + 32yx^3 = 4y(y^3 + 8x^3)$
    $= 4y[y^3 + (2x)^3]$
    $= 4y(y + 2x)(y^2 - 2xy + 4x^2)$

Problem Set 6.9

Factor each of the following expressions.

Expression	Reduction	Factors
1. $2xy^3 - 16x^4 =$	$2x(y^3 - 8x^3)$ $2x(y^3 - (2x)^3)$	$= \underline{2x(y - 2x)(y^2 + 2xy + 4x^2)}$
2. $x^3 - 64 =$		$= $
3. $y^3 - 125 =$		$= $
4. $2x^3 + 16 =$		$= $
5. $6x^3 - 6a^3 =$		$= $
6. $4x^4 - 32x =$		$= $
7. $x^3y - y^4 =$		$= $

8. $5y^3 - 40 =$ = _____

9. $8x^3 - 27 =$ = _____

10. $x^3y + y^4 =$ = _____

11. $16 - 2y^3 =$ = _____

12. $16 + 2y^3 =$ = _____

13. $27x^3 - 1 =$ = _____

14. $54x^4 - 2x =$ = _____

15. $x^6 - 1 =$ = _____

16. $64 - x^6 =$ = _____

17. $(x - y)^3 - (y + 2)^3 =$ = _____

18. $250x^3 - 2 =$ = _____

19. $x^4 - x(y - 1)^3 =$ = _____

20. $27x^3 - (2y + 1)^3 =$ = _____

21. $x^2y^6 - x^8 =$ = _____

In Problems 22–24, simplify the expressions.

Expression		Simplified Form
22. $(2x - y)(4x^2 + 2xy + y^2)$	=	$8x^3 - y^3$
23. $(x + y)(x^2 - xy + y^2)$	=	
24. $(2x - 3)(4x^2 + 6x + 9)$	=	

6.10 Factors by Completing Squares

Consider each of the following examples and see if you can relate the third term on the right side of the equation to the coefficient of x in the second term on the right side of the equation.

1. $(x + 2)^2 = x^2 + 2(2x) + 4$
2. $(x - 3)^2 = x^2 + 2(-3x) + 9$
3. $(x + y)^2 = x^2 + 2(yx) + y^2$
4. $(x - y)^2 = x^2 + 2(-yx) + y^2$
5. $(x + 2z)^2 = x^2 + 2(2zx) + 4z^2$

Do you see that the third term on the right side is the square of one-half the coefficient of x in the second term on the right? We will make heavy use of this observation to rewrite a trinomial as a sum or difference of squares.

Illustrations:
1. *Express $x^2 + 2x + 3$ as a sum or difference of two squares*
   Step 1. Factor out the leading coefficient, if other than 1. In this case, it is 1.
   Step 2. Consider the terms containing the unknowns. That is $x^2 + 2x$. The coefficient of x is 2. Half of 2 is 1. So 1^2 when added to $x^2 + 2x$, will form a perfect square.
   Step 3. Add and subtract the number, obtained in Step 2, to the given expression.
   $$x^2 + 2x + 3 = (x^2 + 2x + \underline{1}) + 3 - \underline{1}$$
   $$= (x + 1)^2 + 2$$
   $$= (x + 1)^2 + (\sqrt{2})^2$$
   We have now rewritten $x^2 + 2x + 3$ as a sum of squares.

2. *Express $x^2 + 6x - 4$ as a sum or difference of two squares*
   Step 1. The leading coefficient is 1.
   Step 2. The coefficient of x is 6. Half of 6 is 3. So the number 3^2 when added to $x^2 + 6x$ will form a perfect square.
   Step 3. Add and subtract 3^2 to the given expression
   $$x^2 + 6x - 4 = x^2 + 6x + \underline{9} - 4 - \underline{9}$$
   $$= (x + 3)^2 - 13$$
   $$= (x + 3)^2 - (\sqrt{13})^2$$

3. *Express $3x^2 - 9x + 4$ as a sum or difference of two squares*
   Step 1. The leading coefficient is 3.
   Therefore, $3x^2 - 9x + 4 = 3\left(x^2 - 3x + \frac{4}{3}\right)$
   We will now change $x^2 - 3x + \frac{4}{3}$ to a sum or a difference of two squares.
   Step 2. The coefficient of x is -3. Half of -3 is $-\frac{3}{2}$. So the number $\left(-\frac{3}{2}\right)^2$ when added to $x^2 - 3x$ will make it a perfect square.
   Step 3. Add and subtract $\left(-\frac{3}{2}\right)^2$ to $x^2 - 3x + \frac{4}{3}$.
   $$x^2 - 3x + \frac{4}{3} = x^2 - 3x + \frac{9}{4} + \frac{4}{3} - \frac{9}{4}$$
   $$= \left(x - \frac{3}{2}\right)^2 - \frac{11}{12}$$
   $$= \left(x - \frac{3}{2}\right)^2 - \left(\sqrt{\frac{11}{12}}\right)^2$$
   Thus, $3x^2 - 9x + 4 = 3\left(x^2 - 3x + \frac{4}{3}\right)$
   $$= 3\left[\left(x - \frac{3}{2}\right)^2 - \left(\sqrt{\frac{11}{12}}\right)^2\right].$$

We observe in Illustrations 1–3 that any trinomial of the form $ax^2 + bx + c$ can be expressed as a sum or a difference of two squares. If it is sum of two squares then it can not be factored further (Example 3(b), Section 6.6) into real factors. If it is a difference of two squares then as before, it can be factored again.

Example 1: Factor $x^2 + 2x - 1$
The square of half the coefficient of x is 1. Therefore,
$$x^2 + 2x - 1 = x^2 + 2x + \underline{1} - 1 - \underline{1}$$
$$= (x + 1)^2 - 2 = (x + 1)^2 - (\sqrt{2})^2$$
$$= (x + 1 + \sqrt{2})(x + 1 - \sqrt{2})$$

Example 2: Factor $2x^2 - 8x - 5$

$$2x^2 - 8x - 5 = 2\left(x^2 - 4x - \frac{5}{2}\right)$$

Consider $x^2 - 4x - \frac{5}{2}$. The square of half the coefficient of x is $(-2)^2 = 4 = (2)^2$. Therefore

$$x^2 - 4x - \frac{5}{2} = x^2 - 4x + \underline{4} - \frac{5}{2} - 4$$
$$= (x - 2)^2 - \frac{13}{2} = (x - 2)^2 - \left(\sqrt{\frac{13}{2}}\right)^2$$
$$= \left(x - 2 - \sqrt{\frac{13}{2}}\right)\left(x - 2 + \sqrt{\frac{13}{2}}\right).$$

Thus, $2x^2 - 8x - 5 = 2\left(x^2 - 4x - \frac{5}{2}\right) = 2\left(x - 2 - \sqrt{\frac{13}{2}}\right)\left(x - 2 + \sqrt{\frac{13}{2}}\right)$

Problem Set 6.10

In Problems 1–8, identify the number to be added and subtracted so that the expression is a sum or difference of two squares.

	Expression	The number		Expression	The number
1.	$x^2 + 6x$	$\left(\frac{6}{2}\right)^2$ 9	2.	$x^2 + 2x$	_____
3.	$x^2 - 2x$	_____	4.	$x^2 - x$	_____
5.	$y^2 + 4y - 1$	_____	6.	$y^2 + 3y - 2$	_____
7.	$2x^2 + 3x - 1$	_____	8.	$5x^2 - 10x + 7$	_____

In Problems 9–20, write the expressions in factored form.

	Expression	Show work here		Factors
9.	$x^2 - 6x - 4$	$x^2 - 6x + 9 - 4 - 9$ $(x - 3)^2 - (\sqrt{13})^2$	9.	$(x - 3 + \sqrt{13})(x - 3 - \sqrt{13})$
10.	$x^2 + 2x - 3$		10.	_____
11.	$x^2 + 3x + 1$		11.	_____
12.	$x^2 + 8x - 1$		12.	_____
13.	$2x^2 + 6x - 7$		13.	_____
14.	$3y^2 + y - 2$		14.	_____

15. $y^2 - 5y + 5$ 15. _____

16. $x^2 + 4x + 2$ 16. _____

17. $2z^2 + 3z - 1$ 17. _____

18. $4z^2 + 5z - 1$ 18. _____

19. $z^2 - 6z + 2$ 19. _____

20. $2b^2 + 8b + 1$ 20. _____

Chapter Summary

1. **Special products**   Products which occur frequently.
2. **Rule 6.1**
$$(x + y)^2 = x + 2xy + y^2$$
$$(x - y)^2 = x^2 - 2xy + y^2$$
$$(x + y)(x - y) = x^2 - y^2$$
3. **Critical integers**   In the expression $ax^2 + bx + c$ if $p = a \cdot c$ and $s = b$, then the integers, if they exist, whose product is p and sum is s are called critical integers.
4. **Factors of $x^2 + sx + p$**   If the critical integers are a and b then $x^2 + sx + p = (x + a)(x + b)$. For $x^2 + 3x + 2$, the CIs are 1 and 2. Therefore, $x^2 + 3x + 2 = (x + 1)(x + 2)$
5. **Factors of $ax^2 + bx + c$**   Find the CIs. Separate the middle terms into two parts using the CIs. Factor the 1st two and the last two terms. At this stage you have a check point, the two terms must have one common factor. Take the common factor out and the factoring is complete.
6. **More on factors**   If there is a factor common in all the terms then it should be factored out first. An expression of the type $ax^2 + bxy + cy^2$ can also be factored using CIs. Factoring is complete only if, in the end, there is no factor that can be factored further.
7. **Rule 6.3** $x^3 - y^3 = (x - y)(x^2 + xy + y^2)$
$x^3 + y^3 = (x + y)(x^2 - xy + y^2)$
8. **Sum and difference of two squares**   The difference of two squares can be factored, viz., $x^2 - y^2 = (x - y)(x + y)$, but the sum of two squares cannot be factored in the real number system.
9. **Completing squares**   Any expression of the type $ax^2 + bx + c$ can be expressed as a sum or a difference of two squares. The main step is to add or subtract the square of half the coefficient of x.

Review Problems

Complete the statements in Problems 1–15.

1. The expression $3x^2 - 4xy + 5y^2 - 7x$ has _____ terms.

2. The literal part of the second term in $5y^3 - 3xy + 4x^2 - 9$ is _____.

3. The coefficient of x^2y in $4x^3 - 3x^2y + 4y^2$ is _____.

4. $(x + y)^2 = x^2 + $ _____ $+ y^2$.

5. $(x - y)^2 = x^2 - $ _____ $+ y^2$.

6. $(2x - y)^2 = $ _____ $+ y^2$.

7. $(x - y)(x + y) = $ _____.

8. $x^3 - y^3 = ($ __ $)(x^2 + xy + $ __$)$.

9. $x^2 - y^2 = ($ __ $)($ __ $)$.

10. $x^2 + y^2 = (x + y)^2$ is _____.
    (true or false)

11. $x^2 - y^2 = (x - y)^2$ is _____.
    (true or false)

12. $x^2 + 3x + 2 = (x + $ __$)(x + $ __$)$.

13. $x^2 - x - 6 = (x - $ __$)(x + $ __$)$.

14. $x^2 + x - 6 = (x - $ __$)(x + $ __$)$.

15. $x^2 - 18x + 72 = (x - $ __$)(x - $ __$)$.

In Problems 16–23, factor completely.

Expression	Factors		Expression	Factors
16. $10x - 30$	$10(x - 3)$	17.	$15x - 18$	_____
18. $ay + ab$	_____	19.	$2x^2 - 4x + 2$	_____
20. $x^3y - y^3x$	_____	21.	$4x^3 - 2x^2$	_____
22. $4x^2 + 4x + 8$	_____	23.	$2x(3x - 4) + 5(3x - 4)$	_____

178

In Problems 24–29, find the integers whose product (p) and sum (s) is given.

	p/s	CIs		p/s	CIs
24.	p = −15 s = −14	−15, 1	25.	p = −15 s = −2	___ , ___
26.	p = 25 s = −10	___ , ___	27.	p = 20 s = −12	___ , ___
28.	p = −20 s = −8	___ , ___	29.	p = −27 s = −6	___ , ___

In Problems 30–45, factor the expressions completely.

	Expressions	p/s	CIs	Step 2 & 3	Factors
30.	$6x^2 + 5x - 4$	p = −24 s = 5	8, −3	$6x^2 + 8x - 3x - 4$ $= 2x(3x + 4) - 1(3x + 4)$ $= (3x + 4)(2x - 1)$	$(3x + 4)(2x - 1)$
31.	$x^2 - x - 12$	p = ___ s = ___	___ , ___		( )( )
32.	$x^2 - 3x - 18$	p = ___ s = ___	___ , ___		( )( )
33.	$x^2 - 16$	p = ___ s = ___	___ , ___		( )( )
34.	$y^2 + 2y - 15$	p = ___ s = ___	___ , ___		( )( )
35.	$A^2 - 6A - 27$	p = ___ s = ___	___ , ___		( )( )
36.	$x^2 + 8x - 20$	p = ___ s = ___	___ , ___		( )( )
37.	$2x^2 - 7x + 3$	p = ___ s = ___	___ , ___		( )( )
38.	$7y^2 + 9y + 2$	p = ___ s = ___	___ , ___		( )( )

39. $2x^2 + 4x - 6$ p = ___ , ___   ( )( )
   s = ___

40. $4x^2 - 4x + 1$ p = ___ , ___   ( )( )
    s = ___

41. $16x^2 + 8x + 1$ p = ___ , ___   ( )( )
    s = ___

42. $5x^2 + 14x - 3$ p = ___ , ___   ( )( )
    s = ___

43. $4x^2 - 3x - 7$ p = ___ , ___   ( )( )
    s = ___

44. $12x^2 - 34x - 6$ p = ___ , ___   ( )( )
    s = ___

45. $16y^2 - 26y + 3$ p = ___ , ___   ( )( )
    s = ___

In Problems 46–54, factor completely.

Expressions	Show work here	Factors
46. $2x^4 - 32$	$2(x^4 - 16) = 2(x^2 - 4)(x^2 + 4)$ $= 2(x - 2)(x + 2)(x^2 + 4)$	$2(x - 2)(x + 2)(x^2 + 4)$
47. $4x^2 - 3xy - 7y^2$		___
48. $12x^2 - 9x - 21$		___
49. $x^4 - 5x^2 + 4$		___
50. $x^3 - y^3$		___
51. $x^3 + 27$		___
52. $2x^4 - 16x$		___

53. $16x^3 - 54y^3$ _____

54. $24 - 3a^3$ _____

In Problems 55–62, mark the correct response.

55. Consider the following statements P and Q:
    P: $(x + 7) + x(x + 7) = (x + 7)(1 + x)$.
    Q: $-(1 - x) = x + 1$.
    Which of the above statements are true?
    (a) P only (b) Q only (c) P and Q (d) None

56. Factor completely the expression: $4x^2 - 8x$
    (a) $4(x^2 - 8x)$ (b) $4(x^2 - 2x)$
    (c) $4x(x - 2x)$ (d) $4x(x - 2)$
    (e) None of these

57. Factor completely the expression: $4x^3 - 16x$
    (a) $4x(x - 2)(x + 2)$ (b) $4x(x^2 - 16x)$
    (c) $4x(x - 4)(x + 4)$ (d) $4(x^3 - 4x)$
    (e) None of these

58. Two numbers whose product is -12 and the sum is -11 are:
    (a) $-4, 3$ (b) $4, -3$ (c) $6, -2$ (d) $-6, 2$ (e) None of these

59. Factor completely the expression: $x^2 + 4x + 5$
    (a) $(x + 4)(x + 5)$ (b) $(x + 1)(x + 4)$
    (c) $(x + 1)(x - 5)$ (d) $(x + 5)(x - 1)$
    (e) None of these

60. Factor completely the expression: $6a^2 - 13a + 6$
    (a) $(6a + 1)(a + 6)$ (b) $(3a + 2)(2a + 3)$
    (c) $(3a - 3)(2a - 2)$ (d) $(3a - 2)(2a - 3)$
    (e) None of these

61. Factor completely the expression: $x^4 - 2x^2 - 8$
    (a) $(x + 1)(x - 1)(x + 8)$ (b) $(x - 2)(x - 2)(x^2 + 2)$
    (c) $(x^2 + 2)(x - 2)(x + 2)$ (d) $(x^2 + 4)(x^2 - 2)$
    (e) None of these

62. Factor completely: $125x^3 + 8$
    (a) $(5x + 2)(25x^2 - 4)$ (b) $(5x + 2)(25x^2 - 10x + 4)$
    (c) $(5x + 2)^3$ (d) $(5x + 2)(25x^2 + 10x + 4)$
    (e) None of these

Chapter Test

Solve each of the following problems and match your answer with the responses.

1. Determine if the entry in column A is less than, greater than, or equal to the entry in column B.
   Given that $x < 0$ and $y < 0$
   Column A   Column B
   $x + y$ [ ] $y - x$
   (a) $>$ (b) $<$ (c) $=$ (d) Cannot be determined

2. What percent of 420 is 140?
   (a) 588   (b) $\frac{7}{2}$ (c) 30   (d) 300   (e) None of these

3. If $x + \frac{7 - 5x}{2} = 1$ then $x =$
   (a) -3 (b) 3   (c) $\frac{5}{3}$ (d) $-\frac{5}{3}$ (e) None of these

4. Smith drove 160 miles in 2 hours. At this rate, how many hours will it take her to travel 440 miles?
   (a) .8 hours   (b) 1.6 hours   (c) 55 hours   (d) 5.5 hours   (e) None of these

5. Consider the following statements P and Q:
   P: $(2a + b)^2 = 4a^2 + b^2$.
   Q: $2x(x + 4) = 2x^2 + 4$.
   Which of the above statements are true?
   (a) P only   (b) Q only   (c) P and Q   (d) None

6. Combine like terms:
   $(4x - 3) - (2x - 3) - (5x - 6) =$
   (a) $-3x$ (b) $-3x + 12$ (c) $-3x - 9$ (d) $-3x - 6$ (e) None of these

7. Simplify: $(3x - 2)(x - 5)$
   (a) $3x^2 - 17x + 10$ (b) $3x^2 - 17x - 10$ (c) $3x^2 - 2x - 10$ (d) None of these

8. Perform the long division $(x^3 - 4) \div (x - 1)$
   The quotient Q   Remainder R is:
   (a) $Q = x^2 - x + 1, R = -3$ (b) $Q = x^2 + x + 1, R = -5$
   (c) $Q = x^2 - x - 1, R = -3$ (d) $Q = x^2 + x + 1, R = -3$
   (e) None of these

9. Consider the following statements P and Q:
   P: $(x + 2)(x - 1) - (x + 2)(x) = (x + 2)(2x - 1)$.
   Q: $-(9 - 11x) = (11x - 9)$.
   Which of the above statements are true?
   (a) P only   (b) Q only   (c) P and Q   (d) None

10. Factor completely the expression: $12x^2 - 144x$
    (a) $12x(x - 12)$ (b) $12(x^2 - 12x)$
    (c) $x(12x - 12)$ (d) $12x(x - 12x)$
    (e) None of these

11. Factor completely the expression: $-3x^4 - 9x^3$
    (a) $-3x^3(x + 3)$
    (b) $3(-x^4 - x^3)$
    (c) $-3x^3(-x + 3)$
    (d) $3x(x^3 + x^2)$
    (e) None of these

12. What factor is common to the two expressions: $5x^2 + 6xy$ and $5xy + 6y^2$
    (a) $5x^2 + 6y^2$
    (b) $5x$
    (c) $6y$
    (d) $5x + 6y$
    (e) None of these

13. Factor completely the expression: $4x^3 - 100x$
    (a) $4(x^2 - 25)$
    (b) $4x(x^2 - 100)$
    (c) $4x(x - 5)(x + 5)$
    (d) $x(4x^2 - 100)$
    (e) None of these

14. Two numbers whose product is -18 and the sum is -7 are:
    (a) $-18, 1$ (b) $9, -2$ (c) $-9, 2$ (d) $6, -3$ (e) None of these

15. Factor completely the expression: $x^2 - 2x - 48$
    (a) $(x - 2)(x - 24)$
    (b) $(x - 8)(x + 6)$
    (c) $(x - 8)(x - 6)$
    (d) $(x + 6)(x - 8)$
    (e) None of these

16. Factor completely the expression: $6a^2 - 13a + 6$
    (a) $(6a + 1)(a + 6)$
    (b) $(3a + 2)(2a + 3)$
    (c) $(3a - 3)(2a - 2)$
    (d) $(3a - 2)(2a - 3)$
    (e) None of these

17. If, for all x, $x^2 - 2x - 15 = (x - R)(x + T)$, then $-R + T$ is
    (a) 2   (b) -2 (c) 3   (d) 5   (e) 7

18. Factor completely the expression: $x^4 - 2x^2 - 8$
    (a) $(x + 1)(x - 1)(x + 8)$
    (b) $(x - 2)(x - 2)(x^2 + 2)$
    (c) $(x^2 + 2)(x - 2)(x + 2)$
    (d) $(x^2 + 4)(x^2 - 2)$
    (e) None of these

19. One of the factors common in the two expressions: $1 + 36x^2 - 12x$, and $-2x + 12x^2$ is
    (a) $1 + 6x^2$ (b) $6x - 1$ (c) $2x$ (d) None of these

20. Factor completely: $8c^3 - 27$
    (a) $(2c - 3)(4c^2 + 6c + 9)$
    (b) $(2c - 3)^3$
    (c) $(2c - 3)(4c^2 - 6c + 9)$
    (d) $(2c - 3)(4c^2 + 9)$
    (e) None of these

Chapter 7

Quadratic Equations

An equation which may be written in the form

$$ax^2 + bx + c = 0, \text{ where } a \neq 0$$

is called a *quadratic equation*. The highest exponent of the unknown in this equation is 2.

In this chapter we discuss

(a) Some basic facts about the solutions,
(b) Solution of quadratic equation by factoring,
(c) Checking the solutions of a quadratic equation,
(d) Solutions of $x^2 = a$,
(e) Solutions of the quadratic equation by the quadratic formula, and
(f) Complex number solutions of $ax^2 + bx + c = 0$.

7.1 Some Basic Facts

Just as a linear equation has only one solution, a quadratic equation has at most two distinct solutions. Consider, for example,

$$x^2 - 1 = 0$$

$x = 1$ and $x = -1$ are the only two solutions of this quadratic equation. Note that

$(1)^2 - 1 = 0$ is true
$(-1)^2 - 1 = 0$ is true

There is no other value of x which satisfies this equation (try to find one). The reason the quadratic equation has only two solutions is that each quadratic equation is a combination of two linear equations. The solutions of those two linear equations are the solutions of the quadratic equation. Consider, for example, the equation

$$(x - 2)(x - 3) = 0 \tag{1}$$

When simplified (1) may be written as
$$x^2 - 5x + 6 = 0 \qquad (2)$$
Therefore, the quadratic Equation (2) and Equation (1) are equivalent, and have the same solutions. But if

$$(x - 2)(x - 3) = 0,$$

then either $x - 2 = 0$ or $\quad x - 3 = 0$
That is, $x = 2$ or $\quad x = 3$

Thus, we see that Equation (1) has exactly two solutions. Therefore, Equation (2), also has exactly two solutions.

Example 1: *Solutions of equations of the type* $(x - a)(x - b) = 0$

(a) $(x - 2)(x + 1) = 0$

Then either
$$\begin{array}{r} x - 2 = 0 \\ +2 \quad +2 \\ \hline x = 2 \end{array}$$
or
$$\begin{array}{r} x + 1 = 0 \\ -1 \quad -1 \\ \hline x = -1 \end{array}$$

Thus, the solutions are 2 and -1.

(b) $(x + 4)(x - 5) = 0$

Then either
$$\begin{array}{r} x + 4 = 0 \\ -4 \quad -4 \\ \hline x = -4 \end{array}$$
or
$$\begin{array}{r} x - 5 = 0 \\ +5 \quad +5 \\ \hline x = 5 \end{array}$$

Thus, the solutions are -4 and 5.

Problem Set 7.1

Complete the statements in Problems 1–5.

1. The equation $2x + 3 = 0$ is called a <u>linear</u> equation and has <u>only one</u> solution.

2. The equation $x^2 + 3x - 5 = 0$ is called a _____ equation and has at most _____ solutions.

3. The equation $x^2 - 2x = 0$ is a _____ equation and has _____ solutions.

4. The equation $x^2 - 9 = 0$ is a _____ equation and has _____ solutions.

5. The equation $x(3 - x) + 5x - 3 = 0$ is a _____ equation and has at most _____ solutions.

In Problems 6–10, find the solutions.

Equation	Linear Equations		Solutions
6. $(x - 2)(x - 7) = 0$	$\begin{array}{r} x - 2 = 0 \\ +2 \quad +2 \\ \hline x = 2 \end{array}$ or $\begin{array}{r} x - 7 = 0 \\ +7 \quad +7 \\ \hline x = 7 \end{array}$		2 7

7. $(2x - 1)(x - 5) = 0$ _____$=0$ or   _____$=0$ ____ ____

    $x =$ _____ $x =$ _____

8. $(3x + 1)(2x - 5) = 0$ _____$=0$ or   _____$=0$ ____ ____

9. $(2 - x)(3x + 4) = 0$ _____$=0$ or   _____$=0$ ____ ____

10. $5(2x - 5)(3x + 2) = 0$ _____$=0$ or   _____$=0$ ____ ____

7.2 Solution by Factoring

In Section 6.5, we discussed the concept of the Critical Integers (CIs) for the expression

$$ax^2 + bx + c.$$

These are two integers whose product is "a · c" and whose sum is "b". We also observed that if the CIs for this expression exist then it can be easily factored. In this Section, we discuss the solutions of the equation

$$ax^2 + bx + c = 0 \qquad (1)$$

where $ax^2 + bx + c$ can be factored by CIs. In those cases where we cannot factor by use of CIs we shall solve the quadratic equation (1) by use of the method to be discussed in Section 7.4.

The procedure for solving a quadratic equation by factoring can be summarized in the following three steps.

Step 1. Simplify the equation so that the right side of the equation is zero.
Step 2. If the expression on the left side is of the form $ax^2 + bx + c$ and it can be factored by use of CIs or some other method, then factor the expression.
Step 3. Separate the equation of Step 2 into two linear equations. The solution of each of these two linear equations together constitute the solution of Equation (1).

Example 1: Solve for x: $x^2 - 3x = 0$

Step 2. $x^2 - 3x = x(x - 3)$
Therefore, the given equation is equivalent to $x(x - 3) = 0$

Step 3. Then either $x = 0$ or $\begin{aligned} x - 3 &= 0 \\ +3 &+3 \\ \hline x &= 3 \end{aligned}$

Thus, the solutions are 0 and 3.

Example 2: Solve $x^2 - 7x + 6 = 0$

Step 2. $p = 6, s = -7$, CIs are -6 and -1
$x^2 - 7x + 6 = (x - 6)(x - 1)$.
Thus, the given equation is equivalent to $(x - 6)(x - 1) = 0$

Step 3. Then either $\begin{aligned} x - 6 &= 0 \\ +6 &+6 \\ \hline x &= 6 \end{aligned}$ or $\begin{aligned} x - 1 &= 0 \\ +1 &+1 \\ \hline x &= 1 \end{aligned}$

The solutions are 6 and 1.

Example 3: Solve for x: $x^2 = 25$

Step 1. $x^2 = 25$ or $x^2 - 25 = 0$ (The left side is the difference of squares.)
Step 2. $(x - 5)(x + 5) = 0$
Step 3. Then either
$$x - 5 = 0 \quad \text{or} \quad x + 5 = 0$$
$$\underline{+5 \quad +5} \quad\quad\quad \underline{-5 \quad -5}$$
$$x = 5 \quad\quad\quad\quad\quad x = -5$$

Thus, the solutions are 5 and -5.

Example 4: Solve $6x^2 + 13x - 5 = 0$

Step 2. $p = 6(-5) = -30$, $s = 13$, CIs are $+15$ and -2
$$\text{Then } 6x^2 + 13x - 5 = 6x^2 + 15x - 2x - 5$$
$$= 3x(2x + 5) - 1(2x + 5)$$
$$= (3x - 1)(2x + 5)$$

Thus, the given equation is equivalent to $(3x - 1)(2x + 5) = 0$

Step 3. Then either
$$3x - 1 = 0 \quad \text{or} \quad 2x + 5 = 0$$
$$\underline{+1 \quad +1} \quad\quad\quad \underline{-5 \quad -5}$$
$$3x = 1 \quad\quad\quad\quad 2x = -5$$
$$x = \frac{1}{3} \quad\quad\quad\quad x = -\frac{5}{2}$$

Thus, the solutions are $\frac{1}{3}$ and $-\frac{5}{2}$.

Example 5: Solve for y: $2y(y - 3) = 7 - y$

Step 1.
$$2y^2 - 6y = 7 - y$$
$$\underline{+y \quad\quad +y}$$
$$2y^2 - 5y = 7$$
$$\underline{-7 \quad -7}$$
$$2y^2 - 5y - 7 = 0.$$

Step 2. $p = -14$, $s = -5$, CIs are -7 and 2
$$\text{Then } 2y^2 - 5y - 7 = 2y^2 - 7y + 2y - 7$$
$$= y(2y - 7) + 1(2y - 7)$$
$$= (2y - 7)(y + 1)$$

The given equation is equivalent to the equation $(2y - 7)(y + 1) = 0$.

Step 3. Then either
$$2y - 7 = 0 \quad \text{or} \quad y + 1 = 0$$
$$\underline{+7 \quad +7} \quad\quad\quad \underline{-1 \quad -1}$$
$$2y = 7 \quad\quad\quad\quad y = -1$$
$$y = \frac{7}{2}$$

Thus, the solutions are $\frac{7}{2}$ and -1.

Problem Set 7.2

Solve the following equations by the factoring method.

	Equation	Linear equations		Solutions	
1.	$x^2 - 3x + 2 = 0$ $(x-2)(x-1) = 0$	$x - 2 = 0$ or $x = 2$	$x - 1 = 0$ $x = 1$	2	1
2.	$x^2 + 4x = 0$		or		
3.	$2x^2 + 3x = 0$		or		
4.	$12x^2 - 15x = 0$		or		
5.	$x^2 - ax = 0$		or		
6.	$bx^2 - cx = 0$		or		
7.	$x^2 - 4 = 0$		or		
8.	$2x^2 - 32 = 0$		or		
9.	$3y^2 - 12 = 0$		or		
10.	$5a^2 - 80 = 0$		or		
11.	$x^2 + 5x + 6 = 0$				
12.	$x^2 - 5x - 6 = 0$				
13.	$x^2 - 6x + 8 = 0$				
14.	$x^2 - 2x - 15 = 0$				
15.	$x^2 - 9x + 14 = 0$				
16.	$y^2 + 4y - 5 = 0$				
17.	$2x^2 - 8 = 0$				
18.	$x^2 - 7x = -6$				
19.	$3x^2 - 16x = -5$				
20.	$4(y^2 + 2y) = -3$				
21.	$15y^2 + 7y - 2 = 0$				
22.	$4x^2 = 5x$				
23.	$x^2 = 3x + 18$				
24.	$6 = 13x - 6x^2$				

25. $12y^2 - 4y = 3 + y$ _____ _____

26. $3(x^2 - 3x) = -x$ _____ _____

27. $4x^2 - 12 = x^2$ _____ _____

28. $x^2 - x = x - 1$ _____ _____

29. $4x^2 - x = 3x - 1$ _____ _____

30. $x - \dfrac{1}{6} = \dfrac{5}{2x}$ _____ _____

7.3 Checking Solutions of $ax^2 + bx + c = 0$

There are two different methods you can use to check the solutions of the equation $ax^2 + bx + c = 0$.

1. The first method is to directly substitute the solutions in the equation and verify that the solutions satisfy the equation.

    Example: $x = 3$, $x = -2$ are the solutions of $x^2 - x - 6 = 0$
    Check: $x = 3$ $\qquad\qquad\qquad\qquad x = -2$
    $\qquad 3^2 - 3 - 6 = 0 \qquad\qquad (-2)^2 - (-2) - 6 = 0$

    But, this method can involve complicated calculations if the solutions are not integers. The second method, described below may provide an easy way of checking solutions.

2. Read the following illustrations carefully. These illustrations explain the relation between the solutions and the coefficients of the left side expression of the equation $ax^2 + bx + c = 0$.

    A. Consider $x^2 - 5x + 6 = 0$ $\qquad$ *Comparing this with $ax^2 + bx + c = 0$, we get*
    or $(x - 2)(x - 3) = 0$ $\qquad a = 1, b = -5, c = 6.$
    Thus, the solutions are $x = 2$, and $x = 3$. Observe that $x^2 - 5x + 6 = 0$ can also be written as
    $x^2 - \underbrace{(2 + 3)}x + \underbrace{(2)(3)} = 0$
    $\qquad\quad\downarrow \qquad\qquad\hookrightarrow$ The product of the solutions
    $\quad$ The sum of the solutions
    Thus, the sum of the solutions $= 5$
    $\qquad\qquad\qquad\qquad\qquad = -(-5) = -\dfrac{-5}{1} = -\dfrac{b}{a}$
    The product of the solution $= 2(3) = 6$
    $\qquad\qquad\qquad\qquad\qquad\quad = \dfrac{6}{1} = \dfrac{c}{a}$

B. Consider $6x^2 - 7x - 3 = 0$ Comparing this with $ax^2 + bx + c = 0$, we get
$(2x - 3)(3x + 1) = 0$ $a = 6, b = -7, c = -3$.

The solutions are $x = \frac{3}{2}$, $x = -\frac{1}{3}$. Observe that $6x^2 - 7x - 3 = 0$ can also be written as

$x^2 - \frac{7}{6}x - \frac{3}{6} = 0$ Divide both sides by 6.

or $x^2 - \underbrace{\left(-\frac{1}{3} + \frac{3}{2}\right)}_{\text{Sum of the solutions}}x + \underbrace{\left(\frac{3}{2}\right)\left(-\frac{1}{3}\right)}_{\text{Product of solutions}} = 0$

Thus, the sum of the solutions $= \frac{3}{2} - \frac{1}{3} = \frac{7}{6}$

$= -\frac{-7}{6} = -\frac{b}{a}$

and the product of the solutions $= \left(\frac{3}{2}\right)\left(-\frac{1}{3}\right)$

$= -\frac{3}{6} = \frac{-3}{6} = \frac{c}{a}$

▶ In order to check the solutions of the equation

$ax^2 + bx + c = 0$

verify that

(1) the sum of the solutions $= -\frac{b}{a}$

(2) the product of the solutions $= \frac{c}{a}$

Sometimes you may find it inconvenient to compute the sum or the product of the solutions, especially when the solutions are irrational or complex numbers as in Sections 7.4 and 7.5. In that case you may obtain a partial check by computing either the sum or the product of the two solutions, whichever you think is easier.

The Sum of Solutions $= -\frac{\text{coefficient of } x}{\text{coefficient of } x^2}$

The Product of Solutions $= \frac{\text{the constant term}}{\text{the coefficient of } x^2}$

In the following Examples we will solve the equations by factoring and we will check the solutions by finding the sum and product of the solutions.

Example 1: Solve $x^2 - 4x = 0$ ($a = 1, b = -4, c = 0$)

Since $x^2 - 4x = x(x - 4)$
The equation is $x(x - 4) = 0$. Then either $x = 0$ or $x - 4 = 0$
$x = 4$

Thus, the solutions are 0 and 4

Check: The sum of solutions $= 4$, $-\frac{b}{a} = -\frac{-4}{1} = 4$

The product of solutions $= 0(4) = 0$, $\frac{c}{a} = \frac{0}{1} = 0$

Example 2: Solve $x^2 - 5x - 14 = 0$ $\quad (a = 1, b = -5, c = -14)$

$$x^2 - 5x - 14 = 0 \qquad \text{(CIs are } -7 \text{ and } 2)$$
$$(x - 7)(x + 2) = 0$$

The solutions are 7 and -2.

Check: The sum of the solutions $= 5$, $\quad -\dfrac{b}{a} = -\dfrac{-5}{1} = 5$

The product of solutions $= 7(-2) = -14$, $\quad \dfrac{c}{a} = \dfrac{-14}{1} = -14$

Example 3: Solve $10y^2 + y - 3 = 0$ $\quad (a = 10, b = 1, c = -3)$

$$p = -30, s = 1, \text{ CIs are } 6 \text{ and } -5$$
$$\begin{aligned}10y^2 + y - 3 &= 10y^2 + 6y - 5y - 3 \\ &= 2y(5y + 3) - 1(5y + 3) \\ &= (5y + 3)(2y - 1)\end{aligned}$$

The given equation is equivalent to

$$(5y + 3)(2y - 1) = 0$$

Then, either

$$5y + 3 = 0 \qquad \text{or} \qquad 2y - 1 = 0$$
$$y = -\dfrac{3}{5} \qquad\qquad\qquad y = \dfrac{1}{2}$$

Thus, the solutions are $-\dfrac{3}{5}$ and $\dfrac{1}{2}$.

Check: The sum of the solutions $= -\dfrac{1}{10}$

$$-\dfrac{b}{a} = -\dfrac{1}{10}$$

The product of the solution $= \left(-\dfrac{3}{5}\right)\left(\dfrac{1}{2}\right) = -\dfrac{3}{10}$

$$\dfrac{c}{a} = \dfrac{-3}{10} = -\dfrac{3}{10}$$

Example 4: Solve $x^2 - 9 = 0$ $\quad (a = 1, b = 0, c = -9)$

$$x^2 - 9 = 0$$
$$(x - 3)(x + 3) = 0$$

The solutions are 3 and -3

Check: In this case you can easily check by inspection that 3 and -3 are solutions.

You could solve the above equation as follows:

$$x^2 - 9 = 0 \text{ or } x^2 = 9$$

The two numbers whose squares are equal to 9 are 3 and -3. Thus, the solutions of $x^2 - 9 = 0$ or of $x^2 = 9$ are $x = 3$ or $x = -3$. These two solutions may be expressed more compactly as $x = \pm 3$. That is, the solutions of $x^2 = 9$ are $x = \pm 3$ or $x = \pm\sqrt{9}$

Example 5: Solve without factoring.

(a) $x^2 - 4 = 0$ or $x^2 = 4$
Then the solutions are $x = \pm\sqrt{4} = \pm 2$

(b) $x^2 - 5 = 0$ or $x^2 = 5$
Then the solutions are $x = \pm\sqrt{5}$

(c) $2x^2 - 32 = 0$ or $2x^2 = 32$ or $x^2 = 16$
Then the solutions are $x = \pm\sqrt{16}$ or ± 4

(d) $(x + 2)^2 = 9$
Then $x + 2 = \pm\sqrt{9}$ or $x + 2 = \pm 3$
Thus, $x = -2 \pm 3$; that is, $x = -2 + 3 = 1$ or $x = -2 - 3 = -5$
Check by inspection.

$\qquad x = 1 \qquad\qquad\qquad x = -5$
$(1 + 2)^2 = 9 \qquad\qquad (-5 + 2)^2 = 9$

(e) $x^2 + 1 = 0$ or $x^2 = -1$
There is no real number whose square is negative. Thus $x^2 + 1 = 0$ has no solution in the real numbers.

Problem Set 7.3

In Problems 1–10, determine which pair of numbers is a solution of the given quadratic equation. *(Remember that: sum $= -\dfrac{b}{a}$ and product $= \dfrac{c}{a}$.)*

1. $x^2 + x - 2 = 0$
   (a) $-1, 2$ (b) $1, -2$ (c) $1, 2$ (d) None of these

2. $y^2 + 7y - 44 = 0$
   (a) $4, -11$ (b) $-4, 11$ (c) $16, -9$ (d) $-16, 9$ (e) None of these

3. $A^2 - A - 2 = 0$
   (a) $1, 2$ (b) $-1, -2$ (c) $-1, 2$ (d) $1, -2$ (e) None of these

4. $x^2 + 2x - 3 = 0$
   (a) $1, 3$ (b) $-1, -3$ (c) $-1, 3$ (d) $1, -3$ (e) None of these

5. $3x^2 + 7x + 2 = 0$
   (a) $2, \frac{1}{3}$ (b) $-2, -\frac{1}{3}$ (c) $-2, -3$ (d) $2, 3$ (e) None of these

6. $6x^2 + 11x - 10 = 0$
   (a) $5, -2$ (b) $-5, -2$ (c) $-\frac{5}{2}, \frac{2}{3}$ (d) $\frac{5}{2}, -\frac{2}{3}$ (e) $-\frac{5}{2}, -\frac{2}{3}$

7. $2x^2 + 5x - 12 = 0$
   (a) $3, -4$ (b) $-4, \frac{3}{2}$ (c) $4, -\frac{3}{2}$ (d) $4, -\frac{2}{3}$ (e) $-4, \frac{2}{3}$

8. $6x^2 - 7x - 20 = 0$
   (a) $\frac{4}{3}, \frac{5}{2}$ (b) $-\frac{4}{3}, -\frac{5}{2}$ (c) $-\frac{4}{3}, \frac{5}{2}$ (d) $\frac{4}{3}, -\frac{5}{2}$ (e) None of these

9. $9a^2 - 42a + 49 = 0$
   (a) $\frac{7}{3}, -\frac{7}{3}$ (b) $\frac{7}{3}, \frac{7}{3}$ (c) $-\frac{7}{3}, -\frac{7}{3}$ (d) None of these

10. $2t^2 - 8 = 0$
    (a) $8, -8$ (b) $4, -4$ (c) $2, -2$ (d) $-2, -2$ (e) $2, 2$

In Problems 11–20, solve the equations by factoring or by taking square roots and check your solutions by using:

$$\text{the sum of the solutions} = -\frac{b}{a}$$

$$\text{and the product of the solutions} = \frac{c}{a}$$

Equation	Solutions	Check
11. $x^2 - x - 2 = 0$ $(x - 2)(x + 1) = 0$ $x = 2, x = -1$	$x = \underline{2}, \underline{-1}$	sum: $1 = -\frac{b}{a}$ product: $-2 = \frac{c}{a}$
12. $x^2 - 36 = 0$	$x = \underline{}, \underline{}$	sum: _____ product: _____
13. $3t^2 - 12t = 0$	$t = \underline{}, \underline{}$	sum: _____ product: _____
14. $6x^2 + 7x - 3 = 0$	$x = \underline{}, \underline{}$	sum: _____ product: _____
15. $2x^2 = 32$	$x = \underline{}, \underline{}$	sum: _____ product: _____
16. $3x^2 - 12 = 0$	$x = \underline{}, \underline{}$	sum: _____ product: _____
17. $t(t + 4) = 5$	$t = \underline{}, \underline{}$	sum: _____ product: _____
18. $\frac{x^2}{2} + \frac{10x}{3} + 2 = 0$	$x = \underline{}, \underline{}$	sum: _____ product: _____

19. $2(x^2 - 5) - 3(x^2 - 6) = 6$ x = ____,____   sum: _____

    product: _____

20. $6x^2 + 10x = 5 - 3x$ x = ____,____   sum: _____

    product: _____

In Problems 21–25, solve the equations and check the solutions by inspection.

Equation	Solutions	Check
21. $(x + 1)^2 - 4 = 0$ $(x + 1)^2 = 4$ $x + 1 = \pm 2$ or $x = 2 - 1$ and $x = -2 - 1$	x = __1__, __−3__	$(1 + 1)^2 - 4 = 0$ $(-3 + 1)^2 - 4 = 0$
22. $(3x - 1)^2 - 16 = 0$	x = ____,____	
23. $2(2x - 5)^2 - 18 = 0$	x = ____,____	
24. $2x^2 - 50 = 0$	x = ____,____	
25. $2x^3 - 5x^2 - 3x = 0$	x = ____,____	

7.4 Solutions by Use of the Quadratic Formula

It was observed in Section 6.10 that sometimes we cannot factor the expression $ax^2 + bx + c$ by use of critical integers. However, we learned that we can always factor such an expression by use of the method of completing squares. In this Section, we will use the method of completing squares to solve quadratic equations.

The following three illustrations will help us to learn how to use the method.

Illustration 1: Solve $x^2 - 2x - 6 = 0$.

To complete the square add and subtract the square of one-half of the coefficient of x.

$$x^2 - 2x + 1 - 1 - 6 = 0$$
$$(x^2 - 2x + 1) - 7 = 0$$
$$(x - 1)^2 - 7 = 0$$
$$(x - 1)^2 = 7$$
$$x - 1 = \pm \sqrt{7}$$
$$x = 1 \pm \sqrt{7}$$

Thus, the solutions of $x^2 - 2x - 6 = 0$ are $1 + \sqrt{7}$ and $1 - \sqrt{7}$.

Illustration 2: Solve $3x^2 + 2x - 2 = 0$.

First we divide both sides of the equation by 3 to produce an equivalent equation which has the coefficient of x^2 equal to one.

$$x^2 + \frac{2}{3}x - \frac{2}{3} = 0$$

Next, we complete the square by adding and subtracting the square of one-half of the coefficient of x.

$$x^2 + \frac{2}{3}x + \frac{1}{9} - \frac{1}{9} - \frac{2}{3} = 0$$

$$\left(x + \frac{1}{3}\right)^2 - \frac{7}{9} = 0$$

$$\left(x + \frac{1}{3}\right)^2 = \frac{7}{9}$$

$$x + \frac{1}{3} = \pm\sqrt{\frac{7}{9}} = \pm\frac{\sqrt{7}}{3}.$$

$$x = -\frac{1}{3} \pm \frac{\sqrt{7}}{3} = \frac{-1 \pm \sqrt{7}}{3}.$$

Thus, the solutions of $3x^2 + 2x - 2 = 0$ are

$$\frac{-1 + \sqrt{7}}{3} \text{ and } \frac{-1 - \sqrt{7}}{3}.$$

Illustration 3: Solve $ax^2 + bx + c = 0$.

First, divide both sides of the equation by a to produce an equivalent equation which has the coefficient of x^2 equal to one.

$$x^2 + \frac{b}{a}x + \frac{c}{a} = 0.$$

Next, add and subtract the square of one-half of the coefficient of x, viz., add and subtract $\left(\frac{b}{2a}\right)^2$.

$$x^2 + \frac{b}{a}x + \frac{b^2}{4a^2} - \frac{b^2}{4a^2} + \frac{c}{a} = 0$$

$$\left(x + \frac{b}{2a}\right)^2 - \frac{b^2}{4a^2} + \frac{c}{a} = 0$$

$$\left(x + \frac{b}{2a}\right)^2 = \frac{b^2}{4a^2} - \frac{c}{a} = \frac{b^2}{4a^2} - \frac{4ac}{4a^2} = \frac{b^2 - 4ac}{4a^2}$$

$$x + \frac{b}{2a} = \pm\sqrt{\frac{b^2 - 4ac}{4a^2}} = \pm\frac{\sqrt{b^2 - 4ac}}{2a}$$

$$x = -\frac{b}{2a} \pm \frac{\sqrt{b^2 - 4ac}}{2a}$$

$$x = \frac{-b \pm \sqrt{b^2 - 4ac}}{2a}$$

Thus, we have derived a general formula for use in solving a quadratic equation.

> **Rule 7.1**
>
> The solutions of the equation $ax^2 + bx + c = 0$ are given by
> $$x = \frac{-b \pm \sqrt{b^2 - 4ac}}{2a}$$

The expression under the radical sign,

$$b^2 - 4ac,$$

is called the *discriminant* of the equation

$$ax^2 + bx + c = 0.$$

When solving any quadratic equation we should first compute $b^2 - 4ac$. The discriminant gives us information about the nature of the solutions.

Case 1. If $b^2 - 4ac = 0$, then the equation has *repeated rational solutions*.

$$x = \frac{-b \pm \sqrt{b^2 - 4ac}}{2a} = \frac{-b \pm 0}{2a} = \frac{-b}{2a}, \frac{-b}{2a}.$$

Example 1: Solve $x^2 - 6x + 9 = 0$
$a = 1, b = -6, c = 9$
$b^2 - 4ac = 0$, and the solutions are
$$x = \frac{6 \pm 0}{2} = \frac{6 + 0}{2}, \frac{6 - 0}{2} = 3, 3.$$

Case 2. If $b^2 - 4ac$ is the *square of a positive integer* (e.g., 4,9,16,etc.) then the equation has two *distinct rational solutions*.

Example 2: Solve $2x^2 + 5x - 3 = 0$
$a = 2, b = 5, c = -3$
$b^2 - 4ac = 49 = 7^2$
The solutions are

$$x = \frac{-5 \pm \sqrt{49}}{2 \cdot 2} = \frac{-5 \pm 7}{4}$$

$$x = \frac{-5 + 7}{4}, \frac{-5 - 7}{4} = \frac{1}{2}, -3$$

Note that the equations discussed in Section 7.2 were of these two types. Thus, whenever the discriminant of an equation is zero or the square of a positive integer then that equation can also be solved by the factoring method. But, in the following two cases, we have no choice but to use the quadratic formula.

Case 3. If $b^2 - 4ac$ is *positive,* but not the square of a positive integer, then the equation has *irrational roots*.

Example 3: Solve $3x^2 + 2x - 2 = 0$
$a = 3, b = 2, c = -2$
$b^2 - 4ac = 28.$
The solutions are

$$x = \frac{-2 \pm \sqrt{28}}{2 \cdot 3} = \frac{-2 \pm \sqrt{28}}{6}$$

$$x = \frac{-2 \pm 2\sqrt{7}}{6} \qquad (\sqrt{28} = 2\sqrt{7})$$

$$x = \frac{-2}{6} + \frac{2\sqrt{7}}{6}, \frac{-2}{6} - \frac{2\sqrt{7}}{6}$$

$$x = -\frac{1}{3} + \frac{\sqrt{7}}{3}, -\frac{1}{3} - \frac{\sqrt{7}}{3}$$

Case 4. If $b^2 - 4ac$ is *negative* then the equation $ax^2 + bx + c = 0$ has complex number solutions. We shall discuss such solutions in Section 7.5 after having a brief discussion of complex numbers in that section.

More Examples.

Example 4: Solve for x: $4x^2 + 4x + 1 = 0$
$a = 4, b = 4, c = 1$
$b^2 - 4ac = 16 - 4 \cdot 4 \cdot 1 = 0$
The solutions are
$$x = \frac{-b \pm \sqrt{b^2 - 4ac}}{2a}$$
$$= \frac{-4 \pm \sqrt{0}}{8} = \frac{-4 \pm 0}{8} = -\frac{1}{2}, -\frac{1}{2}$$

Check: The sum of the solutions $= -\frac{1}{2} - \frac{1}{2} = -1$
$$-\frac{b}{a} = -\frac{4}{4} = -1$$
Thus, the sum $= -\frac{b}{a}$
The product of the solutions $= \frac{1}{4} = \frac{c}{a}$

Example 5. Solve for y: $6y^2 + 13y + 6 = 0$
$a = 6, b = 13, c = 6$
$b^2 - 4ac = 169 - 4 \cdot 6 \cdot 6 = 25$
The solutions are
$$x = \frac{-b \pm \sqrt{b^2 - 4ac}}{2a} = \frac{-13 \pm \sqrt{25}}{2 \cdot 6}$$
$$= \frac{-13 \pm 5}{12} = \frac{-13 + 5}{12}, \frac{-13 - 5}{12}$$
$$= \frac{-8}{12}, \frac{-18}{12} = -\frac{2}{3}, -\frac{3}{2}$$

Check: The sum of the solutions $= -\frac{2}{3} - \frac{3}{2} = -\frac{13}{6} = -\frac{b}{a}$
The product of solutions $= 1 = \frac{6}{6} = \frac{c}{a}$

Example 6: Solve for x: $2x^2 + 3x - 1 = 0$
$a = 2, b = 3, c = -1$
$b^2 - 4ac = 9 - 4(2)(-1) = 17$
The solutions are
$$x = \frac{-3 \pm \sqrt{17}}{4}$$
$$= \frac{-3 + \sqrt{17}}{4}, \frac{-3 - \sqrt{17}}{4}$$
$$= -\frac{3}{4} + \frac{\sqrt{17}}{4}, -\frac{3}{4} - \frac{\sqrt{17}}{4}$$

You might find it too troublesome to check both the sum and product of the roots in this problem. But, the sum is easy to compute and at least you will have a partial check.

Problem Set 7.4

Complete the statements in Problems 1–6.

1. In the equation $2x^2 - 3x + 1 = 0$, the value of a = 2, b = ____, and c = ____.

2. In the equation $3x^2 + 2x = 5 - x$, the value of a = ____, b = ____, and c = ____.

3. The discriminant of $ax^2 + bx + c = 0$ is _____.

4. If $b^2 - 4ac = 0$ then $ax^2 + bx + c = 0$ has _____.

5. If $b^2 - 4ac$ is a square of a positive integer then the equation $ax^2 + bx + c = 0$ has _____.

6. If $b^2 - 4ac$ is positive but not the square of an integer then the equation $ax^2 + bc + c = 0$ has _____.

In Problems 7–15, compute $b^2 - 4ac$ and find the nature of solutions.

Equation	$b^2 - 4ac$	The solutions are
7. $x^2 - 4x + 4$ a = 1, b = −4, c = 4 $b^2 - 4ac = (-4)^2 - 4 \cdot 4$	0	repeated numbers
8. $x^2 - 2x - 24 = 0$ a = 1, b = −2, c = −24 $b^2 - 4ac = 4 - 4(-24)$	100	rational numbers
9. $2x^2 - 3x - 1 = 0$ a = 2, b = −3, c = −1 $b^2 - 4ac = 9 - 4 \cdot 2(-1)$	17	irrational numbers
10. $x^2 - 6x + 9 = 0$		
11. $2x^2 + 4x + 2 = 0$		
12. $3x^2 + 2x - 2 = 0$		
13. $6x^2 + 7x - 3 = 0$		
14. $y(y + 4) = 5$		
15. $3x^2 + 6x - 2 = 0$		

In Problems 16–30, solve by use of the quadratic formula and check your solutions by use of:

the sum of the solutions $= -\dfrac{b}{a}$

and the product of the solutions $= \dfrac{c}{a}$

Equation	Solutions	Check
16. $x^2 - 4x + 4 = 0$ $a = 1, b = -4, c = 4$ $b^2 - 4ac = (-4)^2 - 4 \cdot 1 \cdot 4 = 0$ $x = \dfrac{4 \pm 0}{2} = \dfrac{4}{2}, \dfrac{4}{2}$	2 , 2	Sum $= 4 = -\dfrac{b}{a}$ Product $= 4 = \dfrac{c}{a}$
17. $x^2 - 2x - 24 = 0$ $a = 1, b = -2, c = -24$ $b^2 - 4ac = 100$ $x = \dfrac{2 \pm \sqrt{100}}{2} = 6, -4$	6 , -4	Sum $= 2 = -\dfrac{b}{a}$ Product $= -24 = \dfrac{c}{a}$
18. $2x^2 - 3x - 1 = 0$ $a = 2, b = -3, c = -1$ $b^2 - 4ac = 17$ $x = \dfrac{3 \pm \sqrt{17}}{4}$	$\dfrac{3}{4} + \dfrac{\sqrt{17}}{4}$, $\dfrac{3}{4} - \dfrac{\sqrt{17}}{4}$	Sum $= \dfrac{3}{2} = -\dfrac{b}{a}$ (This is a partial check)
19. $x^2 - 6x + 9 = 0$	_____ , _____	Sum = Product =
20. $2x^2 + 4x + 2 = 0$	_____ , _____	Sum = Product =
21. $3x^2 + 2x - 2 = 0$	_____ , _____	Sum =
22. $6x^2 + 7x - 3 = 0$	_____ , _____	Sum =
23. $3z^2 - z - 4 = 0$	_____ , _____	

24. $7 - y - y^2 = 0$ _____,_____

25. $1 - 6y + 9y^2 = 0$ _____,_____

26. $x(x + 2) = 1 - 2x$ _____,_____

27. $\dfrac{x^2 - 5x}{3} = x - 1$ _____,_____

28. $y(y + 4) = 5$ _____,_____

29. $3x^2 + 6x - 2 = 0$ _____,_____

30. $5x(x + 2) + 8 = 5$ _____,_____

In Problems 31–35, mark the correct response.

Use this space for work

31. What number is a solution of both the equations?
    $x^2 - x - 6 = 0 \quad 2x^2 - 18 = 0$
    (a) -3 (b) -2 (c) 3 (d) 6
    (e) None of these

32. The sum of the solutions of $2x^2 - 3x + 4 = 0$ is
    (a) $-\dfrac{3}{2}$ (b) 2 (c) $\dfrac{3}{2}$ (d) $\dfrac{2}{3}$
    (e) None of these

33. The product of the solutions of $3x^2 - 5x + 12 = 0$ is
    (a) 36 (b) $\dfrac{5}{3}$ (c) $-\dfrac{5}{3}$ (d) 4 (e) -4

34. One of the solutions of $2x^2 - 3x - 4 = 0$ is
    (a) $\frac{-3}{4} + \frac{\sqrt{41}}{4}$ (b) $\frac{-3}{4} - \frac{\sqrt{41}}{4}$
    (c) $\frac{3}{4} - \frac{\sqrt{41}}{4}$ (d) $\frac{3}{4} - \frac{\sqrt{17}}{4}$
    (e) None of these

35. Which of the following pairs of numbers is the solution of $3x^2 - 10x - 8 = 0$
    (a) $\frac{2}{3}, 4$ (b) $-\frac{2}{3}, 4$ (c) $\frac{3}{2}, 4$ (d) $\frac{2}{3}, 4$
    (e) None of these

7.5 Complex Number Solutions of $ax^2 + bx + c = 0$

We observed in Example 5(e) of Section 7.3 that the equation
$$x^2 + 1 = 0 \text{ or } x^2 = -1$$
has no solution in the real numbers because there is no real number whose square is -1. In order to determine solutions of such equations we introduce a new number "i", defined by
$$i = \sqrt{-1}.$$
Any square root of a negative number can be expressed in terms of this number "i".

Example 1:

(a) $\sqrt{-4} = \sqrt{4(-1)} = \sqrt{4}\sqrt{-1} = 2i$
(b) $\sqrt{-9} = \sqrt{9(-1)} = \sqrt{9}\sqrt{-1} = 3i$
(c) $\sqrt{-7} = \sqrt{(7)(-1)} = \sqrt{7}\sqrt{-1} = \sqrt{7}\,i \text{ or } i\sqrt{7}$
(d) $\sqrt{-3} = i\sqrt{3}$

➡ You may note that
$i^2 = -1, i^3 = i^2 \cdot i = -i$
$i^4 = (i^2)^2 = (-1)^2 = 1$

Any number of the type $a + ib$, where a and b are real numbers, is called a *complex number*.

    a is called the *real part* of $a + ib$

and

    b is called the *imaginary part* of $a + ib$

Example 2:

(a) $3 + \sqrt{-2} = 3 + i\sqrt{2},$ Real part = 3
                                             Imaginary part = $\sqrt{2}$
(b) $8 + \sqrt{-4} = 8 + 2i,$ Real part = 8
                                             Imaginary part = 2
(c) $3 - \sqrt{-5} = 3 - i\sqrt{5},$ Real part = 3
                                             Imaginary part = $-\sqrt{5}$

If in a given complex number we change the sign of the imaginary part, we get another complex number called the *conjugate* of the given complex number.

Thus, $a - ib$ is the conjugate of $a + ib$ and $a + ib$ is the conjugate $a - ib$.

Example 3:
(a) The conjugate of $3 + i\sqrt{5}$ is $3 - i\sqrt{5}$.
(b) The conjugate of $3 - i\sqrt{5}$ is $3 + i\sqrt{5}$.

It is interesting to note that the sum and product of complex conjugates is always a real number. In order to verify this, we need to know how to add and multiply complex numbers. For the present, we will restrict ourselves to considering the sum and product of complex conjugates only. We will treat the general problem in Chapter 9. And, we shall merely illustrate these operations in the following example.

Example 4: $2 + i\sqrt{3}$ and $2 - i\sqrt{3}$ are complex conjugates.

(a) $(2 + i\sqrt{3}) + (2 - i\sqrt{3}) = (2 + 2) + (i\sqrt{3} - i\sqrt{3}) = 4$

(b) $(2 + i\sqrt{3})(2 - i\sqrt{3}) = 2(2 - i\sqrt{3}) + i\sqrt{3}(2 - i\sqrt{3})$

$$= 4 - 2i\sqrt{3} + 2i\sqrt{3} - (i\sqrt{3})^2$$

$$= 4 - i^2 \cdot (\sqrt{3})^2$$

$$= 4 - (-1)(3) = 4 + 3 = 7$$

$a + ib$ and $a - ib$ are complex conjugates.

(c) $(a + ib) + (a - ib) = (a + a) + (ib - ib) = 2a$

(d) $(a + ib)(a - ib) = a(a - ib) + ib(a - ib)$
$= a^2 - iab + iab - (ib)^2$
$= a^2 - i^2b^2 = a^2 - (-1)b^2 = a^2 + b^2$.

These relationships will be useful for checking the solutions to quadratic equations which have negative discriminants. From these examples, we can see that

(i) the sum of complex conjugates is twice the real part (i.e., $2a$) of the number, and
(ii) the product of complex conjugates is the sum of the squares of the real and imaginary parts (i.e., $a^2 + b^2$) of the number.

Example 5:
(a) The conjugate of $3 + i\sqrt{5}$ is $3 - i\sqrt{5}$
(b) The conjugate of $5 - 4i$ is $5 + 4i$
(c) $\left(\frac{3}{2} + \frac{i\sqrt{5}}{2}\right)\left(\frac{3}{2} - \frac{i\sqrt{5}}{2}\right) = \left(\frac{3}{2}\right)^2 - \left(\frac{i\sqrt{5}}{2}\right)^2 = \frac{9}{4} + \frac{5}{4} = \frac{14}{4} = \frac{7}{2}$
(d) $(5 - 4i)(5 + 4i) = 5^2 - (4i)^2 = 25 + 16 = 41$
(e) $(3 + i\sqrt{5}) + (3 - i\sqrt{5}) = 6$ . . . (twice the real part)
(f) $(5 - 4i) + (5 + 4i) = 10$ . . . (twice the real part)

Let us now consider solutions of the quadratic equation $ax^2 + bx + c = 0$ when the discriminant $b^2 - 4ac$ is a negative number.

Example 6: Solve $x^2 + x + 1 = 0$

$a = 1, b = 1, c = 1$
$b^2 - 4ac = 1 - 4 = -3$
The solutions are
$$x = \frac{-b \pm \sqrt{b^2 - 4ac}}{2a} = \frac{-1 \pm \sqrt{-3}}{2}$$
$$= \frac{-1 \pm i\sqrt{3}}{2} = -\frac{1}{2} + \frac{i\sqrt{3}}{2}, -\frac{1}{2} - \frac{i\sqrt{3}}{2}$$

Note that these solutions are a pair of conjugate complex numbers—they differ only in the sign of the imaginary part.

Check: Sum = (twice the real part)
$$= -1 = -\frac{1}{1} = -\frac{b}{a}$$
$$\text{Product} = \left(-\frac{1}{2}\right)^2 + \left(\frac{\sqrt{3}}{2}\right)^2 = \frac{1}{4} + \frac{3}{4} = 1 = \frac{c}{a}$$

Example 7: Solve $6x^2 + 8x + 3 = 0$

$a = 6, b = 8, c = 3$
$b^2 - 4ac = 64 - 4(6)(3) = -8$
The solutions are
$$x = \frac{-b \pm \sqrt{b^2 - 4ac}}{2a} = \frac{-8 \pm \sqrt{-8}}{12}$$
$$= \frac{-8 \pm i\,2\sqrt{2}}{12} = -\frac{2}{3} + \frac{i\sqrt{2}}{6}, -\frac{2}{3} - \frac{i\sqrt{2}}{6}$$

Note again that these solutions are conjugate complex numbers.

Check: Sum = (twice the real part)
$$= \left(-\frac{4}{3}\right) = -\frac{8}{6} = \frac{-b}{a}$$
$$\text{Product} = \left(-\frac{2}{3}\right)^2 + \left(\frac{\sqrt{2}}{6}\right)^2 = \frac{4}{9} + \frac{1}{18} = \frac{1}{2} = \frac{c}{a}$$

▶ **Note:** It is always true that if the discriminant of a quadratic is negative then the roots will be a pair of conjugate complex numbers.

Problem Set 7.5

In Problems 1–8, write each number in terms of i and simplify as far as possible.

1. $\sqrt{-9}$ = __$i\sqrt{9}$__ = __$3i$__
2. $\sqrt{-27}$ = __$i\sqrt{27}$__ = __$3i\sqrt{3}$__
3. $\sqrt{-36}$ = _____ = _____
4. $\sqrt{-12}$ = _____ = _____
5. $\sqrt{-28}$ = _____ = _____
6. $-\sqrt{-25}$ = _____ = _____
7. $-\sqrt{-32}$ = _____ = _____
8. $\sqrt{-50}$ = _____ = _____

In Problems 9–12, find the conjugate of the given complex numbers.

	Complex number	Conjugate		Complex number	Conjugate
9.	$-3 + i\sqrt{2}$	$-3 - i\sqrt{2}$	10.	$2 + i\sqrt{3}$	_____
11.	$\frac{3}{2} - \frac{i}{2}$	_____	12.	$-\frac{3}{2} + \frac{i\sqrt{5}}{2}$	_____

In Problems 13–20, find the sum and the product of the given pair of conjugate complex numbers.

	Complex numbers		Sum		Product
13.	$2 + i$ / $2 - i$	$\frac{2+i}{\begin{array}{c}2-i\\\hline 4\end{array}}$	4	$(2+i)(2-i)$ / $2^2 - i^2 =$	$4 + 1 = 5$
14.	$-3 + i\sqrt{2}$ / $-3 - i\sqrt{2}$	$\frac{-3+i\sqrt{2}}{\begin{array}{c}-3-i\sqrt{2}\\\hline -6\end{array}}$	-6		$9 + 2 = 11$
15.	$4 + i\sqrt{3}$ / $4 - i\sqrt{3}$		_____		_____
16.	$-1 \pm \frac{i\sqrt{5}}{4}$		_____		_____
17.	$-\frac{1}{2} \pm \frac{i\sqrt{3}}{2}$		_____		_____
18.	$-\frac{3}{4} \pm \frac{i\sqrt{3}}{2}$		_____		_____
19.	$\frac{5}{8} \pm \frac{i\sqrt{5}}{2}$		_____		_____
20.	$1 \pm \frac{3i}{4}$		_____		_____

In Problems 21–38, solve the equations and check your solutions by use of the sum and the product of the roots.

	Equation	Solutions	Check
21.	$y^2 + 2y + 3 = 0$ $y = \frac{-2 \pm \sqrt{4-12}}{2}$ $= -1 \pm i\sqrt{2}$	$-1 + i\sqrt{2}$ $-1 - i\sqrt{2}$	Sum: $(-2) = -\frac{b}{a}$ Product: $1 + 2 = 3 = \frac{c}{a}$
22.	$x^2 - 2x + 2 = 0$	_____ _____	Sum: Product:

204

23. $x^2 - 3x + 4 = 0$ _____ Sum:

_____ Product:

24. $x^2 + 4 = 0$ _____ Sum:

_____ Product:

25. $2x^2 + 3 = 0$ _____ Sum:

_____ Product:

26. $3x^2 - 2x + 4 = 0$ _____ Sum:

_____ Product:

27. $x^2 + 1 = x$ _____ Sum:

_____ Product:

28. $x^2 + 3 = 3x - x^2$ _____ Sum:

_____ Product:

29. $2x^2 + 3x = 1 - 2x$ _____ Sum:

_____ Product:

30. $3 - 2x - 3x^2 = 0$ _____ Sum:

_____ Product:

31. $2x - x^2 + 4 = 3 + x^2$ _____ Sum:

_____ Product:

32. $4 - x^2 + 2x = 4x$ _____ Sum:

_____ Product:

33. $3t^2 - 5t + 4 = 0$ _____ Sum:

_____ Product:

34. $\frac{x^2}{5} + 2 = x$ _____ Sum:

_____ Product:

35. $x - 2x^2 = \frac{2x}{3} - \frac{1}{2}$ _____ Sum:

_____ Product:

36. $x(4 - x) + 2 = 3 - 2x^2$ _____ Sum:

_____ Product:

37. $x - 2x(5 - x) = \frac{2}{3} - 2x$ _____ Sum:

_____ Product:

38. $\frac{2x}{3}\left(5 - \frac{x}{2}\right) = \frac{5x}{6} - x^2$ _____ Sum:

_____ Product:

Chapter Summary

1. **Quadratic equation** Any equation of the form $ax^2 + bx + c = 0$ or one which can be simplified to this form is called a quadratic equation. The largest exponent of the unknown in this equation is 2.

2. **Solutions** A quadratic equation always has at most two solutions. Solutions of a quadratic equation can be obtained either by factoring or by use of the quadratic formula.

3. **Solutions by factoring** If you can find the Critical Integers of the expression

    $ax^2 + bx + c$

    then the solutions of the equation

    $ax^2 + bx + c = 0$

    can be obtained by equating to zero both factors of

    $ax^2 + bx + c.$

4. **Checking solutions** Solutions of a quadratic equation can be checked either by (a) inspection or (b) by use of the sum and the product of the solutions.

5. **Checking by inspection** If the quadratic equation is not given directly in the form

    $ax^2 + bx + c = 0$

    then the solutions, if convenient, should be checked by inspection. That is, replace the unknown by the solution and see whether the statement is true or not.

6. **Checking by the sum and product method** If the given quadratic equation is of the form
$$a^2 + bx + c = 0$$
then we can check solutions by verifying that

the sum of the solutions $= -\dfrac{b}{a}$

the product of the solutions $= \dfrac{c}{a}$.

7. **Solutions by using the quadratic formula** If critical integers of the expression
$$ax^2 + bx + c$$
do not exist then we should use the quadratic formula for finding the solutions
$$x = \dfrac{-b \pm \sqrt{b^2 - 4ac}}{2a}$$

8. **Discriminant** The expression under the radical sign of the quadratic formula,
$$b^2 - 4ac,$$
is called the discriminant of the equation.

9. **Nature of solutions**
    (a) If $b^2 - 4ac = 0$ then solutions are repeated.
    (b) If $b^2 - 4ac$ is a square of some positive integer then the solutions are distinct rational numbers.
    (c) If $b^2 - 4ac > 0$ but not the square of a whole number, the solutions are irrational numbers.
    (d) If $b^2 - 4ac < 0$ then the solutions are complex numbers.

10. **Complex numbers** Any number of the type $a + ib$ is called a complex number, where
    $i = \sqrt{-1}$ and a, b are real numbers.

    a is called real part of $a + ib$
    b is called imaginary part of $a + ib$.

11. **Conjugate of a complex number** If $a + ib$ is any complex number then $a - ib$ is the conjugate of $a + ib$ and vice versa. Two numbers of the form $a + ib$ and $a - ib$ constitute a pair of conjugate complex numbers.

12. **Complex number solutions** In the equation $ax^2 + bx + c = 0$ if $b^2 - 4ac < 0$, the solutions are complex numbers. In such cases the two solutions will always be a pair of conjugate complex numbers.

    The sum of the solution = twice the real part
    The product of the solutions = the real part squared + the imaginary part squared.

Review Problems

Complete the statements in Problems 1–6.

1. The equation $2x^2 - 5x + 7 = 0$ is called _____ equation and has at most _____ solutions.

2. In the equation $3x^2 - 5x - 9 = 0$, a = ____, b = ____, and c = ____.

3. The CIs of the expression $6x^2 - x - 2$ are ____ and ____ .

4. The number "i" is defined as $i =$ ____ .

5. $i^5 =$ _____ .

6. The conjugate of $2 - 3i$ is _____ .

In Problems 7–15, solve by the factor method.

Equation	Linear equations		Solutions
7. $x^2 + x - 12 = 0$	_____ $x =$	or _____ $x =$	__ __
8. $x^2 - 4 = 0$	_____ $x =$	or _____ $x =$	__ __
9. $2x^2 - 5x = 0$	_____ $x =$	or _____ $x =$	__ __
10. $x^2 - 7x - 8 = 0$	_____ $x =$	or _____ $x =$	__ __
11. $t^2 + 2t = 8$	_____ $t =$	or _____ $t =$	__ __
12. $2y^2 = 9y - 4$	_____ $y =$	or _____ $y =$	__ __
13. $2x^2 - 7x + 3 = 0$	_____ $x =$	or _____ $x =$	__ __
14. $12x^2 - 7x = 12$	_____ $x =$	or _____ $x =$	__ __
15. $2(4y^2 - 3y) = 9$	_____ $y =$	or _____ $y =$	__ __

In Problems 16–20, identify the coefficients, compute the discriminant and indicate the nature of the solutions.

	Equations	a,b,c	Discriminant ($b^2 - 4ac$)	Nature of solutions
16.	$x^2 + 2x + 1 = 0$	1, 2, 1	$2^2 - 4 \cdot 1 \cdot 1 = 0$	equal
17.	$x^2 - 5x + 6 = 0$	__, __, __		
18.	$x^2 - 7x + 12 = 0$	__, __, __		
19.	$x^2 + 2x - 1 = 0$	__, __, __		
20.	$2x^2 + x + 1 = 0$	__, __, __		

In Problems 21–25, find the sum and the product of solutions without solving the equations.

	Equations	Sum of solutions	Product of solutions
21.	$x^2 + 4x + 1 = 0$		
22.	$2x^2 - 3x + 4 = 0$		
23.	$3x^2 - 5 = 0$		
24.	$5x^2 - 6x + 4 = 0$		
25.	$-x^2 - 4x + 1 = 0$		

In Problems 26–31, solve the equations by factoring and check your solutions.

	Equation	Solutions	Check
26.	$x^2 - x - 6 = 0$ CIs __ __ ( )( ) = 0	____ ____	sum $= -\dfrac{b}{a} =$ product $= \dfrac{c}{a} =$
27.	$x^2 - 2x - 24 = 0$ CIs __ __ ( )( ) = 0	____ ____	sum $= -\dfrac{b}{a} =$ product $= \dfrac{c}{a} =$
28.	$2x^2 - 9x + 4 = 0$ CIs __ __ ( )( )	____ ____	sum $= -\dfrac{b}{a} =$ product $= \dfrac{c}{a} =$
29.	$6x^2 + 11x + 3 = 0$ CIs __ __ ( )( )	____ ____	sum $= -\dfrac{b}{a} =$ product $= \dfrac{c}{a} =$

30. $4x^2 - x - 3 = 0$
    CIs ___ ___
    ( )( )

    sum $= -\dfrac{b}{a} =$

    product $= \dfrac{c}{a} =$

31. $2x^2 - 8 = 0$
    CIs ___ ___
    ( )( )

    sum $= -\dfrac{b}{a} =$

    product $= \dfrac{c}{a} =$

In Problems 32–35, solve by using quadratic formula and check your solutions.

    Equation   Solution   Check

32. $x^2 + 2x + 1 = 0$
    $a = \quad , b =$
    $c =$
    $b^2 - 4ac =$

    $x =$ ___ ___

    sum $= -\dfrac{b}{a} =$

    product $= \dfrac{c}{a} =$

33. $x^2 - 3x - 18 = 0$
    $a = \quad , b =$
    $c =$
    $b^2 - 4ac =$

    $x =$ ___ ___

    sum $= -\dfrac{b}{a} =$

    product $= \dfrac{c}{a} =$

34. $x^2 + 2x - 2 = 0$
    $a = \quad , b =$
    $c =$
    $b^2 - 4ac =$

    $x =$ ___ ___

    sum $= -\dfrac{b}{a} =$

    product $= \dfrac{c}{a} =$

35. $x(x + 1) = 1 - 3x$
    $a = \quad , b =$
    $c =$
    $b^2 - 4ac =$

    $x =$ ___ ___

    sum $= -\dfrac{b}{a} =$

    product $= \dfrac{c}{a} =$

In Problems 36–40, write each number in terms of i and simplify as far as possible.

36. $\sqrt{-2} =$ _____   37. $\sqrt{-4} =$ _____

38. $\sqrt{-28} =$ _____   39. $\sqrt{-18} =$ _____

40. $\sqrt{-27} =$ _____

In Problems 41–43, solve and express the solutions in the form $a + ib$.

    Equation   Solutions   Check

41. $x^2 + x + 1 = 0$

    _____

    _____

    sum $= -\dfrac{b}{a} =$

    product $= \dfrac{c}{a} =$

42. $2x^2 + x - 1 = 0$ _____ sum $= -\dfrac{b}{a} =$

_____ product $= \dfrac{c}{a} =$

43. $3x^2 - 4x + 4 = 0$ _____ sum $= -\dfrac{b}{a} =$

_____ product $= \dfrac{c}{a} =$

Mark the correct response in the following questions.

44. One of the solutions of $8y^2 - 4y = 0$ is
    (a) 4 (b) $\dfrac{1}{4}$ (c) $\dfrac{1}{2}$ (d) $\dfrac{-1}{2}$

45. The equation $z^2 + 8z - 9z - 72 = 0$ is equivalent to the equation
    (a) $(z + 18)(z - 4) = 0$ (b) $(z + 6)(z - 12) = 0$ (c) $(z + 8)(z - 9) = 0$
    (d) $(z + 4)(z - 18) = 0$ (e) None of these

46. The solutions of the equation $2x(x - 3) = 7 - x$ are
    (a) $-\dfrac{7}{2}, 1$ (b) $-3, 7$ (c) $3, -7$ (d) $\dfrac{7}{2}, -1$ (e) None of these

47. The critical integers of the equation $x^2 - 8x + 7 = 0$ are
    (a) -1 and -7 (b) 6 and 1 (c) 3 and 4 (d) 7 and 1 (e) None of these

48. The product of the solutions of $8x^2 + 7x - 8 = 0$ is:
    (a) 1 (b) -1 (c) $\dfrac{7}{8}$ (d) $-\dfrac{7}{8}$ (e) None of these

49. The product of the solutions of $3x^2 - 5x + 3 = 0$ is
    (a) $\dfrac{3}{5}$ (b) $\dfrac{5}{3}$ (c) $-\dfrac{3}{5}$ (d) $-\dfrac{5}{3}$ (e) None of these

50. The discriminant of $2x^2 - 5x + 6 = 0$ is:
    (a) -23 (b) $\sqrt{23}$ (c) 73 (d) -73 (e) None of these

51. One of the solutions of $3x^2 - 8x - 3 = 0$ is:
    (a) $\dfrac{1}{3}$ (b) -3 (c) $\dfrac{4}{3} - \dfrac{\sqrt{7}}{6}$ (d) $-\dfrac{1}{3}$ (e) None of these

52. If the discriminant of a quadratic equation equals zero, the equation has
    (a) Distinct rational solutions (b) Repeated solutions (c) Real irrational roots
    (d) Complex number solutions (e) None of these

53. If $b^2 - 4ac$ is positive, but not the square of a positive integer, then the quadratic equation has
    (a) Distinct rational solutions (b) Complex numbers (c) Real irrational solutions (d) None of these

54. The roots of the equation $6p^2 + 13p + 6 = 0$ are
    (a) $\left(-\dfrac{3}{2}, \dfrac{2}{3}\right)$ (b) $\left(-\dfrac{2}{3}, -\dfrac{3}{2}\right)$ (c) $(-2, -3)$ (d) $0, -1$ (e) None of these

55. If $b^2 - 4ac$ is the square of a positive integer then the quadratic equation has one distinct rational solution.
    (a) True   (b) False

56. If $b^2 - 4ac$ is negative then the quadratic equation has
    (a) Repeated solutions   (b) Distinct rational solutions   (c) Real irrational roots   (d) Complex number solutions   (e) None of these

57. What number is a solution of $x^2 - x - 6 = 0$ and $2x^2 - 18 = 0$
    (a) -3 (b) -2 (c) 3   (d) 6   (e) None of these

58. State whether the following statement is true or false: $\sqrt{-16} = -4$
    (a) True   (b) False

59. The conjugate of $3 + i\sqrt{3}$ is $-3 - i\sqrt{3}$
    (a) True   (b) False

60. State whether the following statement is true or false: $i^5 = -i$
    (a) True   (b) False

61. State whether the following statement is true or false: $-\sqrt{-5} = \sqrt{5}$
    (a) True   (b) False

62. The sum of $4 - 5i$ and its conjugate is
    (a) 8   (b) $8 - 10i$ (c) -8 (d) $-8 - 10i$ (e) None of these

63. One of the solutions of $x^2 + 6x + 4 = 0$ is:
    (a) $3 - \sqrt{5}$ (b) $3 - i\sqrt{5}$ (c) $-3 - i\sqrt{5}$ (d) $-3 - \sqrt{5}$ (e) None of these

64. The solutions of the equation $(x + 2)^2 = 8x + 10$ are:
    (a) $2 + \sqrt{10}, 2 - \sqrt{10}$ (b) $2 + 2\sqrt{5}, 2 - 2\sqrt{5}$ (c) $3 + \sqrt{6}, 3 - \sqrt{6}$
    (d) $3 + 2\sqrt{3}, 3 - 2\sqrt{3}$ (e) None of these

Chapter Test

Solve each of the following problems and match your answer with the responses.

1. Simplify: $\sqrt{54} =$
   (a) $-3\sqrt{6}$ (b) $-3\sqrt{3}$ (c) $3\sqrt{6}$ (d) $3\sqrt{3}$ (e) None of these

2. Simplify: $\sqrt{8} - \sqrt{18} - \sqrt{32}$
   (a) $\sqrt{42}$ (b) $5\sqrt{2}$ (c) $-5\sqrt{2}$ (d) $-\sqrt{42}$ (e) $8 - \sqrt{50}$

3. If $z = 3x^2 + 6xy + y^2$ and $x = 2$ and $y = 4$ then $z =$
   (a) 36   (b) 64   (c) 70   (d) 76   (e) 100

4. If $2x + 3\left(2x + \frac{2}{3}\right) = 18$ then $x =$
   (a) 2   (b) -2 (c) 3   (d) -3 (e) None of these

5. Factor completely the expression: $10x^2 - 5x$
   (a) $x(10x - 5x)$ (b) $5(2x^2 - x)$ (c) $10(x^2 - x)$ (d) $5x(2x - 1)$ (e) None of these

6. The two numbers whose sum is 15 and product is 44 are:
   (a) 22,2   (b) 11,4   (c) −11,−4   (d) −22,−2   (e) None of these

7. Factor completely the expression: $x^2 + x - 6$
   (a) $(x - 3)(x + 2)$ (b) $(x + 3)(x - 2)$ (c) $(x + 6)(x - 1)$ (d) $(x + 3)(x + 2)$ (e) Cannot be factored

8. Factor completely the expression: $6p^2 + 11p - 10$
   (a) $(3p + 2)(2p - 5)$ (b) $(2p + 5)(3p + 2)$ (c) $(2p + 5)(3p - 2)$
   (d) $(6p - 1)(p + 10)$ (e) Cannot be factored

9. One of the solutions of $8y^2 - 4y = 0$ is
   (a) 4   (b) $\frac{1}{4}$ (c) $\frac{1}{2}$ (d) $\frac{-1}{2}$

10. The equation $z^2 + 8z - 9z - 72 = 0$ is equivalent to the equation
    (a) $(z + 18)(z - 4) = 0$ (b) $(z + 6)(z - 12) = 0$ (c) $(z + 8)(z - 9) = 0$
    (d) $(z + 4)(z - 18) = 0$ (e) None of these

11. The solutions of the equation $2x(x - 3) = 7 - x$ are
    (a) $\frac{7}{2}, 1$ (b) −3,7   (c) 3,−7   (d) $\frac{7}{2}, -1$ (e) None of these

12. The product of the solutions of $7x^2 + 3x - 3 = 0$ is:
    (a) $-\frac{7}{3}$ (b) $\frac{3}{7}$ (c) $-\frac{3}{7}$ (d) −1   (e) None of these

13. The product of the solutions of $3x^2 - 5x + 3 = 0$ is
    (a) $\frac{3}{5}$ (b) $\frac{5}{3}$ (c) $-\frac{3}{5}$ (d) $-\frac{5}{3}$ (e) None of these

14. The discriminant of $-3x^2 = 5 - 7x$ is:
    (a) 109   (b) −109   (c) −11   (d) $\sqrt{11}$ (e) None of these

15. If $b^2 - 4ac$ is positive, but not the square of a positive integer, then the quadratic equation has
    (a) Distinct rational solutions   (b) Complex numbers   (c) Real irrational solutions   (d) None of these

16. The roots of the equation $6p^2 + 13p + 6 = 0$ are
    (a) $\left(-\frac{3}{2}, \frac{2}{3}\right)$ (b) $\left(-\frac{2}{3}, -\frac{3}{2}\right)$ (c) (−2,−3)   (d) 0,−1   (e) None of these

17. If $b^2 - 4ac$ is the square of a positive integer then the quadratic equation has one distinct rational solution.
    (a) True   (b) False

18. What number is a solution of $x^2 - x - 6 = 0$ and $2x^2 - 18 = 0$
    (a) −3   (b) −2   (c) 3   (d) 6   (e) None of these

19. State whether the following statement is true or false. $-\sqrt{-9} = -3i$
    (a) True   (b) False

20. One of the solutions of $x^2 + 6x + 4 = 0$ is:
    (a) $3 - \sqrt{5}$ (b) $3 - i\sqrt{5}$ (c) $-3 - i\sqrt{5}$ (d) $-3 - \sqrt{5}$ (e) None of these

Chapter 8

Algebraic Fractions

The only difference between numerical fractions and algebraic fractions is that algebraic fractions contain unknowns which can take on any value, *except those values which make the denominator zero,* whereas numerical fractions contain only specified numbers. The process of performing operations on algebraic fractions is therefore similar to what we discussed in Section 1.6. In this chapter we will discuss

- (a) Some basic facts about algebraic fractions,
- (b) Multiplication and division of algebraic fractions,
- (c) Addition of algebraic fractions,
- (d) Simplification of complex fractions, and
- (e) Fractional equations.

Throughout this chapter we will assume that the unknowns involved in the fractions are not assigned any values that make the denominator zero.

8.1 Some Basic Facts

An *algebraic fraction* is a ratio or a quotient of two polynomials. For example,

$$\frac{7x + 5}{2x - 3}, \quad \frac{8x^2 - 3x + 4}{x - 4}, \quad \frac{7}{5x - 3}, \quad \frac{x^3 + 7x - 5}{x^2 - x - 6}$$

are algebraic fractions or rational expressions. The basic rules governing signed numbers and the reduction of arithmetic fractions discussed in Chapter 1 apply to algebraic fractions. We shall review these rules now.

Algebraic fractions can take on a range of values depending upon what values are assigned to the unknown. However, since a fraction is not defined when the denominator is zero, we cannot assign to the unknown any value which makes the denominator zero.

Example 1: *Permissible values of the unknown*

(a) $\dfrac{3x + 2}{x - 1}$ x can have any value except one; $x \neq 1$

(b) $\dfrac{2y - 1}{3y}$ y can have any value except zero; $y \neq 0$

(c) $\dfrac{x^2 + 5}{x^2 - 1}$ x can have any value except 1 or -1; $x \neq 1$, $x \neq -1$

(d) $\dfrac{2x^3 - 3x + 1}{x^2 + 1}$ x can be any real number, since $x^2 + 1 \neq 0$ for any real number.

The rules for assigning signs to algebraic fractions are the same as the rules for signed numbers. We know that

$$\dfrac{-4}{5} = \dfrac{4}{-5} = -\dfrac{4}{5}$$

That is the ratio of two numbers having unlike signs is a negative number. Also the ratio of two numbers with like signs is a positive number and the negative of a negative number is a positive number. For example,

$$\dfrac{-4}{-5} = \dfrac{4}{5}$$
$$-\dfrac{-4}{5} = -\left(-\dfrac{4}{5}\right) = \dfrac{4}{5}$$
$$-\dfrac{4}{-5} = -\left(-\dfrac{4}{5}\right) = \dfrac{4}{5}$$

We follow the same rule even for algebraic fractions.

Example 2: *Fraction signs*

(a) $\dfrac{-4}{x} = -\dfrac{4}{x}$

(b) $\dfrac{-xy}{-z} = \dfrac{xy}{z}$

(c) $\dfrac{-x(x + 1)}{-(x + 2)} = \dfrac{x(x + 1)}{(x + 2)}$

(d) $\dfrac{x}{-y} = -\dfrac{x}{y}$

(e) $-\dfrac{x + 2}{-2x} = -\left(-\dfrac{x + 2}{2x}\right) = \dfrac{x + 2}{2x}$

Another fact used frequently in algebraic fractions is that

$$a - b = -(b - a)$$

For example:

$$9 - 4 = 5$$
$$4 - 9 = -5$$

Therefore,

$$4 - 9 = -(9 - 4)$$

Of course, for addition, we can interchange the numbers we add. That is,

$$a + b = b + a$$
$$4 + 9 = 9 + 4 = 13$$

Example 3: $a - b = -(b - a)$ and $a + b = b + a$

(a) $3 - x = -(x - 3)$
(b) $x - 5 = -(5 - x)$
(c) $3 + x = x + 3$
(d) $x^2 + 2x = 2x + x^2$
(e) $(x)(2 - x) = (x)[-(x - 2)] = -(x)(x - 2)$
(f) $\dfrac{2 - x}{x - 2} = \dfrac{-(x - 2)}{(x - 2)} = -\dfrac{x - 2}{x - 2} = -1$

Fractions should always be left in the most simplified form. We simplify, or reduce numerical fractions, by dividing the numerator and the denominator by the common factors (cancelling common factors). The same process is used for simplifying or reducing algebraic fractions. For example,

$$\frac{8}{12} = \frac{\cancel{2} \cdot \cancel{2} \cdot 2}{\cancel{2} \cdot \cancel{2} \cdot 3} = \frac{2}{3}$$

$$\frac{x^3}{x^2 y} = \frac{\cancel{x} \cdot \cancel{x} \cdot x}{\cancel{x} \cdot \cancel{x} \cdot y} = \frac{x}{y}$$

$$\frac{x(x - 1)^2}{x^2(x - 1)} = \frac{\cancel{x}(\cancel{x - 1})(x - 1)}{\cancel{x} \cdot x(\cancel{x - 1})} = \frac{x - 1}{x}$$

Example 4: *Reducing fractions*

(a) $\dfrac{x^2 y^3}{x^3 y} = \dfrac{x \cdot x \cdot y \cdot y \cdot y}{x \cdot x \cdot x \cdot y} = \dfrac{y \cdot y}{x} = \dfrac{y^2}{x}$

(b) $\dfrac{x(x - 1)}{1 - x} = \dfrac{x(x - 1)}{-(x - 1)} = \dfrac{x}{-1} = -\dfrac{x}{1} = -x$

(c) $\dfrac{x(x - 1)}{y(1 - x)} = \dfrac{x(x - 1)}{-y(x - 1)} = \dfrac{x}{-y} = -\dfrac{x}{y}$

The numerator or the denominator may not always be given in the factored form. If this is the case, then before we reduce fractions, the expressions in the numerator and the denominator must be written in factored form. For example,

$$\frac{5 + 10x}{5x} = \frac{5(1 + 2x)}{5x} = \frac{1 + 2x}{x}$$

Always keep in mind whenever we reduce fractions

➡ *WE CANCEL ONLY FACTORS AND NOT TERMS.*

Note that $\dfrac{4x + 16x^2}{4x} \neq 1 + 16x^2$

A common error is to cancel the terms $4x$ in the numerator and denominator to obtain $1 + 16x^2$. The correct way is

$$\frac{4x + 16x^2}{4x} = \frac{4x(1 + 4x)}{4x} = 1 + 4x.$$

Example 5: *Reduce the fractions*

(a) $\dfrac{4x^2y}{4x^2 - 2x} = \dfrac{4x^2y}{2x(2x - 1)} = \dfrac{2 \cdot 2 \cdot x \cdot x \cdot y}{2x(2x - 1)} = \dfrac{2xy}{2x - 1}$

(b) $\dfrac{2x^2 - 8}{x^2 - 3x + 2} = \dfrac{2(x - 2)(x + 2)}{(x - 1)(x - 2)} = \dfrac{2(x + 2)}{x - 1}$

(c) $\dfrac{4x^3 - 36x}{3x^2 + 13x + 12} = \dfrac{4x(x - 3)(x + 3)}{(3x + 4)(x + 3)} = \dfrac{4x(x - 3)}{3x + 4}$

(d) $\dfrac{6x^2 + 5x - 4}{1 - 4x^2} = \dfrac{(3x + 4)(2x - 1)}{(1 + 2x)(1 - 2x)}$

$= \dfrac{(3x + 4)(2x - 1)}{-(1 + 2x)(2x - 1)}$

$= -\dfrac{3x + 4}{1 + 2x}$

$\begin{aligned}6x^2 + 5x - 4 &= 6x^2 + 8x - 3x - 4\\&= 2x(3x + 4) - 1(3x + 4)\\&= (3x + 4)(2x - 1)\end{aligned}$

and $\begin{aligned}1 - 4x^2 &= 1^2 - (2x)^2\\&= (1 - 2x)(1 + 2x)\end{aligned}$

Problem Set 8.1

In Problems 1–6, write the values of the unknown for which the given expression is not defined.

	Expression	Not defined for x =		Expression	Not defined for x =
1.	$\dfrac{3x + 4}{x}$	0	2.	$\dfrac{2x}{x - 1}$	
3.	$\dfrac{3x^2 + 4x - 1}{x + 1}$		4.	$\dfrac{x^2 + 1}{x(x + 1)}$	
5.	$\dfrac{2x - 1}{x^2 - 1}$		6.	$\dfrac{x^3 + 1}{x^2 - 3x + 2}$	

In Problems 7–16, simplify the expressions as in Examples 2 and 3.

	Expression	Simplified form		Expression	Simplified form
7.	$-\dfrac{3}{-4} =$	$\dfrac{3}{4}$	8.	$-\dfrac{-8}{-9} =$	
9.	$-\dfrac{-4}{9} =$		10.	$-\dfrac{-x}{-y} =$	
11.	$-\dfrac{x + 2}{-x} =$		12.	$-(x - 1) =$	
13.	$-\dfrac{x + 1}{1 - x} =$		14.	$\dfrac{3 + x}{x + 3} =$	
15.	$-\dfrac{x - 2}{2 - x} =$		16.	$\dfrac{x - 3}{3 - x} =$	

In Problems 17–46, simplify the expressions as in Examples 4 and 5.

	Expression	Simplified form		Expression	Simplified form
17.	$\dfrac{16}{24}$	$= \dfrac{2\cdot 2\cdot 2\cdot 2}{3\cdot 2\cdot 2\cdot 2} = \dfrac{2}{3}$	18.	$\dfrac{x^2 + x}{x^2 - 1} =$	$\dfrac{x(x+1)}{(x-1)(x+1)} = \dfrac{x}{x-1}$

19. $\dfrac{x^2 yz}{xy^2} =$

20. $\dfrac{xy^3}{z^2} =$

21. $\dfrac{24xy^3 z^5}{20yz^2} =$

22. $\dfrac{-x^2 y^3}{x^2 y} =$

23. $\dfrac{4x^2 y^3 z}{-2xy^3} =$

24. $\dfrac{-20xy^2 z}{4x^2 yz} =$

25. $\dfrac{-45x(x-1)}{15(1-x)} =$

26. $\dfrac{x^2 - 1}{x - 1} =$

27. $\dfrac{2x^2 - 8}{x^2 - 2x} =$

28. $\dfrac{x^3 + 4x^2}{x^2 - 16} =$

29. $\dfrac{10x^3 - 40x^2}{x^3 - 4x} =$

30. $\dfrac{ax + ab}{x + b} =$

31. $\dfrac{x^2 y - xy}{x^2 - 1} =$

32. $\dfrac{ab - ac}{5a} =$

33. $\dfrac{2a^2 b - ab}{4ab} =$

34. $\dfrac{x^2 + x}{2x^2 - 2} =$

35. $\dfrac{4x^2 - 4x}{2x - 2x^2} =$

36. $\dfrac{5x^2 - 5x}{2x - 2} =$

37. $\dfrac{3x^3 - 81}{6x - 18} =$

38. $\dfrac{x^2 - x}{1 - x} =$

39. $\dfrac{y^2 - 4y}{4 - y} =$

40. $\dfrac{2x^2 - x - 6}{x^2 - 4} =$

41. $\dfrac{x^2 - x - 6}{x^2 - 9} =$

42. $\dfrac{2 - 2x^2}{x^2 + x - 2} =$

43. $\dfrac{15x^3 - 15x}{1 - x} =$

44. $\dfrac{a^2 - b^2}{5b - 5a} =$

45. $\dfrac{x^3 - 27}{3 - x} =$

46. $\dfrac{6x^2 + xy - 35y^2}{9x^2 - 6xy - 35y^2} =$

Mark the correct response to the following items.

47. Consider the statements P and Q

    P: $x - 1 = -(1 - x)$ Q: $-\dfrac{2 - x}{x - 2} = -1$

    Which of the above statements is true?
    (a) P only   (b) Q only   (c) P and Q   (d) none

48. The expression $\dfrac{2x^2 - 8}{x^2 - 3x + 2}$ when simplified equals

    (a) $\dfrac{2x + 2}{x + 1}$ (b) $\dfrac{2x + 2}{x - 1}$ (c) $\dfrac{2x + 4}{x - 1}$ (d) none of these

49. The expression $\dfrac{6x^2 + x - 1}{6x^2 - 5x + 1}$ when simplified equals

    (a) $\dfrac{3x + 1}{3x - 1}$ (b) $\dfrac{3x - 1}{3x + 1}$ (c) $\dfrac{2x - 1}{2x + 1}$ (d) $\dfrac{2x + 1}{2x - 1}$ (e) none of these

8.2 Multiplication and Division

Recall that to multiply two or more numerical fractions you multiply the numerators and denominators. The numerator and the denominator are then expressed in factored form and the factors which are common to both are cancelled to get the final result as a completely reduced fraction. For example,

$$\left(\frac{3}{5}\right) \cdot \left(\frac{10}{9}\right) = \frac{3 \cdot 10}{5 \cdot 9} = \frac{\cancel{3} \cdot 2 \cdot \cancel{5}}{\cancel{5} \cdot \cancel{3} \cdot 3} = \frac{2}{3}$$

We follow exactly the same procedure for multiplying algebraic fractions. Read carefully the following examples.

Example 1: *Perform the indicated multiplication*

$\dfrac{6x^2y}{5xz} \times \dfrac{10z^2}{24xy^2}$

$= \dfrac{(6x^2y)(10z^2)}{(5xz)(24xy^2)}$ Multiply the numerators and denominators

$= \dfrac{(2 \cdot 3)(2 \cdot 5)x \cdot x \cdot y \cdot z \cdot z}{5 \cdot 2 \cdot 2 \cdot 2 \cdot 3 \cdot x \cdot z \cdot x \cdot y \cdot y}$ Factor the numerator and the denominator

$= \dfrac{2 \cdot 2}{2 \cdot 2} \cdot \dfrac{3 \cdot 5}{3 \cdot 5} \cdot \dfrac{x \cdot x}{x \cdot x} \cdot \dfrac{y}{y} \cdot \dfrac{z}{z} \cdot \dfrac{z}{2y}$ Reduce the fraction

$= 1 \cdot 1 \cdot 1 \cdot 1 \cdot 1 \cdot \dfrac{z}{2y} = \dfrac{z}{2y}$

Example 2: *Multiply*

$\dfrac{x^2 - 1}{x^2 - 5x + 6} \cdot \dfrac{2x^2 - 7x + 3}{2x^2 + x - 1}$

$= \dfrac{(x - 1)(x + 1)}{(x - 2)(x - 3)} \cdot \dfrac{(2x - 1)(x - 3)}{(2x - 1)(x + 1)}$ Factor each trinomial

Multiply the numerators and the denominators. Arrange the factors in the denominator in such a manner that the factors common in the numerator and the denominator are right below each other. This arrangement will make it convenient to identify the common factors.

$= \dfrac{(x - 1)\cancel{(x + 1)}\cancel{(2x - 1)}\cancel{(x - 3)}}{(x - 2)\cancel{(x + 1)}\cancel{(2x - 1)}\cancel{(x - 3)}} = \dfrac{x - 1}{x - 2}$

Example 3: *Divide*

$$\frac{2x^2 - 8}{2x^2 + x - 6} \div \frac{3x^2 - 5x - 2}{6x^2 - 7x - 3}$$

Just as with numerical fractions, we change the division problem to a multiplication problem. Multiply the dividend by the reciprocal of the divisor. Recall that the fraction before the division symbol is the dividend and the fraction after the division symbol is the divisor.

$$\frac{2x^2 - 8}{2x^2 + x - 6} \times \frac{6x^2 - 7x - 3}{3x^2 - 5x - 2}$$

$$= \frac{2(x - 2)(x + 2)}{(2x - 3)(x + 2)} \cdot \frac{(3x + 1)(2x - 3)}{(x - 2)(3x + 1)} \qquad \text{Factoring all the expressions}$$

Multiply the numerators and the denominators in such a manner that factors in the denominator are written directly under identical factors in the numerator.

$$= \frac{2(x - 2)(x + 2)(3x + 1)(2x - 3)}{(x - 2)(x + 2)(3x + 1)(2x - 3)} = 2.$$

Example 4: *Simplify*

$$\frac{\dfrac{2x}{2x^2 - 18}}{\dfrac{4}{x + 3}} = \frac{2x}{2x^2 - 18} \cdot \frac{x + 3}{4} = \frac{2 \cdot x}{2(x - 3)(x + 3)} \cdot \frac{x + 3}{4}$$

$$= \frac{2 \cdot x \cdot (x + 3)}{4 \cdot 2(x - 3)(x + 3)} = \frac{x}{4(x - 3)}$$

Problem Set 8.2

In Problems 1–29, perform the indicated operations and reduce.

Answer

1. $\dfrac{x^2}{y} \cdot \dfrac{y^2}{x^3} =$ $\dfrac{y}{x}$

2. $\dfrac{a^2}{b} \cdot \dfrac{c}{a} =$ _____

3. $\dfrac{2x + 1}{x^2} \cdot \dfrac{x}{2x + 1} =$ _____

4. $\dfrac{x + 4}{x - 2} \cdot \dfrac{x - 2}{x + 4} =$ _____

5. $\dfrac{2x - 4}{x + 1} \cdot \dfrac{x^2 - 1}{x - 2} =$ _____

6. $\dfrac{5x - 10}{x^2} \cdot \dfrac{x^2 + 2x}{x^2 - 4} =$ _____

7. $\dfrac{x + y}{x^2 - y^2} \cdot \dfrac{x^2 - xy}{x} =$ _____

8. $\dfrac{a^2 - ab}{a} \cdot \dfrac{a + b}{(a - b)^2} =$ _____

9. $\dfrac{x^2 + 5x + 4}{x^2 + 8x + 15} \cdot \dfrac{x^2 + 5x + 6}{x^2 + 7x + 12} =$

10. $\dfrac{3x^2 - 6x}{x^3 + 3x^2} \cdot \dfrac{x^2 - 4x}{6x - 12} =$

11. $\dfrac{12x + 4}{6x - 8} \cdot \dfrac{20x^2 - 5x}{8x + 24x^2} =$

12. $\dfrac{x^2 - 2x - 3}{12x^2y^3} \cdot \dfrac{4xy^3}{x^2 - 9x + 18} =$ 12. _____

13. $\dfrac{2x^2 - 3x}{x + 4} \cdot \dfrac{x^2 + x}{2x^3 - 3x^2} =$ 13. _____

14. $\dfrac{x^2 + x - 2}{x^2y^2} \cdot \dfrac{x^3y^4}{x^2 - 4} =$ 14. _____

15. $\dfrac{a^2 - a}{a - 1} \cdot \dfrac{a + 1}{a} =$ 15. _____

16. $\dfrac{x + 3}{x^3 + 3x^2} \cdot \dfrac{x^3}{x - 3} =$ 16. _____

17. $\dfrac{2x^2 + 4x}{12x^2y} \cdot \dfrac{6x}{x^2 + 6x + 8} =$ 17. _____

18. $\dfrac{x^2 - 6x + 9}{x^2 - x - 6} \cdot \dfrac{x^2 + 2x}{x^2 + 2x - 15} =$ 18. _____

19. $\dfrac{6b^2c + 5bc^2 - c^3}{6b^2 + 11bc - 2c^2} \cdot \dfrac{2b^2 + 7bc + 6c^2}{b^2c - c^3} =$ 19. _____

20. $\dfrac{x}{5} \div \dfrac{x^2}{25} = \dfrac{x}{5} \cdot \dfrac{25}{x^2} = \dfrac{x}{x} \cdot \dfrac{5}{5} \cdot \dfrac{5}{x} = 1 \cdot 1 \cdot \dfrac{5}{x}$ 20. $\dfrac{5}{x}$

21. $\dfrac{x + y}{12} \div \dfrac{2(x + y)}{3} =$ 21. _____

22. $\dfrac{\frac{4x^2}{5y}}{\frac{16x}{15y^2}} =$ 22. _____

23. $\dfrac{5y}{9} \div 10y =$ 23. _____

24. $\dfrac{10x^3}{x^2 - 1} \div \dfrac{5x}{x - 1} =$ 24. _____

25. $-6x^3 \div \dfrac{4x^2}{3} =$ 25. _____

26. $\dfrac{x^2 + 4x + 3}{6x^2 + x - 1} \div \dfrac{2x^2 + 5x - 3}{4x^2 - 1} =$ 26. _____

27. $\dfrac{2x^2 + 5x - 3}{4x^2 - 1} \div \dfrac{x^2 + 4x + 3}{6x^2 + x - 1} =$ 27. _____

28. $\dfrac{\frac{12}{5}(x^2 - x - 20)}{\frac{3x + 15}{4}} =$

28. _____

29. $\dfrac{\frac{9x + 18}{4 - 9x^2}}{\frac{(x + 2)^2}{3x - 2}} =$

29. _____

8.3 Addition and Subtraction of Fractions

All that you need for adding algebraic fractions is, (1) the skill to factor algebraic expressions, and (2) the skill to add numerical fractions. Just like other operations, we add algebraic fractions in a way similar to the way we add numerical fractions. Read the following examples carefully and observe the similarity.

Example 1: *Addition of fractions with like denominators*

Numerical fractions

(a) $\dfrac{3}{5} + \dfrac{4}{5} = \dfrac{3 + 4}{5} = \dfrac{7}{5}$ $\dfrac{y}{x} + \dfrac{z}{x} = \dfrac{y + z}{x}$

(b) $\dfrac{5}{3} - \dfrac{4}{3} = \dfrac{5 - 4}{3} = \dfrac{1}{3}$ $\dfrac{x}{y} - \dfrac{z}{y} = \dfrac{x - z}{y}$

(c) $\dfrac{x}{x - y} + \dfrac{y}{x - y} = \dfrac{x + y}{x - y}$

(d) $\dfrac{y}{x^2} - \dfrac{z}{x^2} = \dfrac{y - z}{x^2}$

(e) $\dfrac{x}{(x - 1)} + \dfrac{1}{(x - 1)} = \dfrac{x + 1}{x - 1}$

(f) $\dfrac{x}{x - 1} - \dfrac{1}{x - 1} = \dfrac{(x - 1)}{(x - 1)} = 1$

(g) $\dfrac{2x - 3}{x - 2} + \dfrac{x}{x - 2} = \dfrac{(2x - 3) + x}{x - 2}$
$= \dfrac{3x - 3}{x - 2}\ \dfrac{3(x - 1)}{x - 2}$

(h) $\dfrac{x - 1}{x^2 - 1} + \dfrac{3}{x + 1} = \dfrac{(x - 1)}{(x - 1)(x + 1)} + \dfrac{3}{x + 1}$
$= \dfrac{1}{x + 1} + \dfrac{3}{x + 1}$ (same denominator)
$= \dfrac{1 + 3}{x + 1} = \dfrac{4}{x + 1}$

(i) $\dfrac{3x}{2x + 1} + \dfrac{x + 1}{2x^2 + 3x + 1}$
$= \dfrac{3x}{(2x + 1)} + \dfrac{(x + 1)}{(2x + 1)(x + 1)}$ Factoring the denominator
$= \dfrac{3x}{2x + 1} + \dfrac{1}{2x + 1}$ Same denominator
$= \dfrac{3x + 1}{2x + 1}$

If the denominators of fractions involved in addition are not the same then as in the case of numerical fractions we perform the following steps. Recall, Rule 1.6.

Step 1. *Find the Least Common Denominator (LCD).*
Step 2. *Change each fraction to an equivalent fraction with the LCD as the common denominator.*
Step 3. *Perform the addition as in Example 1.*

In Illustrations 1–3, we discuss the procedure for finding the Least Common Denominator which is similar to what we discussed in Section 1.6.

Illustrations: *TO FIND THE LCD*
(Step 1)

1. Consider the fractions $\frac{3}{8}, \frac{8}{12}, \frac{5}{18}$

   Express each denominator as a product of prime factors.

   $8 = 2 \cdot 2 \cdot 2$
   $12 = 2 \cdot 2 \cdot 3$
   $18 = 2 \cdot 3 \cdot 3$

   Then the LCD $= 2 \cdot 2 \cdot 2 \cdot 3 \cdot 3 = 2^3 \cdot 3^2$

   Note that there are only two *distinct* factors, 2 and 3, but 2 is repeated three times in 8 and 3 is repeated two times in 18. In general as in Rule 1.5 we adopt the following procedure.

Rule 8.1

Finding the LCD

If x, y and z are any three *distinct* prime factors in the denominators then the

$$LCD = x^a \cdot y^b \cdot z^c$$

where a = the maximum number of times factor "x" is repeated in any single denominator,
b = the maximum number of times factor "y" is repeated in any single denominator,
c = the maximum number of times factor "z" is repeated in any single denominator.

---

2. Consider the fractions $\frac{4}{x^2}, \frac{5x}{x^2 + x}, \frac{3x}{x^2 - 1}$.

   Factor each denominator.

   $x^2 = x \cdot x$
   $x^2 + x = x(x + 1)$
   $x^2 - 1 = (x - 1)(x + 1)$

   There are three distinct factors namely, x, x + 1, and x − 1 and factor x is repeated *TWO TIMES* in the first denominator. Thus, the LCD $= x^2(x + 1)(x - 1)$

3. Consider the fractions $\frac{2x-1}{x^3-4x}, \frac{5}{x^3-2x^2}, \frac{3x-4}{(x-2)^2}$.

   Factor each denominator.

   $x^3 - 4x = x(x^2 - 4) = x(x - 2)(x + 2)$
   $x^3 - 2x^2 = x^2(x - 2) = x \cdot x(x - 2)$
   $(x - 2)^2 = (x - 2)(x - 2)$

   There are three distinct factors: x, $x - 2$, and $x + 2$. The factor x is repeated twice in the second denominator and the factor $x - 2$ is repeated twice in the third denominator. Thus, the LCD $= x^2(x - 2)^2(x + 2)$.

In Illustrations 4–6, we will discuss the procedure for changing the given fractions to equivalent fractions with the LCD as the common denominator.

Illustrations: *To Change Fractions to Equivalent Fractions*
(Step 2)

4. In Illustration 1, we obtained $2^3 \cdot 3^2$ or 72 as the LCD of fractions $\frac{3}{8}, \frac{7}{12}, \frac{5}{18}$. The second step for adding these fractions is to change these fractions to equivalent fractions with $2 \cdot 2 \cdot 2 \cdot 3 \cdot 3$ as the denominator.

   $\frac{3}{8} = \frac{3}{2 \cdot 2 \cdot 2} = \frac{3}{2 \cdot 2 \cdot 2} \cdot \frac{3 \cdot 3}{3 \cdot 3} = \frac{27}{72}$

   $\frac{7}{12} = \frac{7}{2 \cdot 2 \cdot 3} = \frac{7}{2 \cdot 2 \cdot 3} \cdot \frac{2 \cdot 3}{2 \cdot 3} = \frac{42}{72}$

   $\frac{5}{18} = \frac{5}{2 \cdot 3 \cdot 3} = \frac{5}{2 \cdot 3 \cdot 3} \cdot \frac{2 \cdot 2}{2 \cdot 2} = \frac{20}{72}$

5. In Illustration 2, we obtained $x^2(x + 1)(x - 1)$ as the LCD of the fractions $\frac{4}{x^2}, \frac{5x}{x^2 + x}, \frac{3x}{x^2 - 1}$. The fractions equivalent to these with the LCD as the denominator are

   $\frac{4}{x^2} = \frac{4}{x \cdot x} = \frac{4}{x \cdot x} \cdot \frac{(x + 1)(x - 1)}{(x + 1)(x - 1)} = \frac{4(x + 1)(x - 1)}{x^2(x + 1)(x - 1)}$

   $\frac{5x}{x^2 + x} = \frac{5x}{x(x + 1)} = \frac{5x}{x(x + 1)} \cdot \frac{x(x - 1)}{x(x - 1)} = \frac{5x^2(x - 1)}{x^2(x + 1)(x - 1)}$

   $\frac{3x}{x^2 - 1} = \frac{3x}{(x + 1)(x - 1)} = \frac{3x}{(x + 1)(x - 1)} \cdot \frac{x^2}{x^2} = \frac{3x^3}{x^2(x + 1)(x - 1)}$

6. In Illustration 3, we obtained $x^2(x - 2)^2(x + 2)$ as the least common denominator of the fractions $\frac{2x - 1}{x^3 - 4x}, \frac{5}{x^3 - 2x^2}$, and $\frac{3x - 4}{(x - 2)^2}$. The fractions equivalent to these with the LCD as the denominator are:

   $\frac{2x - 1}{x^3 - 4x} = \frac{2x - 1}{x(x - 2)(x + 2)}$
   $= \frac{2x - 1}{x(x - 2)(x + 2)} \cdot \frac{x - 2}{x - 2} \cdot \frac{x}{x} = \frac{x(2x - 1)(x - 2)}{x^2(x - 2)^2(x + 2)}$

   $\frac{5}{x^3 - 2x^2} = \frac{5}{x^2(x - 2)} = \frac{5}{x^2(x - 2)} \cdot \frac{x + 2}{x + 2} \cdot \frac{x - 2}{x - 2} = \frac{5(x + 2)(x - 2)}{x^2(x - 2)^2(x + 2)}$

   $\frac{3x - 4}{(x - 2)^2} = \frac{3x - 4}{(x - 2)^2} \cdot \frac{x^2}{x^2} \cdot \frac{x + 2}{x + 2} = \frac{(3x - 4)x^2(x + 2)}{x^2(x - 2)^2(x + 2)}$

We now use the procedures we learned above to add fractions.

Illustrations: *Addition of Fractions with Unlike Denominators*
(Step 3)

7. Compute: $\dfrac{3}{8} + \dfrac{7}{12} + \dfrac{5}{18}$

   LCD = 72   Step 1 (Illustration 1)

   $\dfrac{27}{72} + \dfrac{42}{72} + \dfrac{20}{72}$ Step 2 (Illustration 4)

   $= \dfrac{27 + 42 + 20}{72} = \dfrac{89}{72}$

8. Simplify: $\dfrac{4}{x^2} + \dfrac{5x}{x^2 + x} + \dfrac{3x}{x^2 - 1}$

   LCD = $x^2(x + 1)(x - 1)$ Step 1 (Illustration 2)

   $\dfrac{4(x + 1)(x - 1)}{x^2(x + 1)(x - 1)} + \dfrac{5x^2(x - 1)}{x^2(x + 1)(x - 1)} + \dfrac{3x^3}{x^2(x + 1)(x - 1)}$ Step 2 (Illustration 5)

   $= \dfrac{4(x + 1)(x - 1) + 5x^2(x - 1) + 3x^3}{x^2(x + 1)(x - 1)}$

   $= \dfrac{4(x^2 - 1) + 5x^3 - 5x^2 + 3x^3}{x^2(x + 1)(x - 1)} = \dfrac{8x^3 - x^2 - 4}{x^2(x + 1)(x - 1)}$

9. Add: $\dfrac{2x - 1}{x^3 - 4x} + \dfrac{5}{x^3 - 2x^2} + \dfrac{3x - 4}{(x - 2)^2}$

   LCD = $x^2(x - 2)^2(x + 2)$ Step 1 (Illustration 3)

   $\dfrac{x(2x - 1)(x - 2)}{x^2(x - 2)^2(x + 2)} + \dfrac{5(x + 2)(x - 2)}{x^2(x - 2)^2(x + 2)} + \dfrac{x^2(3x - 4)(x + 2)}{x^2(x - 2)^2(x + 2)}$ Step 2 (Illustration 6)

   $= \dfrac{x(2x - 1)(x - 2) + 5(x + 2)(x - 2) + x^2(3x - 4)(x + 2)}{x^2(x - 2)^2(x + 2)}$

   $= \dfrac{3x^4 + 4x^3 - 8x^2 + 2x - 20}{x^2(x - 2)^2(x + 2)}$

Example 1: Add $\dfrac{5}{12} + \dfrac{7}{20}$

Step 1.   Find the LCD

$12 = 2 \cdot 2 \cdot 3$
$20 = 2 \cdot 2 \cdot 5$
LCD = $2^2 \cdot 3 \cdot 5 = 60$

Step 2.   Change each fraction to an equivalent fraction with the LCD as the denominator.

$\dfrac{5}{12} = \dfrac{5}{12} \cdot \dfrac{5}{5} = \dfrac{25}{60}$

$\dfrac{7}{20} = \dfrac{7}{20} \cdot \dfrac{3}{3} = \dfrac{21}{60}$

Step 3.   Perform the addition

$\dfrac{5}{12} + \dfrac{7}{20} = \dfrac{25}{60} + \dfrac{21}{60} = \dfrac{25 + 21}{60} = \dfrac{46}{60} = \dfrac{23}{30}$

Example 2: Add $\dfrac{x+1}{x-1} + \dfrac{x}{x+1}$

Step 1. LCD $= (x-1)(x+1)$

Step 2. $\dfrac{x+1}{x-1} = \dfrac{(x+1)}{(x-1)} \cdot \dfrac{(x+1)}{(x+1)}$

$\dfrac{x}{x+1} = \dfrac{x}{(x+1)} \cdot \dfrac{(x-1)}{(x-1)}$

Step 3. $\dfrac{x+1}{x-1} + \dfrac{x}{x+1}$

$= \dfrac{(x+1)(x+1)}{(x-1)(x+1)} + \dfrac{x(x-1)}{(x+1)(x-1)} = \dfrac{(x+1)(x+1) + x(x-1)}{(x-1)(x+1)}$

$= \dfrac{x^2 + 2x + 1 + x^2 - x}{(x-1)(x+1)} = \dfrac{2x^2 + x + 1}{(x-1)(x+1)}$

Example 3: Perform the indicated operation: $\dfrac{x-1}{x^2-4} - \dfrac{2}{x^2-2x}$

Step 1. Find the LCD

$x^2 - 4 = (x-2)(x+2)$
$x^2 - 2x = x(x-2)$
LCD $= x(x-2)(x+2)$

Step 2. Change each fraction to an equivalent fraction.

$\dfrac{x-1}{x^2-4} = \dfrac{x-1}{(x-2)(x+2)} = \dfrac{x-1}{(x-2)(x+2)} \cdot \dfrac{x}{x}$

$\dfrac{2}{x^2-2x} = \dfrac{2}{x(x-2)} = \dfrac{2}{x(x-2)} \cdot \dfrac{(x+2)}{(x+2)}$

Step 3. Perform the addition

$\dfrac{x-1}{x^2-4} - \dfrac{2}{x^2-2x} = \dfrac{x(x-1)}{x(x-2)(x+2)} - \dfrac{2(x+2)}{x(x-2)(x+2)}$

$= \dfrac{x(x-1) - 2(x+2)}{x(x-2)(x+2)} = \dfrac{x^2 - x - 2x - 4}{x(x-2)(x+2)} = \dfrac{x^2 - 3x - 4}{x(x-2)(x+2)}$

Example 4: Perform the indicated operations.

$4 + \dfrac{2x}{x+1} - \dfrac{5}{x^2 - 2x - 3}$

Note that $4 = \dfrac{4}{1}$

Step 1. Find the LCD

$1 = 1$
$x + 1 = x + 1$
$x^2 - 2x - 3 = (x-3)(x+1)$
LCD $= 1(x+1)(x-3) = (x+1)(x-3)$

Steps 2 and 3.

$$\frac{4}{1} + \frac{2x}{x+1} - \frac{5}{x^2 - 2x - 3}$$
$$= \frac{4}{1} \cdot \frac{(x+1)(x-3)}{(x+1)(x-3)} + \frac{2x}{x+1} \cdot \frac{x-3}{x-3} - \frac{5}{(x-3)(x+1)}$$
$$= \frac{4(x+1)(x-3) + 2x(x-3) - 5}{(x+1)(x-3)}$$
$$= \frac{4(x^2 - 2x - 3) + 2x^2 - 6x - 5}{(x+1)(x-3)}$$
$$= \frac{6x^2 - 14x - 17}{(x+1)(x-3)}$$

Example 5: Perform the indicated operations and simplify

$$\frac{x+1}{x^2 - 7x + 10} - \frac{3x}{x^2 - 6x + 8}$$
$$= \frac{x+1}{(x-2)(x-5)} - \frac{3x}{(x-2)(x-4)}$$

The LCD is $(x-2)(x-5)(x-4)$

$$= \frac{x+1}{(x-2)(x-5)} \cdot \frac{(x-4)}{(x-4)} - \frac{3x}{(x-2)(x-4)} \cdot \frac{(x-5)}{(x-5)}$$
$$= \frac{(x+1)(x-4) - 3x(x-5)}{(x-2)(x-5)(x-4)}$$
$$= \frac{x^2 - 3x - 4 - 3x^2 + 15x}{(x-2)(x-5)(x-4)} = \frac{-2x^2 + 12x - 4}{(x-2)(x-5)(x-4)}$$

Example 6: Simplify $\frac{2x}{x-1} - \frac{1}{1-x}$

We have seen in Section 8.1 that

$$-\frac{1}{1-x} = -\frac{1}{-(x-1)} = \frac{1}{x-1}$$

Therefore,

$$\frac{2x}{x-1} - \frac{1}{1-x} = \frac{2x}{x-1} + \frac{1}{x-1} = \frac{2x+1}{x-1}$$

Problem Set 8.3

In Problems 1–7, find the Least Common Denominator of the fractions.

Fractions	Factors of the Denominator	LCD
1. $\frac{5}{x^3 - x^2}, \frac{4x}{x^2 - 3x + 2}$	$x^3 - x^2 = x \cdot x \cdot (x-1)$ $x^2 - 3x + 2 = (x-2)(x-1)$	$x \cdot x \cdot (x-1)(x-2)$
2. $\frac{4}{15}, \frac{5}{9}$	$15 =$ $9 =$	_____
3. $\frac{4}{x^2} - \frac{5}{3x}$	$x^2 =$ $3x =$	_____

4. $\dfrac{8}{x-1} - \dfrac{9}{x-3}$ $x - 1 =$
                                       $x - 3 =$ _____

5. $\dfrac{5x-1}{x^2} - \dfrac{2x+3}{x-1}$ $x^2 =$
                                       $x - 1 =$ _____

6. $\dfrac{x}{x^2-9}, \dfrac{2x-1}{(x-3)^2}, \dfrac{5}{2x+6}$ $x^2 - 9 =$
                                                    $(x-3)^2 =$ _____
                                                    $2x + 6 =$

7. $\dfrac{x-1}{4x^2+8x-5}, \dfrac{7x-3}{4x^2-1}$ $4x^2 + 8x - 5 =$
                                                   $4x^2 - 1 =$ _____

In Problems 8–14, find the missing numerators.

8. $\dfrac{2x-1}{(x-1)} = \dfrac{}{(x-1)(x+1)} = \dfrac{}{(x-1)(x+1)}$

9. $\dfrac{5}{9} = \dfrac{}{9 \cdot 4} = \dfrac{}{9 \cdot 4}$

10. $\dfrac{y}{x+4} = \dfrac{}{(x+4)(x+1)} = \dfrac{}{(x+4)(x+1)}$

11. $\dfrac{2x}{x-5} = \dfrac{}{(x-5)(x-3)} = \dfrac{}{(x-5)(x-3)}$

12. $\dfrac{x^2-4}{x(x+1)} = \dfrac{}{x(x+1)(x+2)} = \dfrac{}{x(x+1)(x+2)}$

13. $\dfrac{x+1}{x} = \dfrac{}{x^2(x+1)(x+2)} = \dfrac{}{x^2(x+1)(x+2)}$

14. $\dfrac{4}{(x+1)(x^2+4)} = \dfrac{}{x(x-1)(x+1)(x^2+4)} = \dfrac{}{x(x-1)(x+1)(x^2+4)}$

In Problems 15–39, perform the indicated operations and simplify.

Problem	Show work here (Steps 2 and 3)	Answer
15. $\dfrac{5}{4x^2} + \dfrac{3x-4}{x^2-x} =$ $4x^2 = 2 \cdot 2 \cdot x \cdot x$ $x^2 - x = x(x-1)$ $\text{LCD} = 2 \cdot 2 \cdot x \cdot x(x-1)$ $\phantom{\text{LCD}} = 4x^2(x-1)$	$\dfrac{5(x-1)}{4x^2(x-1)} + \dfrac{4x(3x-4)}{4x^2(x-1)}$ $= \dfrac{5x - 5 + 12x^2 - 16x}{4x^2(x-1)}$	$\dfrac{12x^2 - 11x - 5}{4x^2(x-1)}$
16. $\dfrac{5x-1}{x-2} + \dfrac{7}{x-2} =$		_____

228

17. $\dfrac{9x}{7x-1} - \dfrac{5x}{7x-1} =$ _____

18. $\dfrac{2x}{2x-1} - \dfrac{1}{1-2x} =$ _____
    (be careful)

19. $\dfrac{x}{x^2-1} - \dfrac{1}{x^2-1} =$ _____

20. $\dfrac{x}{x-1} + \dfrac{1}{2(x-1)} =$ _____

21. $\dfrac{2x-1}{x^2} + \dfrac{5}{x} =$ _____

22. $\dfrac{2y}{3x} + \dfrac{1}{x^2} =$ _____

23. $\dfrac{4}{x} - \dfrac{5}{y} =$ _____

24. $\dfrac{3}{x} - \dfrac{y}{2x} =$ _____

25. $\dfrac{3y}{2x} + \dfrac{2x}{3y} =$ _____
    LCD =

26. $\dfrac{5}{x} + \dfrac{4}{x-1} =$
    LCD =

27. $\dfrac{2x}{x^2-1} - \dfrac{x}{x^2+x} =$
    $x^2 - 1 =$
    $x^2 + x =$
    LCD =

28. $\dfrac{2y-1}{y^3-2y^2} - \dfrac{4}{y^2-4} =$
    $y^3 - 2y^2 =$
    $y^2 - 4 =$
    LCD =

29. $\dfrac{5}{x} + \dfrac{2x}{x^2+x} - \dfrac{x}{x^2-1} =$
    $x^2 + x =$
    $x^2 - 1 =$
    LCD =

30. $\dfrac{x}{x+1} - \dfrac{2x}{x^2+3x+2} =$
    $x^2 + 3x + 2 =$
    LCD =

31. $\dfrac{3}{x-1} - \dfrac{4}{1-x} =$
    (see example 6)

32. $\dfrac{x}{1-x} + \dfrac{1}{x-1} =$

33. $2 + \dfrac{3}{x} - \dfrac{5}{x^2} =$
    LCD =

34. $2x - 1 + \dfrac{4}{x-1} =$

35. $\dfrac{x+1}{x^2+7x+10} - \dfrac{3x}{x^2+6x+5} =$
    $x^2 + 7x + 10 =$
    $x^2 + 6x + 5 =$
    LCD =

36. $\dfrac{4}{x^2(x-3)} + \dfrac{7}{x(x-3)^2} =$
    LCD =

37. $\dfrac{x+2}{x^2+x} + \dfrac{3}{x^3+2x^2} - \dfrac{3x-2}{x^2+3x+2} =$
    $x^2 + x =$
    $x^3 + 2x^2 =$
    $x^2 + 3x + 2 =$
    LCD =

38. $\dfrac{2y}{x^2-y^2} + \dfrac{x}{x^2+2xy+y^2} + \dfrac{1}{x+y} =$
    $x^2 - y^2 =$
    $x^2 + 2xy + y^2 =$
    LCD =

39. $\dfrac{x^2-4}{x^2-5x+6} - \dfrac{5}{x-3} =$

8.4 Complex Fractions

We define a *complex fraction* as an algebraic expression for which the numerator or the denominator or both contain fractions. For example

$$\frac{\frac{3}{x} + \frac{5}{x-1}}{\frac{7}{x-1} + \frac{2}{x}}$$

is a complex fraction. We discuss two methods for simplifying such expressions.

Method 1. The procedure is described in the following steps.
Step 1. Simplify the numerator to a single fraction as in Section 8.3.
Step 2. Simplify the denominator to a single fraction as in Section 8.3.
Step 3. Divide the fraction of Step 1 by the fraction of Step 2 and simplify as in Section 8.2.

Method 2. The procedure is described in the following steps.
Step 1. Find the LCD of all the fractions that appear in the numerator and the denominator.
Step 2. Multiply the numerator and the denominator by the LCD obtained in Step 1.
Step 3. Simplify the numerator and the denominator to obtain an equivalent fraction.
Step 4. Reduce the fraction of Step 3.

We will use both of these methods in Examples 1 and 2.

Example 1: Simplify

$$\frac{2 + \frac{3}{x}}{4 - \frac{1}{x}}$$

Method 1.
Step 1. The numerator $2 + \frac{3}{x} = \frac{2}{1} + \frac{3}{x} = \frac{2x}{1 \cdot x} + \frac{3}{x} = \frac{2x+3}{x}$

Step 2. The denominator $4 - \frac{1}{x} = \frac{4}{1} - \frac{1}{x} = \frac{4}{1} \cdot \frac{x}{x} - \frac{1}{x} = \frac{4x-1}{x}$

Step 3. $\left(\frac{2x+3}{x}\right) \div \left(\frac{4x-1}{x}\right) = \frac{2x+3}{x} \cdot \frac{x}{4x-1} = \frac{2x+3}{4x-1} \cdot \frac{x}{x} = \frac{2x+3}{4x-1}$

Method 2.
Step 1. The LCD of all the fractions that appear in the numerator and the denominator is "x"

Step 2. $\dfrac{x\left(2 + \frac{3}{x}\right)}{x\left(4 - \frac{1}{x}\right)} = \dfrac{2x+3}{4x-1}$

Since $\dfrac{2x+3}{4x-1}$ cannot be simplified further, we get

$$\frac{2 + \frac{3}{x}}{4 - \frac{1}{x}} = \frac{2x+3}{4x-1}$$

Example 2: Simplify

$$\frac{\frac{1}{x} + \frac{x}{2x-3}}{\frac{x-1}{x}}$$

Method 1.

Step 1. The numerator $\frac{1}{x} + \frac{x}{2x-3} = \frac{1}{x} \cdot \frac{2x-3}{2x-3} + \frac{x}{2x-3} \cdot \frac{x}{x}$

$$= \frac{(2x-3) + x^2}{x(2x-3)} = \frac{x^2 + 2x - 3}{x(2x-3)} = \frac{(x+3)(x-1)}{x(2x-3)}$$

Step 2. The denominator $\frac{x-1}{x}$ is already a single fraction

Step 3. $\frac{(x+3)(x-1)}{x(2x-3)} \div \frac{x-1}{x} = \frac{(x+3)(x-1)}{x(2x-3)} \cdot \frac{x}{x-1}$

$$= \frac{x+3}{2x-3} \cdot \frac{x}{x} \cdot \frac{x-1}{x-1} = \frac{x+3}{2x-3}$$

Method 2.

Step 1. The LCD of the numerator and the denominator is $x(2x-3)$

Step 2. $\frac{x(2x-3)\left(\frac{1}{x} + \frac{x}{2x-3}\right)}{x(2x-3)\left(\frac{x-1}{x}\right)}$

Step 3. $\frac{2x - 3 + x^2}{(2x-3)(x-1)} = \frac{x^2 + 2x - 3}{(2x-3)(x-1)}$

Step 4. $\frac{x^2 + 2x - 3}{(2x-3)(x-1)} = \frac{(x-1)(x+3)}{(x-1)(2x-3)} = \frac{x+3}{2x-3}$

Problem Set 8.4

Simplify the fractions in the following problems.

Fraction	Show work here	Simplified form
1. $\dfrac{\frac{2}{x} - \frac{3}{x^2}}{\frac{2x-3}{x}} =$	$\dfrac{x^2\left(\frac{2}{x} - \frac{3}{x^2}\right)}{x^2 \cdot \frac{2x-3}{x}} = \dfrac{2x-3}{x(2x-3)}$	$\dfrac{1}{x}$
2. $\dfrac{2 + \frac{3}{4}}{4 - \frac{5}{2}}$		

233

3. $\dfrac{x - \dfrac{1}{x}}{x - 1}$

4. $\dfrac{\dfrac{2}{x} - \dfrac{4}{x}}{1 - \dfrac{5}{x}}$

5. $\dfrac{\dfrac{2}{x} + \dfrac{x}{x-1}}{\dfrac{2x-1}{x^2 - x}}$

6. $\dfrac{1 - \dfrac{1}{3x}}{1 - \dfrac{1}{9x^2}}$

7. $\dfrac{\dfrac{x^2}{y} - y}{\dfrac{y}{x} - \dfrac{x}{y}}$

8. $\dfrac{x - 2 - \dfrac{12}{x-1}}{\dfrac{x}{x-1} - \dfrac{5}{x-1}}$

9. $\dfrac{\dfrac{x}{y} - 2 + \dfrac{y}{x}}{\dfrac{x}{y} - \dfrac{y}{x}}$

10. $\dfrac{\dfrac{x-1}{x^2 + 4x - 12}}{1 + \dfrac{3x}{x^2 - 4}}$

11. $\dfrac{\dfrac{m^2 + m - 6}{m^2}}{\dfrac{m^3}{m^2 - 9}}$

12. $\dfrac{\dfrac{3}{x} - \dfrac{4}{x^2 - x}}{\dfrac{2x}{x^2 - 1} + \dfrac{1}{x}}$

13. $\dfrac{a - \dfrac{a^2}{a - b}}{b + \dfrac{b^2}{a - b}}$

14. $1 - \dfrac{1}{1 + \dfrac{2}{x + 2}}$

8.5 Rational Equations

By a *rational equation* we mean an equation in which algebraic fractions appear such as

$$\frac{2x}{x - 2} = 3 - \frac{4}{x}$$

The steps used to solve such an equation are described below.

Step 1. Find the LCD of all the fractions.
Step 2. Multiply both sides of the equation by the LCD and simplify.
Step 3. Solve the resulting equation. In this book, the equation in Step 2 will be either a linear equation or a quadratic equation. Solve this equation by any of the methods discussed in this book.
Step 4. Check the solutions. Checking the solutions of rational equations is very important, because sometimes you may obtain a solution which makes the denominator of a rational term zero in the given equation. Such solutions, called *extraneous solutions,** are to be disregarded.

> Make sure the solutions that you indicate are not extraneous solutions.

Example 1: Solve

$$\frac{3}{x} + \frac{1}{3} = \frac{4}{x}$$

Step 1:   LCD = $3x$

---

*Extraneous solutions arise because of Step 2. One of the factors of the LCD is zero when the unknown equals the extraneous solution.

Step 2: $\quad 3x\left(\dfrac{3}{x} + \dfrac{1}{3}\right) = 3x\left(\dfrac{4}{x}\right)$ \hfill multiply by LCD

or $\quad 3x\left(\dfrac{3}{x}\right) + 3x\left(\dfrac{1}{3}\right) = 3x\left(\dfrac{4}{x}\right)$

$$9 + x = 12$$
$$x = 12 - 9$$
$$x = 3$$

Check: $\quad \dfrac{3}{x} + \dfrac{1}{3} = \dfrac{3}{3} + \dfrac{1}{3} = \dfrac{4}{3}$

$$\dfrac{4}{x} = \dfrac{4}{3}$$

Example 2: Solve

$$\dfrac{x}{x-4} = \dfrac{4}{x-4} + 2$$

Step 1. $\quad$ LCD $= (x - 4)$

Step 2. $\quad (x-4)\dfrac{x}{(x-4)} = (x-4)\left[\dfrac{4}{x-4} + 2\right]$

$$\dfrac{x-4}{x-4} \cdot x = (x-4) \cdot \dfrac{4}{x-4} + (x-4) \cdot 2$$

$$x = 4 + 2(x-4) = 4 + 2x - 8 = 2x - 4$$

Thus, $x = 2x - 4$

Step 3. $\quad$ The solution is $x = 4$

Step 4. $\quad x = 4$ makes the denominators in the given equation zero. Therefore, $x = 4$ is an extraneous solution.

Answer: *No Solution*

Example 3: Solve

$$\dfrac{1}{x^2} + \dfrac{6}{x} + 8 = 0$$

Step 1. $\quad$ LCD $= x^2$

Step 2. $\quad x^2\left[\dfrac{1}{x^2} + \dfrac{6}{x} + 8\right] = x^2 \cdot 0$

$$1 + 6x + 8x^2 = 0 \quad \text{or} \quad 8x^2 + 6x + 1 = 0$$

Step 3. $\quad (4x + 1)(2x + 1) = 0$

$$\begin{array}{ll} 4x + 1 = 0 & \text{or} \quad 2x + 1 = 0 \\ \underline{-1 \quad -1} & \underline{-1 \quad -1} \\ 4x = -1 & 2x = -1 \\ x = -\dfrac{1}{4} & x = -\dfrac{1}{2} \end{array}$$

Step 4. $\quad$ Neither of these solutions is an extraneous solution.

Thus, the solutions are $x = -\dfrac{1}{4}$, $x = -\dfrac{1}{2}$.

Example 4: Solve

$$\frac{3}{x+1} + \frac{4}{x} = \frac{12}{x+2}$$

Step 1.   LCD $= x(x+1)(x+2)$

Step 2. $x(x+1)(x+2)\left[\frac{3}{x+1} + \frac{4}{x}\right] = x(x+1)(x+2) \cdot \frac{12}{x+2}$

$\frac{3(x)(x+1)(x+2)}{(x+1)} + \frac{4x(x+1)(x+2)}{x} = \frac{12(x)(x+1)(x+2)}{(x+2)}$

$3x(x+2) + 4(x+1)(x+2) = 12x(x+1)$
$3x^2 + 6x + 4(x^2 + 3x + 2) = 12x^2 + 12x$
$3x^2 + 6x + 4x^2 + 12x + 8 = 12x^2 + 12x$
$7x^2 - 12x^2 + 18x - 12x + 8 = 0$
$-5x^2 + 6x + 8 = 0$ or $5x^2 - 6x - 8 = 0$

Step 3. $(5x+4)(x-2) = 0$
   $5x + 4 = 0$ or $x - 2 = 0$
   $5x = -4$ $x = 2$
   $x = -\frac{4}{5}$

Step 4.   Neither of these solutions is an extraneous solution.
Thus, the solutions are $x = -\frac{4}{5}$, $x = 2$

Problem Set 8.5

Solve the following equations. Check your solutions. Indicate extraneous solutions, if any.

Equation	Show work here	Solution	Cal. Check
1. $\frac{5}{2x} + \frac{3}{4} = \frac{1}{x}$	LCD $= 4x$ $4x \cdot \frac{5}{2x} + 4x \cdot \frac{3}{4} = 4x \cdot \frac{1}{x}$ $10 + 3x = 4$, $3x = -6$ $x = -2$	-2	LHS $= -.5$ RHS $= -.5$
2. $\frac{x}{3} - \frac{4}{5} = 1 - \frac{3x}{5}$		_____	LHS $=$ RHS $=$
3. $\frac{y}{4} - \frac{2y}{5} = \frac{1}{10}$		_____	LHS $=$ RHS $=$
4. $\frac{3}{x} - \frac{2}{3} = -\frac{1}{x}$		_____	LHS $=$ RHS $=$

5. $\dfrac{4}{x} - \dfrac{3}{4} = 2 - \dfrac{5}{2x}$ _____ LHS =
RHS =

6. $\dfrac{1}{2}\left(\dfrac{3}{x} - \dfrac{1}{3}\right) = 1 - \dfrac{2}{x}$ _____ LHS =
RHS =

7. $\dfrac{3}{x+1} - 2 = \dfrac{5}{x+1}$ _____ LHS =
RHS =

8. $\dfrac{2}{x-1} = \dfrac{1}{x-1} - 2$ _____ LHS =
RHS =

9. $\dfrac{3}{y+3} - 2 = \dfrac{-y}{y+3}$ _____ LHS =
RHS =

10. $\dfrac{1}{x} + \dfrac{2}{x^2} - 3 = 0$ _____ LHS =
RHS =

11. $\dfrac{4}{x} - \dfrac{3}{x-1} = \dfrac{10}{x}$ _____ LHS =
RHS =

12. $\dfrac{3}{x+1} + x = 3 - \dfrac{2}{x+1}$ _____ LHS =
RHS =

13. $\dfrac{2}{x+2} - \dfrac{5}{x-2} = \dfrac{7}{x^2-4}$ _____   LHS =
  RHS =

14. $\dfrac{4}{x} - \dfrac{3}{x-2} = \dfrac{5}{x^2-2x}$ _____   LHS =
  RHS =

15. $\dfrac{2}{x-1} + \dfrac{2x}{(x+4)(x-1)} = \dfrac{2}{x+4} + \dfrac{2}{(x+4)(x-1)}$ _____   LHS =
  RHS =

8.6 Equations with More Than One Unknown

It is quite common to have relationships between two or more unknown quantities. These relationships are often called formulas. For example,

1. $F = \dfrac{9}{5}C + 32$ is an expression for the relationship between temperature in Fahrenheit (F) and temperature in Celsius (C).
2. $V = \pi r^2 h$ is an expression for the volume (V) of a right circular cylinder with height h and base with radius r.
3. $F = \dfrac{GmM}{d^2}$ is an expression for the force F due to the gravitational attraction between two bodies with masses m and M and which are d units of distance apart. G is a constant, called the gravitational constant.

It is often the case that we need to express one of the unknowns in terms of the others. We now illustrate this in the following examples.

Example 1: Solve $F = \dfrac{9}{5}C + 32$ for C.

Step 1:   Simplify parentheses or fractions.

$$F = \dfrac{9}{5}C + 32$$

LCD = 5

$$5F = 9C + 160$$

Step 2:   Isolate the terms involving the unknown for which we intend to solve.

$$5F - 160 = 9C$$

Step 3:   Solve for the unknown.

$$\dfrac{5F - 160}{9} = C$$

or $C = \dfrac{5}{9}(F - 32)$

Example 2: Solve $V = \pi r^2 h$ for r.

Step 1. There are no parentheses or fractions.
Step 2. Isolate the term involving the unknown.
$$r^2 = \frac{V}{\pi h}$$
Step 3. $r = \sqrt{\dfrac{V}{\pi h}}$ (take square root)

Note: We choose the positive square root as the answer because r is a length, and we usually take length to be nonnegative.

Example 3: Solve $F = \dfrac{GmM}{d^2}$ for M.

$$F = \frac{GmM}{d^2}$$
$$d^2 F = GmM \quad \text{(multiply by } d^2\text{)}$$
$$\frac{d^2 F}{Gm} = M \quad \text{(divide by Gm)}$$

Example 4: Solve $\dfrac{1}{R} = \dfrac{1}{r} + \dfrac{1}{s}$ for r.

In this example it is more convenient to interchange Steps 1 and 2.

$$\frac{1}{R} - \frac{1}{s} = \frac{1}{r}$$
$$\frac{s}{sR} - \frac{R}{sR} = \frac{1}{r}$$
$$\frac{s - R}{sR} = \frac{1}{r}$$
$$\frac{sR}{s - R} = r \quad \text{(take reciprocal of both sides)}$$

Example 5: Solve $\dfrac{k}{N}(N - M) = N(N + M)$ for M.

$$k - \frac{kM}{N} = NM + N^2 \quad \text{(distributive law)}$$
$$kN - kM = N^2 M + N^3 \quad \text{(multiply by N)}$$
$$kN - N^3 = kM + N^2 M \quad \text{(isolate terms involving M)}$$
$$kN - N^3 = M(k + N^2) \quad \text{(factor out M)}$$
$$M = \frac{kN - N^3}{k + N^2} \quad \text{(divide by } k + N^2\text{)}$$

Problem Set 8.6

In Problems 1–15, solve the equation for the indicated unknown.

Equation	Solution
1. $C = 2\pi r$ for r	$r =$
2. $C = 5 - 2a$ for a	$a =$
3. $E = mc^2$ for m	$m =$

4. $A = P + I$ for I $\qquad$ I =

5. $ay + b = 0$ for y $\qquad$ y =

6. $d = rt$ for t $\qquad$ t =

7. $y = ax + b$ for x $\qquad$ x =

8. $3x - 4y = 5$ for y $\qquad$ y =

9. $3x - 4y = 5$ for x $\qquad$ x =

10. $C = \frac{5}{9}(F - 32)$ for F $\qquad$ F =

11. $A = P(1 + rt)$ for r $\qquad$ r =

12. $P = M - Mdt$ for d $\qquad$ d =

13. $5M(L - N) = N(2L - 3M)$ for L $\qquad$ L =

14. $y = \frac{2x - 5}{x - 3}$ for x $\qquad$ x =

15. $\frac{1}{R} = \frac{1}{R_1} + \frac{1}{R_2}$ for R_2 $\qquad$ R_2 =

Chapter Summary

1. **Algebraic fraction** A ratio of two polynomials.
2. **Some basic facts**
   (1) $b - a = -(a - b)$
   (2) $\frac{x - 1}{1 - x} = \frac{-(1 - x)}{1 - x} = -1$
   (3) $\frac{1}{1 - x} = \frac{1}{-(x - 1)} = -\frac{1}{x - 1}$
   (4) When reducing fractions we cancel only the factors and not terms.
3. **Operations on fractions** The procedure of performing addition or multiplication on algebraic fractions is similar to the procedure for performing these operations on numerical fractions.
4. **Simplifying complex fractions** There are two methods for simplifying complex fractions.

   Method 1. Simplify the numerator and the denominator to single fractions and then perform the division of the two fractions.

   Method 2. Multiply the numerator and the denominator by the LCD of all the fractions and simplify.
5. **Rational equations** Equations containing fractions.
6. **Extraneous solution** A value for the unknown which is obtained by algebraic methods, and which is a candidate for a solution; but, it is not a solution because when substituted in the original expression it makes a denominator zero.

Review Problems

In Problems 1–10, mark True or False.

1. $x - 2 = 2 - x$
   (a) True   (b) False

2. $3 + x = x + 3$
   (a) True   (b) False

3. $\dfrac{x - 2}{2 - x} = -1$
   (a) True   (b) False

4. $\dfrac{x + 2}{2 + x} = -1$
   (a) True   (b) False

5. $\dfrac{x}{1 - x} = -\dfrac{x}{x - 1}$
   (a) True   (b) False

6. $\dfrac{2 - 2x}{1 - x} = 2$
   (a) True   (b) False

7. $\dfrac{x + x^2}{x} = 1 + x^2$
   (a) True   (b) False

8. $\dfrac{x + x^2}{x} = 2x$
   (a) True   (b) False

9. $\dfrac{x(2 + x)}{x} = 2 + x$
   (a) True   (b) False

10. $\dfrac{x(2 + x)}{2 + x} = x$
    (a) True   (b) False

In Problems 11–24, reduce the fractions to lowest terms.

	Fraction	Reduced form		Fraction	Reduced form
11.	$\dfrac{x^2 - 4x + 4}{x^2 - 2x} = \dfrac{(x - 2)(x - 2)}{x(x - 2)} =$	$\dfrac{x - 2}{x}$	12.	$\dfrac{12}{32} =$	_____
13.	$\dfrac{-4}{12} =$	_____	14.	$\dfrac{-20}{-35} =$	_____
15.	$\dfrac{-2x}{-x} =$	_____	16.	$\dfrac{-6x}{-3} =$	_____
17.	$\dfrac{12}{-4} =$	_____	18.	$\dfrac{-4x}{x} =$	_____
19.	$\dfrac{-4x^2}{2x} =$	_____	20.	$\dfrac{2x - 2}{x^2 - x} =$	_____
21.	$\dfrac{x - 1}{x^2 - 1} =$	_____	22.	$\dfrac{x^2 - 2x}{x^2 - 4} =$	_____
23.	$\dfrac{x^2 - x - 6}{x^2 - 9} =$	_____	24.	$\dfrac{a^3 - 4a}{a^2 - 5a + 6} =$	_____

Perform the indicated operations in Problems 25–69.

25. $\dfrac{x-1}{x-x^2} \cdot \dfrac{x^2}{x+1} = \dfrac{x-1}{x(1-x)} \cdot \dfrac{x \cdot x}{x+1}$

    $\dfrac{x \cdot x \cdot (x-1)}{-x(x-1)(x+1)}$

    25. $-\dfrac{x}{x+1}$

26. $\dfrac{25}{120} \cdot \dfrac{24}{5} =$ _____

27. $\dfrac{2x^2}{3y} \cdot \dfrac{6y^2}{8x} =$ _____

28. $\dfrac{4x^2y}{6z} \cdot \dfrac{9z^2}{4y} =$ _____

29. $\dfrac{x+1}{x^2-1} \cdot \dfrac{x-1}{5} =$ _____

30. $(2x - 3) \cdot \dfrac{5x}{2x-3} =$ _____

31. $\dfrac{x}{x-1} \cdot \dfrac{(x-1)^2}{x} =$ _____

32. $\dfrac{(x-1)^2}{(x-2)(x-3)} \cdot \dfrac{x-3}{x-1} =$

    32. _____

33. $\dfrac{(x-2)^2}{x^2-3x+2} \cdot \dfrac{x^2-4x+3}{x^2-5x+6} =$

    33. _____

34. $\dfrac{5x}{x^2+6x+8} \cdot (x+2) =$

    34. _____

35. $\dfrac{b}{2y-4} \cdot \dfrac{y^2-4}{by-2b} =$

    35. _____

36. $\dfrac{2a^2-11a-21}{3a^2+11a-4} \cdot \dfrac{9a^2-1}{3a^2-20a-7} =$

    36. _____

37. $\dfrac{12}{5} \div \dfrac{9}{10} =$ _____

38. $\dfrac{24}{9} \div \dfrac{15}{16} =$ _____

39. $\dfrac{x}{y} \div \dfrac{x}{y} =$ _____

40. $\dfrac{x^2}{y^2} \div \dfrac{y}{x} =$ _____

41. $\dfrac{x}{x-1} \div \dfrac{2x^2}{x^2-1} =$ _____

42. $\dfrac{5}{9} \div y =$ _____

43. $3x \div \dfrac{4x}{x-1} =$ _____

44. $\frac{15}{28} \div 45 =$

45. $\frac{2x^2 - 18}{x + 3} \div (x - 3) =$

46. $\frac{4}{5} \cdot \frac{15}{12} \div \frac{12}{20} =$

47. $\frac{3}{12} \div \frac{6}{25} \cdot \frac{9}{100} =$

48. $\frac{2a^2 - 50}{a^2 - 8a + 15} \div \frac{a^2 + 3a - 10}{4a^2 - 9} =$

49. $\frac{2}{3} + \frac{4}{5} = \frac{10}{15} + \frac{12}{15} =$ _____ 50. $\frac{2}{x} + \frac{3}{x} =$ _____

51. $\frac{2x - 1}{x + 1} + \frac{x}{x + 1} =$ _____ 52. $\frac{x + 1}{x^2 - 1} - \frac{3}{x - 1} =$ _____

53. $\frac{3x + 4}{2x - 3} - \frac{x - 1}{2x - 3} =$ _____ 54. $\frac{3}{x} + \frac{y}{x} - \frac{5}{x} =$ _____

55. $\frac{4}{x - 3} + \frac{x}{x - 3} - \frac{x - 1}{x - 3} =$ _____ 56. $\frac{x - 1}{x + 2} - \frac{x}{x^2 + 2x} =$ _____

57. $\frac{2}{2x} - \frac{5}{x} =$ _____ 58. $\frac{x}{y} + \frac{3}{y^2} =$ _____

59. $\frac{1}{x - 1} + \frac{x}{1 - x} =$ _____ 60. $\frac{1}{(x - 1)^2} - \frac{2x}{(x - 1)^3} =$ _____

61. $\frac{3}{x} - \frac{4}{y} =$ _____ 62. $\frac{3}{x - 1} - \frac{y}{y - 1} =$ _____

63. $\frac{2}{x - 1} + \frac{3}{x + 2} = \frac{2(x + 2)}{(x - 1)(x + 2)} + \frac{3(x - 1)}{(x + 2)(x - 1)}$ $\frac{5x + 1}{(x - 1)(x + 2)}$
$= \frac{2x + 4 + 3x - 3}{(x - 1)(x + 2)} = \frac{5x + 1}{(x - 1)(x + 2)}$
LCD $= (x - 1)(x + 2)$

64. $\dfrac{3x}{x^2-1} - \dfrac{5}{x+1} =$

   LCD =

65. $\dfrac{3x+4}{x^2+x-2} - \dfrac{x-2}{x^2+x-2} =$

66. $5 + \dfrac{4}{x-1} =$

67. $\dfrac{x}{2x-1} - (x+1) =$

68. $\dfrac{x-1}{x^2-4} - \dfrac{x-1}{x^2+4x+4} + \dfrac{4}{x-2}$

   $x^2 - 4 =$

   $x^2 + 4x + 4 =$

   $x - 2 =$

   LCD =

69. $\dfrac{\dfrac{1}{x} - \dfrac{3}{x^2}}{\dfrac{9}{x^2} - 1}$

In Problems 70–73, solve the given equation and check your answer.

| Equation | Show work here | Solution | Check |

70. $\dfrac{2}{x} = 1$ LHS: ____   RHS: ____

71. $\dfrac{2}{x-1} = \dfrac{1}{x}$ LHS: ____   RHS: ____

72. $2 + \dfrac{2}{x-1} = \dfrac{3x}{x-1}$ LHS: ____   RHS: ____

73. $\dfrac{x}{x-1} + \dfrac{2}{x^2-1} = 1 - \dfrac{5}{x+1}$ _____ LHS: _____

RHS: _____

Solve each of the following equations for the indicated unknown.

74. $2a(b - 3c) = a - 4c$ for c   c =

75. $\dfrac{x - 2y}{x} = 2 - y$ for x   x =

76. $x^2 + y^2 = a^2$ for x   x =

77. $ax + by + c = 0$ for y   y =

78. $\dfrac{x}{a} + \dfrac{y}{b} = 1$ for y   y =

79. $\dfrac{x - y}{x + y} = a$ for y   y =

80. $y = mx + b$ for m   m =

Chapter Test

Solve each of the following problems and match your answer with the given responses.

1. What factor is common to the two expressions: $5x^2 + 6xy$ and $5xy + 6y^2$
   (a) $5x^2 + 6y^2$ (b) $5x$ (c) $6y$ (d) $5x + 6y$ (e) None of these

2. Factor completely the expression: $4x^3 - 16x$
   (a) $4x(x - 2)(x + 2)$ (b) $4x(x^2 - 16x)$ (c) $4x(x - 4)(x + 4)$ (d) $4(x^3 - 4x)$
   (e) None of these

3. One of the solutions of $5x^2 + 9x - 2 = 0$ is:
   (a) $-\dfrac{1}{5}$ (b) 2 (c) $\dfrac{1}{5}$ (d) $\dfrac{-9 + \sqrt{41}}{10}$ (e) None of these

4. One of the solutions of $x^2 + 3x - 5 = 0$ is:
   (a) $5 - i$ (b) $1 - i\sqrt{5}$ (c) $\dfrac{3}{2} + i\sqrt{29}$ (d) $-\dfrac{3}{2} + \dfrac{i\sqrt{29}}{2}$ (e) None of these

5. The expression $\dfrac{4x^2 + 5x - 1}{x - 2}$ is not defined for x =
   (a) 1 (b) 2 (c) -2 (d) 0 (e) None of these

6. Simplify: $\dfrac{15a^2b^3}{35ab} =$
   (a) $\dfrac{5a^3b^4}{7}$ (b) $\dfrac{3ab}{7b}$ (c) $\dfrac{3ab^2}{7}$ (d) None of these

7. Divide: $\dfrac{4xy}{3} \div \dfrac{x}{y} =$

   (a) $\dfrac{4y^2}{3}$ (b) $\dfrac{4x^2}{3}$ (c) $\dfrac{3x^2}{4}$ (d) None of these

8. Find the missing numerator: $\dfrac{5x}{x+6} = \dfrac{?}{4(x-5)(x+6)}$.

   (a) $20x^2 - 100$ (b) $4x - 20$ (c) $20x^2 + 100$ (d) $20x - 100$ (e) None of these

9. The LCD of the fractions $\dfrac{1}{x^2+6x+9}, \dfrac{x+3}{2(x^2-9)}$ is:

   (a) $2(x+3)^2(x-3)$ (b) $(x+3)^2(x-3)$ (c) $2(x+3)(x-3)$ (d) $2(x+3)^2$ (e) None of these

10. Simplify: $\dfrac{2x-2}{2x^2-x} - \dfrac{2}{x} =$

    (a) $\dfrac{x^2}{x^2(2x-1)}$ (b) $\dfrac{-2}{2x-1}$ (c) $\dfrac{x-2}{2x^2-x}$ (d) None of these

11. Simplify: $\dfrac{x-1}{x^2-4} - \dfrac{2}{x^2-2x} =$

    (a) $\dfrac{x^2-3x-4}{x(x-2)(x+2)}$ (b) $\dfrac{x-3}{x(x^2-4)}$ (c) $\dfrac{x^2-3x-4}{(x-2)(x+2)}$ (d) None of these

12. Simplify: $\dfrac{x-1}{1-\dfrac{1}{x}} =$

    (a) $-\dfrac{1}{x}$ (b) x (c) $1-x$ (d) $x-1$ (e) None of these

13. Simplify: $\dfrac{\dfrac{x+1}{x}}{\dfrac{1}{x}+1} =$

    (a) $\dfrac{1}{x}$ (b) $1+x$ (c) x (d) 1 (e) None of these

14. Solve for x: $\dfrac{5}{x-3} = \dfrac{x+1}{10}$

    (a) 3 (b) $1 + 3\sqrt{6}, 1 - 3\sqrt{6}$ (c) $2 + \sqrt{6}, 2 - \sqrt{6}$ (e) None of these

15. Which of the following pairs is a solution of $\dfrac{2}{x+1} - \dfrac{3}{x-1} = \dfrac{1}{x-4}$.

    (a) $\dfrac{7}{2}, -3$ (b) $\dfrac{7}{2}, 3$ (c) $-\dfrac{7}{2}, 3$ (d) $-\dfrac{7}{2}, -3$ (e) None of these

Chapter 9

Exponents and Radicals

The basic rules of exponents and the concept of square root were discussed in Chapter 2. In those rules the exponents were integers and the discussion on simplifying expressions with radicals or fractional exponents was postponed until this chapter.

In this chapter we discuss

(a) Review of exponent rules,
(b) Fractional exponents and radicals,
(c) Simplifying expressions with rational number exponents,
(d) Simplifying radicals,
(e) Basic operations on radicals,
(f) Complex numbers.

9.1 Review

Recall that in Sections 2.2 and 2.3 we discussed the following rules. We assume that variables are not assigned any value which makes the denominator zero.

Rule #	Rule
2.3	$x^m \cdot x^n = x^{m+n}$
2.4	$(xy)^n = x^n y^n$
	$\left(\dfrac{x}{y}\right)^n = \dfrac{x^n}{y^n}$
2.5	$(x^m)^n = x^{m \cdot n}$
2.6	$\dfrac{x^m}{x^n} = x^{m-n}$
2.7	$x^0 = 1,\ x \neq 0$
2.8	$x^{-n} = \dfrac{1}{x^n}$
	$\dfrac{1}{x^{-n}} = x^n$

In all these rules the exponents m and n were assumed to be positive integers.

Example 1: *Use of exponent rules 2.3–2.8*

(a) $\dfrac{(2x^2)^3(3x^2)}{12x^5} = \dfrac{8x^6 \cdot 3x^2}{12x^5}$
$= \dfrac{24x^8}{12x^5} = 2x^3$

(b) $\dfrac{x^{-2}(3x^0)y^3}{x^4 y^{-2}} = \dfrac{3y^3 \cdot y^2}{x^4 \cdot x^2} = \dfrac{3y^5}{x^6}$

(c) $\dfrac{(x^{-2}y^0 z^3)^{-2}}{(2x)^0 z^6 x^{-4}} = \dfrac{x^4 y^0 z^{-6}}{1 \cdot z^6 x^{-4}} = \dfrac{x^4 \cdot x^4}{z^6 \cdot z^6} = \dfrac{x^8}{z^{12}}$

Problem Set 9.1

Complete the statements in Problems 1–6.

1. $x^0 = 1$ for all x except when _____.

2. The base for the exponent 4 in $5x^4$ is _____.

3. $3x^0 =$ _____.

4. $2x^{-2} =$ _____.

5. $\dfrac{x^{-2}}{x^2} = x^?$.

6. $\dfrac{x^{-2}}{x^2} = \dfrac{1}{x^?}$.

In Problems 7–15, use the exponent rules to simplify the expressions. There should be no negative exponents in the answer.

Expression	Answer	Expression	Answer
7. $(2x)^2 x^5 = 4x^2 x^5 =$	$4x^7$	8. $\dfrac{(2x^2)^3}{4x^4} =$	
9. $4x^{-2} =$		10. $2x^0 y^{-2} =$	

11. $(3x^2 y^3)^2(-2x)^2 =$ 11. _____

12. $\dfrac{(-3x^0)(-2x^4)}{4x^{-3}} =$ 12. _____

13. $\dfrac{1}{(-2x)^{-4}} =$ 13. _____

14. $\left(\dfrac{-a^{-1}}{b}\right)^{-3} =$ 14. _____

15. $(x^{-5})^{-4}(y^{-3})^2 =$ 15. _____

9.2 Fractional Exponents and Radicals

In this section, we shall discuss the meaning of fractional exponents and how expressions which involve exponents can be written in radical form.

$$\text{Let } b = a^n = \underbrace{a \cdot a \cdot a \cdot \ldots \cdot a}_{n \text{ factors}}$$

Then, a is said to be an N^{TH} ROOT of b and we also write

$$a = b^{1/n}$$

Thus, if $a^n = b$, then $a = b^{1/n}$.

You may also interpret $b^{1/n}$ as a number whose n^{th} power is b.

Example 1: *To find n^{th} roots*

(a) $9^{1/2} = 3$, because $3^2 = 9$.

(b) $8^{1/3} = 2$, because $2^3 = 8$.
    $8^{1/3}$ is read as a 'cube root' of 8.

(c) $16^{1/4} = 2$, because $2^4 = 16$.
    $16^{1/4}$ is read as a 'fourth root' of 16.

(d) $32^{1/5} = 2$, because $2^5 = 32$.
    $32^{1/5}$ is called a 'fifth root' of 32.

Instead of using a fractional power to express an nth root we often use a *radical sign*.

Example 2: *An nth root expressed as a radical*

(a) $8^{1/3}$ may be written as $\sqrt[3]{8}$

(b) $16^{1/4}$ may be written as $\sqrt[4]{16}$

(c) $32^{1/5}$ may be written as $\sqrt[5]{32}$

In the symbol $\sqrt[n]{b}$, the number n is referred to as the *index*, b is called the *radicand*, and the symbol $\sqrt{}$ is called the *radical*.

▶ If there is no index then it is understood that the index is 2. That is $\sqrt[2]{9}$ or $\sqrt{9}$ are the same thing.

Above we gave the definition of a root of a number as a solution to an equation. That is, given b, we seek a number "a" which satisfies the equation

$$x^n = b.$$

That is, $a^n = b$ and if n is even, then $(-a)^n$ is also equal to b. Thus, if n is even, then the equation $x^n = b$ will have two real solutions. That is to say, there will be two real values of the nth root of b, one positive and one negative.

We will reserve the radical symbol for the positive real root, if there is one, and we will also use the radical symbol for the negative real root when that is the only real root.

The following examples may help to point out the various situations which can arise.

Example 3: *Different types of radicands and indices*

(a) $\sqrt{4} = 2$ since $2^2 = 4$. But we also have $(-2)^2 = 4$. Thus -2 is also a square root of 4. But we will designate -2 as $-\sqrt{4}$ because $\sqrt{}$ is reserved for the positive root when one exists.

(b) $\sqrt{-4}$ is not defined in the reals, since there is no real number whose square is -4.

(c) $\sqrt[3]{8} = 2$ since $2^3 = 8$.

(d) $\sqrt[3]{-8} = -2$ since $(-2)^3 = -8$.

We see in these examples that we get unique real numbers as n^{th} roots if n is odd. But when the index is even, then either there are two real roots or there are none. In the future, by a root we shall always mean the *principal root* which we define as follows:

1. The principal n^{th} root of a positive number is positive.
2. The principal n^{th} root of a negative number is negative if n is odd.
3. The principal n^{th} root of a negative number when n is even is not defined in the real numbers.
4. The principal n^{th} root of 0 is 0.
5. The symbol $\sqrt[n]{b}$ is reserved for the principal nth root of b.

▶ Henceforth by a root we shall always mean the principal root.

Example 4: *Principal roots*

(a) $\sqrt{16} = 4$.

(b) $\sqrt[3]{8} = 2$, that is $\sqrt[3]{2^3} = 2$

(c) $\sqrt[3]{-8} = -2$, since $(-2)^3 = -8$, that is $\sqrt[3]{(-2)^3} = -2$

If the radicand for an even root is a literal number, then we have to be more careful in taking the principal root.

For example, it would not be correct to say that

$$\sqrt{(-3)^2} = -3$$

because $(-3)^2 = 9$ and $\sqrt{9} = 3$. We have agreed that the symbol $\sqrt{}$ is reserved for the principal square root. By analogy, it would not be correct to say

$$\sqrt{x^2} = x$$

unless you know $x \geq 0$. In fact, if $x < 0$ you must write

$$\sqrt{x^2} = -x$$

in order to get a positive answer.

Thus, we have

$$\sqrt{x^2} = \begin{cases} x \text{ if } x \geq 0 \\ -x \text{ if } x < 0 \end{cases}$$

or, put more simply,

$$\sqrt{x^2} = |x|. \quad \text{Recall that } |x| \text{ stands for the absolute value of x which is always nonnegative.}$$

In a similar manner we must write

$$\sqrt[4]{x^4} = |x|.$$

The situation for an odd root is simple. The odd root of a positive number is positive and the odd root of a negative number is negative. For example,

$$\sqrt[3]{x^3} = x \quad \text{where x is positive or negative.}$$

Example 5: *Roots of literal numbers*

(a) $\sqrt{x^2} = |x|$
(b) $\sqrt[3]{x^3} = x$
(c) $\sqrt[4]{x^4} = |x|$
(d) $\sqrt[3]{x^6} = \sqrt[3]{(x^2)^3} = x^2$
(e) $\sqrt[4]{x^8} = \sqrt[4]{(x^2)^4} = x^2$ since $|x^2| = x^2$.
(f) $\sqrt{x^2 y^4} = |x| y^2$

For the purpose of discussion in this chapter we shall assume that unknowns involved with even roots are positive. Thus, we have two restrictions now for performing operations with radicals.

1. The value of a root means its principal value.
2. The unknown in a radicand with an even root is positive.

With these restrictions there is no need for using absolute value symbols in (a), (c) and (f) of Example 5.

Problem Set 9.2

Complete the statements in Problems 1–6.

1. If $y = x^n$ then $x =$ _____ .

2. The expression $a^{1/n}$ can be expressed in radical form as _____ .

3. In $\sqrt[n]{y}$, n is called the _____ .

4. In $\sqrt[n]{y}$, y is called the _____ .

5. An even root of a negative number is _____ in reals.

6. The principal odd root of a negative number is _____ .

In Problems 7–20, find the indicated root.

7. $\sqrt{36x^4} = \sqrt{(6x^2)^2} =$ __6x²__
8. $\sqrt{9x^4} =$ _____
9. $\sqrt[3]{x^3} =$ _____
10. $\sqrt[4]{x^4} =$ _____
11. $\sqrt[3]{-27} =$ _____
12. $\sqrt[4]{16} =$ _____
13. $\sqrt[3]{-27x^3} =$ _____
14. $\sqrt[4]{16x^4 y^8} =$ _____
15. $\sqrt[5]{-32} =$ _____
16. $\sqrt[5]{-32x^5 y^{10}} =$ _____
17. $\sqrt[6]{64x^6 y^{18}} =$ _____
18. $\sqrt[5]{-243x^{10}} =$ _____
19. $\sqrt[3]{1000x^3 y^{18}} =$ _____
20. $\sqrt[3]{-125x^{21}} =$ _____

9.3 Expressions with Rational Exponents

In Section 9.2, the nth root of a number "x" was represented symbolically in two different ways;

$x^{1/n}$ and $\sqrt[n]{x}$.

Since by a root we mean principal root, we have

$$x^{1/n} = \sqrt[n]{x}$$

Recall $x^{1/n}$ or $\sqrt[n]{x}$ represent numbers whose n^{th} power is x. The base and the radicand in the two forms are the same.

Example 1: *Change to radical form and find the value, if possible.*

(a) $16^{1/2} = \sqrt[2]{16} = \sqrt{16} = 4$, since $4^2 = 16$
(b) $8^{1/3} = \sqrt[3]{8} = 2$, since $2^3 = 8$
(c) $(-8)^{1/3} = \sqrt[3]{-8} = -2$, since $(-2)^3 = -8$
(d) $8^{-1/3} = \dfrac{1}{8^{1/3}} = \dfrac{1}{\sqrt[3]{8}} = \dfrac{1}{2}$
(e) $(x^5 - 4)^{1/3} = \sqrt[3]{x^5 - 4}$
(f) $(x^7)^{1/4} = \sqrt[4]{x^7}$
(g) $(x^2 + y^2)^{1/2} = \sqrt{x^2 + y^2}$

Example 2: *Change to exponent form*

(a) $\sqrt{9} = 9^{1/2} = (3^2)^{1/2} = 3$
(b) $\sqrt{11} = 11^{1/2}$
(c) $\sqrt[5]{x^3} = (x^3)^{1/5} = x^{3/5}$

In Example 2(c) we have

$\sqrt[5]{x^3} = x^{3/5}$

It must be true then that

$x^{3/5} = \sqrt[5]{x^3}$

This leads to a general rule. If n is any positive integer and m is any integer, then

$$\boxed{\text{Rule 9.1} \\ x^{m/n} = \sqrt[n]{x^m}}$$

➡ Note that the denominator of the exponent goes as the index and the numerator of the exponent goes as the exponent of the radicand

Example 3: *Change to radical form*

(a) $4^{2/3} = \sqrt[3]{4^2} = \sqrt[3]{16}$

(b) $(-8)^{2/3} = \sqrt[3]{(-8)^2} = \sqrt[3]{64} = 4$

(c) $(x)^{3/7} = \sqrt[7]{x^3}$

(d) $27^{-2/3} = \dfrac{1}{27^{2/3}} = \dfrac{1}{\sqrt[3]{27^2}} = \dfrac{1}{\sqrt[3]{729}} = \dfrac{1}{9}$

Note that whenever the exponent is negative, we first use Rule 2.8 to change the expression so that all the exponents are positive.

Example 4: *Simplify*

(a) $8^{-1/3} = \dfrac{1}{8^{1/3}} = \dfrac{1}{\sqrt[3]{8}} = \dfrac{1}{2}$

(b) $\dfrac{1}{(-8)^{-1/3}} = (-8)^{1/3} = \sqrt[3]{-8} = -2$

Example 5: *Change to exponent form and simplify whenever possible.*

(a) $\sqrt[3]{2^6} = (2^6)^{1/3} = 2^2 = 4$

(b) $\sqrt[3]{-27} = \sqrt[3]{(-3)^3} = [(-3)^3]^{1/3} = -3$

(c) $\sqrt[5]{x^3} = (x^3)^{1/5} = x^{3/5}$

(d) $(\sqrt{x})^6 = (x^{1/2})^6 = x^3$

(e) $\sqrt[4]{x^{12}} = (x^{12})^{1/4} = x^3$

(f) $\sqrt[4]{x^4 + y^4} = (x^4 + y^4)^{1/4}$

(g) $\sqrt[5]{xy^3} = (xy^3)^{1/5}$

The Exponent Rules 2.3–2.8 are applicable even when the exponents are fractions.

Example 6: *Simplify by use of the exponent rules and express the answer with positive exponents*

(a) $x^{2/3} x^{1/6} = x^{(2/3 + 1/6)} = x^{5/6}$

(b) $\dfrac{x^{4/5}}{x^{2/5}} = x^{(4/5 - 2/5)} = x^{2/5}$

(c) $(x^{2/3})^{1/4} = x^{2/3 \cdot 1/4} = x^{1/6}$

(d) $(x^{2/3})^{-6} = x^{-4} = \dfrac{1}{x^4}$

or $(x^{2/3})^{-6} = \dfrac{1}{(x^{2/3})^6} = \dfrac{1}{x^4}$

(e) $(x^{1/2} y^{-1/3})^{-12} = x^{-6} y^4 = \dfrac{y^4}{x^6}$

(f) $\left(\dfrac{x^{5/6} y^{1/3}}{z^{2/3}}\right)^{-6} = \dfrac{x^{-5} y^{-2}}{z^{-4}} = \dfrac{z^4}{x^5 y^2}$

Problem Set 9.3

Complete the statements in Problems 1–5.

1. $x^{1/5}$ expressed in radical form equals _____ .

2. $\sqrt[3]{x^2}$, when expressed in exponent form equals _____ .

3. If $\sqrt[5]{x^4} = x^{m/n}$, then m = ____ , n = ____ .

4. If $y^{3/4} = \sqrt[n]{y^m}$, then m = ____ , n = ____ .

5. An odd root of a negative number is always _____ .

Mark the correct response in Problems 6–10.

6. $\sqrt[3]{-27} =$
   (a) -9 (b) 3   (c) -3 (d) not defined in reals   (e) none

7. $\sqrt[4]{-16} =$
   (a) -4 (b) 2   (c) -2 (d) not defined in reals   (e) none

8. $x^{3/7} = \sqrt[3]{x^7}$
   (a) True   (b) False

9. Consider the following statements P and Q
   P: $(5)^{-1/3} = \dfrac{1}{\sqrt[3]{5}}$ Q: $\sqrt{x^2 + y^2} = x + y$
   Which of the above statements is true?
   (a) P only   (b) Q only   (c) P and Q   (d) none

10. Consider the following statements P and Q
    P: An odd root of a negative number is not defined in real numbers
    Q: An even root of a negative number is a negative number
    Which of the above statements is true?
    (a) P only   (b) Q only   (c) P and Q   (d) none

Find the value of the expressions in Problems 11–20.

Expression	The value	Expression	The value
11. $(-8)^{-1/3} =$ $= \dfrac{1}{(-8)^{1/3}}$	$-\dfrac{1}{2}$	12. $25^{1/2} =$	_____
13. $\sqrt[3]{-125} =$	_____	14. $25^{(-1/2)} =$	_____
15. $\left(\dfrac{1}{27}\right)^{1/3} =$	_____	16. $(2^3)^{1/3} =$	_____

17. $(2^3)^{(-1/3)} =$ _____   18. $\left(\dfrac{1}{8}\right)^{(-1/3)} =$ _____

19. $\left(\dfrac{1}{27}\right)^{(-2/3)} =$ _____   20. $\left(-\dfrac{1}{32}\right)^{-1/5} =$ _____

In Problems 21–30, rewrite the expressions in exponent form.

	Radical Form	Exponent Form		Radical form	Exponent Form
21.	$\sqrt[6]{x^7 y^{18}} = (x^7 y^{18})^{1/6}$	$x^{7/6} y^3$	22.	$\sqrt{x} =$	_____
23.	$\sqrt{x^3} =$	_____	24.	$\sqrt[3]{y^9} =$	_____
25.	$\sqrt[7]{x^5} =$	_____	26.	$x^2 \sqrt{x} =$	_____
27.	$\dfrac{1}{\sqrt[5]{x^2}} =$	_____	28.	$\dfrac{1}{\sqrt[7]{x^3}} =$	_____
29.	$\dfrac{1}{\sqrt[3]{-8}} =$	_____	30.	$\sqrt[4]{x}\ \sqrt[9]{y} =$	_____

In Problems 31–43, simplify by use of the Exponent Rules. Express the answer with positive exponents.

	Expression	Show work here	Simplified form
31.	$x^{2/3} x^{-4/3} =$	$x^{(2/3 - 4/3)} = x^{-2/3} =$	$\dfrac{1}{x^{2/3}}$
32.	$x^{1/5} x^{3/5} =$		_____
33.	$x^2 \cdot x^{-1/3} \cdot x^{2/3} =$		_____
34.	$(x^{2/3})^6 \cdot x^{1/4} =$		_____
35.	$(x^{1/3} \cdot y^2)^6 =$		_____
36.	$x^{-1/3} \cdot x^{2/3} \cdot x^{-5} =$		_____
37.	$(x^{1/2} x^{-1/3})^6 =$		_____
38.	$\left(\dfrac{x^{1/2} y^{2/3}}{z^{-1/2}}\right)^{-12} =$		_____

39. $(x^{-2} \cdot y^{1/3})^{-3} =$ _____

40. $(x^{1/3} \cdot y^{-2/5})^{-15} =$ _____

41. $(x^5 \cdot y^{-2})^{-1/20} =$ _____

42. $(8^{-2} \cdot y^3)^{-1/6} =$ _____

43. $\dfrac{x^{2/3} y^{-1/3} z}{x^{-1/3} y^{2/3}} =$ _____

9.4 Simplifying Radicals

Simplifying an expression containing radicals means:
1. pulling out of the radicand as many factors as possible
2. reducing the index completely
3. in the simplified form, the radicand is not a fraction and there is no radical in the denominator.

Before we discuss procedures for simplifying expressions containing radicals, we need to develop some rules used in these methods.

$$(\sqrt[n]{x} \cdot \sqrt[n]{y})^n = (x^{1/n} \cdot y^{1/n})^n$$
$$= (x^{1/n})^n (y^{1/n})^n = xy$$

Thus, $(\sqrt[n]{x} \cdot \sqrt[n]{y})^n = x \cdot y$

We know by definition that

if $b^n = a$, then $b = \sqrt[n]{a}$.

Therefore,

if $(\sqrt[n]{x} \cdot \sqrt[n]{y})^n = xy$ then $\sqrt[n]{x} \sqrt[n]{y} = \sqrt[n]{xy}$

which is the same thing as

$$\sqrt[n]{xy} = \sqrt[n]{x} \sqrt[n]{y}$$

In a similar way we can prove that

$$\sqrt[n]{\dfrac{x}{y}} = \dfrac{\sqrt[n]{x}}{\sqrt[n]{y}}$$

We restate these two results in the following rule.

Rule 9.2
$$\sqrt[n]{xy} = \sqrt[n]{x} \sqrt[n]{y}$$
$$\sqrt[n]{\dfrac{x}{y}} = \dfrac{\sqrt[n]{x}}{\sqrt[n]{y}}$$

Example 1: *Simplify by use of Rule 9.2*

(a) $\sqrt{12} = \sqrt{4 \cdot 3} = \sqrt{4} \cdot \sqrt{3} = 2\sqrt{3}$

(b) $\sqrt{x^2 y^2} = \sqrt{x^2} \sqrt{y^2} = x \cdot y$

(c) $\sqrt{8} \sqrt{18} = \sqrt{8 \cdot 18} = \sqrt{144} = 12$

(d) $\sqrt[3]{x^3 y^6} = \sqrt[3]{x^3} \cdot \sqrt[3]{y^6} = x \cdot \sqrt[3]{(y^2)^3} = xy^2$

Rule 9.2 is quite convenient for pulling out the maximum number of factors from the radicand.

Illustration 1:
Simplify

$$\sqrt[4]{24 \cdot 54 \cdot 5}$$

Identify which of the factors in the radicand is repeated four times because it can then be expressed as a single factor with exponent 4.

$$\sqrt[4]{24 \cdot 54 \cdot 5} = \sqrt[4]{2 \cdot 2 \cdot 2 \cdot 3 \cdot 2 \cdot 3 \cdot 3 \cdot 3 \cdot 5}$$
$$= \sqrt[4]{(2 \cdot 2 \cdot 2 \cdot 2)(3 \cdot 3 \cdot 3 \cdot 3) \cdot 5}$$
$$= \sqrt[4]{2^4 \cdot 3^4 \cdot 5}$$
$$= \sqrt[4]{2^4} \cdot \sqrt[4]{3^4} \cdot \sqrt[4]{5} \qquad \text{Rule 9.2}$$
$$= 2 \cdot 3 \cdot \sqrt[4]{5} = 6 \sqrt[4]{5}$$

Example 2: *Simplify by use of Rule 9.2*

(a) $\sqrt[3]{5^3 \cdot 7^3} = \sqrt[3]{5^3} \cdot \sqrt[3]{7^3} = 5 \cdot 7$

(b) $\sqrt[6]{x^6 y^6} = \sqrt[6]{x^6} \cdot \sqrt[6]{y^6} = xy$

(c) $\sqrt[5]{x^5 \cdot y^5 \cdot z} = \sqrt[5]{x^5} \cdot \sqrt[5]{y^5} \cdot \sqrt[5]{z} = xy \cdot \sqrt[5]{z}$

Illustration 2: Simplify

$$\sqrt[4]{x^4 y^{12} z^2}$$

$$\sqrt[4]{x^4 y^{12} z^2} = \sqrt[4]{x^4} \cdot \sqrt[4]{y^{12}} \cdot \sqrt[4]{z^2}$$

We have

$$\sqrt[4]{x^4} = x$$
$$\sqrt[4]{y^{12}} = (y^{12})^{1/4} = y^{12 \cdot 1/4} = y^3$$
$$\sqrt[4]{z^2} = z^{2/4} = z^{1/2} = \sqrt{z}$$

Thus,

$$\sqrt[4]{x^4 y^{12} z^2} = \sqrt[4]{x^4} \cdot \sqrt[4]{y^{12}} \cdot \sqrt[4]{z^2} = xy^3 \sqrt{z}$$

From these illustrations we make the following observations that are very helpful in simplifying radicals.

To simplify $\sqrt[n]{x^m}$, reduce completely the fraction

$$\frac{m}{n} \quad \begin{array}{l} \rightarrow \text{exponent in the radicand} \\ \rightarrow \text{index} \end{array}$$

Case 1. If it is a whole number, say p then

$$\sqrt[n]{x^m} = x^p.$$

Case 2. If it is a proper fraction say $\frac{p}{q}$, then

$$\sqrt[n]{x^m} = \sqrt[q]{x^p}$$

Case 3. If it is an improper fraction then express it as a sum of a whole number and a proper fraction say, $p + \dfrac{q}{r}$.

Then $\sqrt[n]{x^m} = x^p \sqrt[r]{x^q}$
since $\sqrt[n]{x^m} = x^{m/n} = x^{p + (q/r)} = x^p \cdot x^{q/r} = x^p \sqrt[r]{x^q}$

Example 3: *Simplify the radicals*

(a) $\sqrt[3]{x^6 y^{12}} = \sqrt[3]{x^6} \cdot \sqrt[3]{y^{12}}$
$= x^2 \cdot y^4$ since $\dfrac{6}{3} = 2$, and $\dfrac{12}{3} = 4$

(b) $\sqrt[4]{x^{10}} = x^2 \cdot \sqrt[2]{x^1}$ since $\dfrac{10}{4} = \dfrac{5}{2} = 2 + \dfrac{1}{2}$
$= x^2 \sqrt{x}$

(c) $\sqrt[6]{x^{18} y^{15}} = \sqrt[6]{x^{18}} \cdot \sqrt[6]{y^{15}}$
$= x^3 y^2 \cdot \sqrt[2]{y^1}$ since $\dfrac{18}{6} = 3$ and $\dfrac{15}{6} = \dfrac{5}{2} = 2 + \dfrac{1}{2}$
$= x^3 y^2 \sqrt{y}$

(d) $\sqrt[5]{32 x^{10} y^8} = \sqrt[5]{32} \cdot \sqrt[5]{x^{10}} \cdot \sqrt[5]{y^8}$
$= 2 x^2 y^1 \cdot \sqrt[5]{y^3}$ since $32 = 2^5$, $\dfrac{10}{5} = 2$ and $\dfrac{8}{5} = 1 + \dfrac{3}{5}$

Sometimes the radicand is a fraction. When required to simplify such radical expressions, the final answer must not have radicals in the denominator. The procedure to simplify such radical expressions is described in the following illustration.

Illustration 3: Simplify

$$\sqrt[c]{\dfrac{x^a}{y^b}}$$

Case 1. If b is a multiple of c, say $b = mc$ then

$$\sqrt[c]{\dfrac{x^a}{y^b}} = \dfrac{\sqrt[c]{x^a}}{\sqrt[c]{y^{mc}}} = \dfrac{\sqrt[c]{x^a}}{y^m} \qquad \left(\dfrac{mc}{c} = m\right)$$

Case 2. If b is not a multiple of c then find a number, say b_1, such that $b + b_1$ is a multiple of c. Let $b + b_1 = mc$; then we multiply and divide the radicand by y^{b_1}. We shall call such a factor a *rationalizing factor* (RF). This factor helps remove all the radicals from the denominator.

$$\sqrt[c]{\dfrac{x^a}{y^b}} = \sqrt[c]{\dfrac{x^a \, y^{b_1}}{y^b \, y^{b_1}}} = \sqrt[c]{\dfrac{x^a y^{b_1}}{y^{b + b_1}}}$$
$$= \sqrt[c]{\dfrac{x^a y^{b_1}}{y^{mc}}} = \dfrac{\sqrt[c]{x^a y^{b_1}}}{\sqrt[c]{y^{mc}}}$$
$$= \dfrac{\sqrt[c]{x^a y^{b_1}}}{y^m} \qquad \text{since } \dfrac{mc}{c} = m$$

This process of simplifying an expression is called rationalizing the denominator.

Example 4: *Simplify and leave no radicals in the denominator.*

(a) $\sqrt[6]{\dfrac{x^4}{y^6}} = \dfrac{\sqrt[6]{x^4}}{\sqrt[6]{y^6}} = \dfrac{\sqrt[3]{x^2}}{y} \quad \left(\dfrac{4}{6} = \dfrac{2}{3}, \dfrac{6}{6} = 1\right)$
(b is a multiple of c)

(b) $\sqrt[5]{\dfrac{A^{10}}{B^3}}$

Here, the exponent of the denominator of the radicand is 3 and is not a multiple of the index 5. Since $3 + 2 = 5$, the RF is B^2.

Thus $\sqrt[5]{\dfrac{A^{10}}{B^3}} = \sqrt[5]{\dfrac{A^{10}}{B^3} \cdot \dfrac{B^2}{B^2}} = \sqrt[5]{\dfrac{A^{10}B^2}{B^5}} = \dfrac{A^2}{B} \sqrt[5]{B^2} \quad \left(\dfrac{10}{5} = 2; \dfrac{5}{5} = 1\right)$

(c) $\sqrt[3]{\dfrac{5}{2x^5y}} = \sqrt[3]{\dfrac{5}{2^1 x^5 y^1}}$

Since the index in this case is 3, the exponent of each factor in the denominator of the radicand must be made a multiple of 3. The RF = $2^2 x^1 y^2$

Then $\sqrt[3]{\dfrac{5}{2x^5y}} = \sqrt[3]{\dfrac{5 \cdot 2^2 \cdot x^1 \cdot y^2}{2^3 x^6 y^3}} = \dfrac{\sqrt[3]{20xy^2}}{2x^2y} \quad \left(\dfrac{3}{3} = 1; \dfrac{6}{3} = 2; \dfrac{3}{3} = 1\right)$

Example 5: *Simplify the radicals*

(a) $\sqrt[3]{\dfrac{5}{3}} = \sqrt[3]{\dfrac{5}{3^1}},$ $\quad$ RF = 3^2

$= \sqrt[3]{\dfrac{5 \cdot 3^2}{3^1 \cdot 3^2}} = \sqrt[3]{\dfrac{45}{3^3}} = \dfrac{\sqrt[3]{45}}{3}$

(b) $\sqrt[4]{x^{10}y^6} = \sqrt[4]{x^{10}} \sqrt[4]{y^6}$

$= x^2 \cdot \sqrt[4]{x^1} \cdot y \cdot \sqrt[4]{y^1} \quad \dfrac{10}{4} = \dfrac{5}{2} = 2 + \dfrac{1}{2}$ and $\dfrac{6}{4} = \dfrac{3}{2} = 1 + \dfrac{1}{2}$

$= x^2 y \sqrt{x} \sqrt{y}$
$= x^2 y \sqrt{xy}$ $\quad$ Rule 9.2

(c) $\sqrt[3]{\dfrac{-x^3 y^4}{z}}$

The odd root of a negative number is negative. The RF = z^2

Thus, $\sqrt[3]{\dfrac{-x^3 y^4}{z}} = -\sqrt[3]{\dfrac{x^3 y^4 z^2}{z^3}} = -\dfrac{\sqrt[3]{x^3 y^4 z^2}}{\sqrt[3]{z^3}} = \dfrac{-xy}{z} \sqrt[3]{yz^2}$

(d) $\sqrt[4]{\dfrac{x^{-4} y^2}{z}} = \sqrt[4]{\dfrac{y^2}{x^4 z}}$

$= \sqrt[4]{\dfrac{y^2 z^3}{x^4 z^4}} \quad$ The RF is z^3

$= \dfrac{\sqrt[4]{y^2 z^3}}{xz}$

Problem Set 9.4

In Problems 1–5, complete the statements.

1. The nth root of x^2 is expressed in radical form as _____ .

2. If we change $x^{m/n}$ to radical form, then the index is _____ and the exponent of the radicand is _____ .

3. If we change $\sqrt[11]{y^5}$ to the exponent form then the denominator of the exponent is _____ .

4. The Rationalizing Factor (RF) needed to simplify the radical $\sqrt{\dfrac{5}{7}}$ is _____ .

5. If a proper fraction $\dfrac{m}{n}$ is equivalent to its reduced form $\dfrac{p}{q}$ then $\sqrt[n]{x^m} = \sqrt[?]{x^?}$

In Problems 6–10, mark the correct response.

6. Consider the following statements P and Q
   P: $\sqrt[3]{27} = 9$ Q: $\sqrt[2]{x^4} = x^2$
   Which of the above statements is true?
   (a) P only (b) Q only (c) P and Q (d) None

7. Consider the following statements P and Q
   P: $\sqrt{x^{-2}} = -x$ Q: $\sqrt[3]{-x^3} = -x$
   Which of the above statements is true?
   (a) P only (b) Q only (c) P and Q (d) None

8. Simplify $\sqrt[3]{x^5 y^4} =$
   (a) $x^3 y^3 \sqrt[3]{x^2 y}$ (b) $x^3 y^3 \sqrt{x^2 y^2}$ (c) $xy\sqrt{x^2 y}$ (d) $xy \sqrt[3]{x^2 y}$ (e) None of these

9. $\sqrt{x^2 + y^2} = x + y$
   (a) True (b) False

10. $\sqrt[6]{27} =$
    (a) $\dfrac{27}{6}$ (b) $\dfrac{6}{27}$ (c) 3^2 (d) $\sqrt{3}$ (e) None of these

In Problems 11–20, simplify the radicals.

	Radicals	Answer		Radicals	Answer
11.	$\sqrt{9} = \sqrt{3^2} =$	3	12.	$\sqrt[3]{x^3} =$	
13.	$\sqrt{x^2 y^2} =$		14.	$\sqrt[3]{x^6} =$	
15.	$\sqrt[3]{x^6 y^9} =$		16.	$\sqrt{\dfrac{x^2}{y^2}} =$	
17.	$\sqrt[3]{\dfrac{x^3}{y^6}} =$		18.	$\sqrt[3]{8 y^{12}} =$	
19.	$\sqrt[5]{-32 y^{10}} =$		20.	$\sqrt[3]{\dfrac{1}{8}} =$	

In Problems 21–26, simplify the radicals. Reduce the index whenever possible.

	Radicals	Answer		Radicals	Answer
21.	$\sqrt[4]{x^2} =$ since $\dfrac{2}{4} = \dfrac{1}{2}$	$\sqrt{x}$	22.	$\sqrt[5]{x^{10}} =$	
23.	$\sqrt[15]{x^5}$		24.	$\sqrt[10]{x^5} =$	
25.	$\sqrt[7]{x^5}$		26.	$\sqrt[12]{x^4 y^8} =$	

In Problems 27–37 simplify the radicals. *Remember if $\frac{m}{n}$ for $\sqrt[n]{x^m}$ is an improper fraction and equals $p + \frac{q}{r}$ then $\sqrt[n]{x^m} = x^p \sqrt[r]{x^q}$.*

	Radical	Show work here		Answer
27.	$\sqrt[6]{27x^{15}} =$	$\sqrt[6]{3^3 x^{15}} = \sqrt[6]{3^3} \sqrt[6]{x^{15}} = \sqrt{3} \, x^2 \sqrt{x}$	27.	$x^2 \sqrt{3x}$
28.	$\sqrt{20} =$		28.	
29.	$\sqrt[3]{54} =$		29.	
30.	$\sqrt[3]{x^{11}} =$		30.	
31.	$\sqrt[3]{24x^6} =$		31.	
32.	$\sqrt{x^5 y^4} =$		32.	
33.	$\sqrt[5]{x^{21} y^{15}} =$		33.	
34.	$\sqrt[3]{81 y^3 z^5} =$		34.	
35.	$\sqrt[6]{x^2 y^4} = \sqrt[6]{x^2} \cdot \sqrt[6]{y^4} = \sqrt[3]{x} \, \sqrt[3]{y^2}$		35.	$\sqrt[3]{xy^2}$
36.	$\sqrt[6]{27 x^3 y^3 z^3} =$		36.	
37.	$\sqrt[12]{x^{18} y^{30}} =$		37.	

In Problems 38–46, simplify the radicals. (RF stands for Rationalizing Factor.)

	Radical	RF	Show work here	Answer
38.	$\sqrt[3]{\dfrac{x^4 z^5}{y^2}}$	y	$\sqrt[3]{\dfrac{x^4 z^5 y}{y^3}} = \dfrac{\sqrt[3]{x^4 z^5 y}}{\sqrt[3]{y^3}} =$	$\dfrac{xz}{y} \cdot \sqrt[3]{xyz^2}$
39.	$\sqrt{\dfrac{1}{y}}$	___		
40.	$\sqrt[4]{\dfrac{x^6}{y^3}}$	___		
41.	$\sqrt[4]{\dfrac{x}{z^4}}$	___		
42.	$\sqrt[3]{\dfrac{x}{yz^2}}$	___		
43.	$\sqrt{\dfrac{y^3}{x^9}}$	___		

44. $\sqrt[6]{\dfrac{9x^2}{y^2}}$ _____ _____

45. $\sqrt[5]{\dfrac{1}{5}}$ _____ _____

46. $\sqrt[3]{\dfrac{-8x^4y^6}{z^7}}$ _____ _____

9.5 Basic Operations with Radicals

The addition of expressions involving radicals is similar to the addition of two algebraic expressions. The terms having the same index and the same radicand, called *LIKE TERMS*, can be combined into one term just as we do in algebraic expressions. For example,

$5\sqrt{x} + 2\sqrt{x}$ can be simplified as $(5 + 2)\sqrt{x}$ or $7\sqrt{x}$.

Sometimes the radical terms in an expression do not appear to be like terms but can be made like terms after simplification. For example,

$$\sqrt{18} + 5\sqrt{2}$$

Although $\sqrt{18}$ and $5\sqrt{2}$ cannot be combined in this form, these terms can be combined after simplification.

Thus, $\sqrt{18} = \sqrt{2 \cdot 9} = \sqrt{2}\sqrt{9} = 3\sqrt{2}$
$\sqrt{18} + 5\sqrt{2} = 3\sqrt{2} + 5\sqrt{2} = (3 + 5)\sqrt{2} = 8\sqrt{2}$

Example 1: *Adding expressions containing radicals*

(a) $(2\sqrt{3} - 5\sqrt{2}) + (3\sqrt{2} - 5\sqrt{3}) = 2\sqrt{3} - 5\sqrt{2} + 3\sqrt{2} - 5\sqrt{3}$
$= 2\sqrt{3} - 5\sqrt{3} - 5\sqrt{2} + 3\sqrt{2}$
$= -3\sqrt{3} - 2\sqrt{2}$

(b) $(3\sqrt{5} - 2\sqrt[3]{4}) - (5\sqrt[3]{4} - 2\sqrt{5}) = 3\sqrt{5} - 2\sqrt[3]{4} - 5\sqrt[3]{4} + 2\sqrt{5}$
$= 3\sqrt{5} + 2\sqrt{5} - 2\sqrt[3]{4} - 5\sqrt[3]{4}$
$= 5\sqrt{5} - 7\sqrt[3]{4}$

(c) $\sqrt{27} + \sqrt{12} - \sqrt{3a^2} = 3\sqrt{3} + 2\sqrt{3} - a\sqrt{3}$
$= (3 + 2 - a)\sqrt{3} = (5 - a)\sqrt{3}$

(d) $\sqrt[6]{x^8} - 4\sqrt[3]{x} + \sqrt[3]{8x^4} = x\sqrt[3]{x} - 4\sqrt[3]{x} + \sqrt[3]{8}\sqrt[3]{x^4}$
$= x\sqrt[3]{x} - 4\sqrt[3]{x} + 2x\sqrt[3]{x}$
$= (x - 4 + 2x)\sqrt[3]{x} = (3x - 4)\sqrt[3]{x}$

We can multiply two radical terms with the same index by using Rule 9.2. In case the index of the two terms to be multiplied is not the same, we can obtain equivalent forms of the two terms so that they have the same index. This common index is normally the Least Common Multiple (LCM) of the indices of the given two terms. We obtain the product of the given two terms by multiplying their equivalent forms as in Example 2. For example consider the product,

$$\sqrt{x} \, \sqrt[3]{y} = \sqrt[2]{x} \, \sqrt[3]{y}$$

The LCM of 2 and 3 is 6

$$\sqrt{x} = x^{1/2} = x^{3/6} = \sqrt[6]{x^3}$$
$$\sqrt[3]{y} = y^{1/3} = y^{2/6} = \sqrt[6]{y^2}$$

Thus, $\sqrt{x} \sqrt[3]{y} = \sqrt[6]{x^3} \sqrt[6]{y^2} = \sqrt[6]{x^3 y^2}$

➡ Note this is just like changing the fractions $\frac{1}{2}, \frac{1}{3}$ to equivalent fractions $\frac{3}{6}, \frac{2}{6}$ having the same denominator.

Example 2: *Multiply the radical terms with the same index*

(a) $\sqrt{8} \sqrt{2} = \sqrt{8 \cdot 2} = \sqrt{16} = 4$

(b) $\sqrt[3]{a} \sqrt[3]{a^2} = \sqrt[3]{a \cdot a^2} = \sqrt[3]{a^3} = a$

(c) $\sqrt[3]{x^2 y} \sqrt[3]{xy} = \sqrt[3]{x^2 y \cdot yx} = \sqrt[3]{x^3 y^2} = x \sqrt[3]{y^2}$

(d) $\sqrt[4]{x^2 y^5} \sqrt[4]{\frac{x^3}{y}} = \sqrt[4]{x^2 y^5 \cdot \frac{x^3}{y}} = \sqrt[4]{x^5 y^4}$
$= \sqrt[4]{x^5} \sqrt[4]{y^4} = x \sqrt[4]{x} \cdot y = xy \sqrt[4]{x}$

Example 3: *Multiply the terms with different indexes*

(a) $\sqrt{2} \sqrt[3]{9} = \sqrt[6]{2^3} \sqrt[6]{9^2} = \sqrt[6]{8 \cdot 81} = \sqrt[6]{648}$

(b) $\sqrt[3]{a^2} \sqrt[4]{b^3} = \sqrt[12]{(a^2)^4} \sqrt[12]{(b^3)^3} = \sqrt[12]{a^8 b^9}$ Rule 9.2

We can multiply two expressions involving radicals using the distributive law the same way we multiply two algebraic expressions.

Example 4: *Multiply the expressions and simplify.*

(a) $\sqrt{2}(2\sqrt{3} - \sqrt{2}) = 2\sqrt{2} \sqrt{3} - \sqrt{2} \sqrt{2} = 2\sqrt{6} - 2$

(b) $(\sqrt{3} - \sqrt{2})(2\sqrt{3} - 5\sqrt{2}) = \sqrt{3}(2\sqrt{3} - 5\sqrt{2}) - \sqrt{2}(2\sqrt{3} - 5\sqrt{2})$
$= 2 \cdot 3 - 5\sqrt{6} - 2\sqrt{6} + 5 \cdot 2$
$= 6 + 10 - 7\sqrt{6} = 16 - 7\sqrt{6}$

(c) $(\sqrt{5} + \sqrt{11})(\sqrt{5} - \sqrt{11}) = (\sqrt{5})^2 - (\sqrt{11})^2$ Rule 6.2
$= 5 - 11 = -6$

We can divide two radical terms with the same index by using Rule 9.2

$$\frac{\sqrt[n]{x}}{\sqrt[n]{y}} = \sqrt[n]{\frac{x}{y}}.$$

But remember, the final answer should be expressed in the simplified form. That is, the denominator should be rationalized.

Example 5: *Divide the radical terms with the same index.*

(a) $\dfrac{\sqrt{8}}{\sqrt{2}} = \sqrt{\dfrac{8}{2}} = \sqrt{4} = 2$

(b) $\dfrac{\sqrt{x}}{\sqrt{y}} = \sqrt{\dfrac{x}{y}} = \sqrt{\dfrac{x \cdot y}{y^2}}$ RF is x

$= \dfrac{\sqrt{xy}}{y}$

(c) $\dfrac{\sqrt[3]{x^2}}{\sqrt[3]{y}} = \sqrt[3]{\dfrac{x^2}{y}}$ RF $= y^2$

$= \sqrt[3]{\dfrac{x^2 y^2}{y^3}} = \dfrac{\sqrt[3]{x^2 y^2}}{y}$

We can *rationalize the denominator* when the denominator has two terms involving radicals. The Rationalization becomes very simple when the two terms in the denominator contain square roots only, such as

$$3 - \sqrt{5}, \text{ or } \sqrt{3} + \sqrt{2}$$

In this case just multiply and divide the fraction by the expression that makes the denominator a difference of two squares.

Recall that $(a + b)(a - b) = a^2 - b^2$.

Example 6: *Rationalize the denominator*

(a) $\dfrac{3}{2 - \sqrt{3}} = \dfrac{3}{2 - \sqrt{3}} \cdot \dfrac{2 + \sqrt{3}}{2 + \sqrt{3}}$

$= \dfrac{3(2 + \sqrt{3})}{(2 - \sqrt{3})(2 + \sqrt{3})} = \dfrac{6 + 3\sqrt{3}}{4 - 3} = 6 + 3\sqrt{3}$

(b) $\dfrac{\sqrt{2}}{\sqrt{2} - \sqrt{3}} = \dfrac{\sqrt{2}}{\sqrt{2} - \sqrt{3}} \cdot \dfrac{\sqrt{2} + \sqrt{3}}{\sqrt{2} + \sqrt{3}}$

$= \dfrac{\sqrt{2}(\sqrt{2} + \sqrt{3})}{(\sqrt{2} - \sqrt{3})(\sqrt{2} + \sqrt{3})} \cdot \dfrac{2 + \sqrt{6}}{2 - 3}$

$= \dfrac{2 + \sqrt{6}}{-1} = -2 - \sqrt{6}$

(c) $\dfrac{2\sqrt{3} - \sqrt{2}}{3\sqrt{2} - \sqrt{5}} = \dfrac{2\sqrt{3} - \sqrt{2}}{3\sqrt{2} - \sqrt{5}} \cdot \dfrac{3\sqrt{2} + \sqrt{5}}{3\sqrt{2} + \sqrt{5}} = \dfrac{(2\sqrt{3} - \sqrt{2})(3\sqrt{2} + \sqrt{5})}{(3\sqrt{2} - \sqrt{5})(3\sqrt{2} + \sqrt{5})}$

$= \dfrac{6\sqrt{6} + 2\sqrt{15} - 6 - \sqrt{10}}{(3\sqrt{2})^2 - (\sqrt{5})^2}$

$= \dfrac{2\sqrt{15} - \sqrt{10} + 6\sqrt{6} - 6}{13}$

Problem Set 9.5

In Problems 1–8, simplify.

Expression	Show work here	Answer
1. $5\sqrt{x} - 3\sqrt{x} + 2\sqrt{x} =$	$(5 - 3 + 2)\sqrt{x}$	$4\sqrt{x}$
2. $\sqrt{20} - \sqrt{5} =$		

3. $3\sqrt{3} - \sqrt{27} =$ _____

4. $2\sqrt{x^3} + 4\sqrt{x} - \sqrt{9x} =$ _____

5. $\sqrt{18} + \sqrt{50} + \sqrt{72} =$ _____

6. $\sqrt[3]{-16} + 2\sqrt[3]{54} =$ _____

7. $\sqrt[3]{a^4} - \sqrt[6]{a^8} + \sqrt[3]{27a^4} =$ _____

8. $\sqrt{ba^2} - a\sqrt{4b} + \sqrt{9b^3} =$ _____

In Problems 9–15, perform the multiplication and express the answer in the most simplified form.

Product	Show work here	Answer
9. $\sqrt{3}\sqrt{12} =$	$\sqrt{3 \cdot 12} = \sqrt{36}$	6
10. $\sqrt[3]{x}\sqrt[3]{x^2} =$		
11. $\sqrt{2x^3}\sqrt{18x} =$		
12. $\sqrt{x}\sqrt[3]{y}$		$\sqrt[6]{}$
13. $\sqrt[3]{x^2}\sqrt[4]{y}$		$\sqrt[12]{}$
14. $3x\sqrt[3]{2xy^2}\sqrt{2x} =$		
15. $\sqrt{a}\sqrt[3]{b^2}\sqrt[4]{c} =$		

In Problems 16–20, perform the indicated operations and simplify.

Answer

16. $(\sqrt{3} - 2)(\sqrt{2} - 4\sqrt{3}) =$ $-12 - 2\sqrt{2} + 8\sqrt{3} + \sqrt{6}$
$\sqrt{3}(\sqrt{2} - 4\sqrt{3}) - 2(\sqrt{2} - 4\sqrt{3}) =$
$\sqrt{6} - 4 \cdot 3 - 2\sqrt{2} + 8\sqrt{3}$

17. $\sqrt{3}(\sqrt{2} - 4\sqrt{3}) =$ _____

18. $(\sqrt{3} - 2)(\sqrt{3} + 2) =$ _____

19. $(\sqrt{3} - \sqrt{5})^2 =$ _____

20. $(\sqrt{3} + \sqrt{2})(\sqrt{3} - 2\sqrt{2}) =$ _____

In Problems 21–25, simplify.

Problem	Show work here	Answer
21. $\dfrac{\sqrt{6}}{\sqrt{2}} =$	$\sqrt{\dfrac{6}{2}} = \sqrt{3}$	$\sqrt{3}$
22. $\dfrac{\sqrt{3}}{\sqrt{y}} =$		
23. $\dfrac{\sqrt[3]{x}}{\sqrt[3]{2y}} =$		
24. $\dfrac{\sqrt[3]{5}}{\sqrt[2]{2}} =$		
25. $\dfrac{\sqrt{4x}}{\sqrt{2y}} =$		

In Problems 26–30, rationalize the denominator.

Fraction	Show work here	Answer
26. $\dfrac{3}{3 + \sqrt{2}} =$	$\dfrac{3}{3 + \sqrt{2}} \cdot \dfrac{3 - \sqrt{2}}{3 - \sqrt{2}} = \dfrac{3(3 - \sqrt{2})}{9 - 2}$	$\dfrac{9 - 3\sqrt{2}}{7}$
27. $\dfrac{4}{\sqrt{5} + 1} =$		
28. $\dfrac{x}{\sqrt{y} - 1} =$		

29. $\dfrac{2 - \sqrt{3}}{2 + \sqrt{3}} =$ _____

30. $\dfrac{\sqrt{2} - \sqrt{3}}{2\sqrt{2} + \sqrt{3}} =$ _____

9.6 Complex Numbers

Up to this point we have not discussed any problems where the radicand is negative and the index is an even integer. For example, we have not considered expressions of the form

$$\sqrt{-4} \text{ or } \sqrt[4]{-2}.$$

The reason for this is that all such numbers are not real numbers. We call such numbers complex numbers. Complex numbers were introduced earlier in Section 7.5.

Before we discuss how to perform the basic operations when the numbers are not real, let us review in brief what we discussed in Section 7.5.

1. The number "i" is defined as

$$\boxed{i = \sqrt{-1}}$$

   Therefore,

$$\boxed{i^2 = -1}$$

   Also, note that $i^3 = -i$, $i^4 = 1$, $i^5 = i$.

2. In general any number of the type "a + bi" where "a" and "b" are real numbers is called a complex number.

3. In a complex number a + bi
   a = Real part of a + bi
   b = Imaginary part of a + bi

4. Given a complex number a + bi then a number obtained by changing the sign of i is called the *conjugate* of a + bi

   a + bi is a conjugate of a − bi
   a − bi is a conjugate of a + bi

   and the pair (a ± bi) is called a conjugate pair of complex numbers.

5. The sum and product of a conjugate pair of numbers are real numbers.

   (a + bi) + (a − bi) = 2a
    = 2(real part)
   (a + bi)(a − bi) = $a^2 - i^2 b^2 = a^2 + b^2$

6. *Observation 1.* If a complex number is zero then its real and imaginary parts are separately zero. That is, if $a + bi = 0$, then $a = 0$ and $b = 0$. To see this, suppose $a + bi = 0$. Then $a = -ib$ and we have

   $a^2 = (-ib)^2$ or $a^2 = i^2 b^2$
   Hence, $a^2 = -b^2$ or $a^2 + b^2 = 0$

   The sum of squares of two real numbers is zero. This is possible only if $a = 0$ and $b = 0$.

7. *Observation 2.* If $a + bi = c + di$ then $a = c$ and $b = d$. You can prove it by using Observation 1.

Example 1: *Review from Section 7.5*
  (a) $i = \sqrt{-1}$, $i^2 = -1$
  (b) The real part of the complex number $-4 + i\sqrt{5}$ is -4, and its imaginary part is $\sqrt{5}$
  (c) The conjugate of $-3 - 5i$ is $-3 + 5i$
  (d) $(5 + 4i) + (5 - 4i) = 10$
  (e) $(5 + 4i)(5 - 4i) = 25 + 16 = 41$

We add, subtract and multiply complex numbers just as we perform these operations with algebraic expressions. While doing that we treat "i" as a literal part of the term and we keep in mind that

$$i^n = \pm 1 \text{ or } \pm i \quad \text{depending upon the value of n.}$$

Example 2: *Add complex numbers*
  (a) $(2 + 3i) + (5 - 7i)$
     $= (2 + 5) + (3 - 7)i = 7 - 4i$
  (b) $(2i^2 + 3i - 4) - (5 - 4i + i^2)$
     $i^2 = -1$
     Therefore, $2i^2 + 3i - 4 = -2 + 3i - 4 = -6 + 3i$
     and $5 - 4i + i^2 = 5 - 4i - 1 = 4 - 4i$
     Thus, $(2i^2 + 3i - 4) - (5 - 4i + i^2) =$
     $(-6 + 3i) - (4 - 4i) = -6 + 3i - 4 + 4i = -10 + 7i$

Example 3: *Multiply complex numbers*
  (a) $(4 + 3i)(4 - 3i) = 16 - 9i^2 = 16 + 9 = 25$
  (b) $(3 - 4i)(2 + i) = 3(2 + i) - 4i(2 + i)$
     $= 6 + 3i - 8i - 4i^2 = 6 + 3i - 8i - 4(-1)$
     $= 10 - 5i$
  (c) $(3 + 2i - 4i^2)(i^3 - 5i^5)$
     $-4i^2 = -4(-1) = 4$
     $i^3 = i^2 \cdot i = -i$
     $i^5 = i^4 \cdot i = (i^2)^2 \cdot i = (-1)^2 i = i$
     Therefore, $(3 + 2i - 4i^2)(i^3 - 5i^5) = (3 + 2i + 4)(-i - 5i)$
     $= (7 + 2i)(-6i)$
     $= 7(-6i) + (2i)(-6i)$
     $= -42i - 12i^2$
     $= -42i + 12$
     $= 12 - 42i$

Note that all the answers in Examples 2 and 3 are expressed in the form $a + bi$. That is the way it should be whenever we perform operations on complex numbers. But when we divide two complex numbers, we need to do some extra work to express the answer in the form $a + bi$. The procedure for doing this is described in the following illustration.

Illustration: *Divide $2i^3 - 4i^2 + 5$ by $i^3 - 4$*

Step 1. Express the divisor and the dividend in the form $a + bi$

$$2i^3 - 4i^2 + 5 = 2i^2 i - 4(-1) + 5$$
$$= -2i + 4 + 5$$
$$= 9 - 2i$$
$$i^3 - 4 = i^2 i - 4$$
$$= -i - 4$$
$$= -4 - i$$

Step 2: Express the division as a fraction and rationalize the denominator

$$\frac{2i^3 - 4i^2 + 5}{i^3 - 4} = \frac{9 - 2i}{-4 - i} = \frac{9 - 2i}{-4 - i} \cdot \frac{-4 + i}{-4 + i}$$

▶ Note that the rationalizing factor in complex numbers is the conjugate of the denominator.

Step 3: Simplify the fraction of Step 2.

$$\frac{(9 - 2i)(-4 + i)}{(-4 - i)(-4 + i)} = \frac{(9 - 2i)(-4 + i)}{(-4)^2 - i^2}$$
$$= \frac{-34 + 17i}{17} = \frac{-34}{17} + \frac{17i}{17} = -2 + i$$

Example 4: *Simplify the fractions involving complex numbers.*

(a) $\dfrac{2}{1 - i} = \dfrac{2(1 + i)}{(1 - i)(1 + i)} = \dfrac{2 + 2i}{1 - i^2} = \dfrac{2 + 2i}{2} = 1 + i$

(b) $\dfrac{2 - i}{3 - 2i} = \dfrac{(2 - i)(3 + 2i)}{(3 - 2i)(3 + 2i)} = \dfrac{8 + i}{9 - 4i^2} = \dfrac{8 + i}{13} = \dfrac{8}{13} + \dfrac{1}{13}i$

(c) $\dfrac{3i^3 + 2i^2 - 4}{i^2 + 2i - 2} = \dfrac{3i^2 i + 2i^2 - 4}{-1 + 2i - 2} = \dfrac{-3i - 2 - 4}{-3 + 2i} = \dfrac{-6 - 3i}{-3 + 2i}$
$$= \dfrac{(-6 - 3i)(-3 - 2i)}{(-3 + 2i)(-3 - 2i)} = \dfrac{12 + 21i}{(-3)^2 - (2i)^2}$$
$$= \dfrac{12 + 21i}{13} = \dfrac{12}{13} + \dfrac{21}{13}i$$

Problem Set 9.6

Complete the statements in Problems 1–5.

1. The real part of the complex number $3 + 4i$ is _____ and its imaginary part is _____ .

2. The sum of two complex conjugate numbers is twice its _____ .

3. Given a complex number $3 + 4i$, the product of this number and its conjugate is _____ .

4. If $i(i - 4) = a + ib$ then $a = $ _____ and $b = $ _____ .

5. If $i^3 + 2i^2 - i + 4 = a + ib$ then $a = $ _____ and $b = $ _____ .

Mark the correct response in Problems 6–8.

6. Consider the statements P and Q
   P: $-\dfrac{1}{i} = i$ Q: $i^3 = -i$
   Which of the above statements is true?
   (a) P only   (b) Q only   (c) P and Q   (d) None of these

7. Consider the following statements P and Q
   P: $\sqrt{-5} = -\sqrt{5}$ Q: $(2 + i\sqrt{3})(2 - i\sqrt{3}) = 7$
   Which of the above statements is true?
   (a) P only   (b) Q only   (c) P and Q   (d) None of these

8. $\dfrac{2}{1 - i} = $
   (a) $1 - i$ (b) $1 + i$ (c) $2 + i$ (d) $1 + 2i$ (e) None of these

In Problems 9–20, perform the indicated operations and express the answers in the form $a + bi$.

Problem	Show work here	Answer
9. $(2i - 3)(4 - i) = $	$-2i^2 + 11i - 12 = $ $-2(-1) + 11i - 12$	9. $-10 + 11i$
10. $4i^2 - 3i + 4 = $		10. _____
11. $i^4 - 4i^3 - 2i^2 = $		11. _____
12. $(3 + i) + (3 - i) = $		12. _____
13. $(\sqrt{3} + \sqrt{2}i) + (\sqrt{3} - \sqrt{2}i) = $		13. _____
14. $2i(4 - i) = $		14. _____
15. $3i^2(5 + 2i) = $		15. _____
16. $(3 - i)(3 + i) = $		16. _____

17. $(\sqrt{2} - i\sqrt{3})(\sqrt{2} + i\sqrt{3}) =$ 17. _____

18. $(2 + 5i)(3 - i) =$ 18. _____

19. $(2i^2 - 3)(i^4 - 3i) =$ 19. _____

20. $(4 - 3i^4)(3 - 2i^5) =$ 20. _____

In Problems 21–25, perform the division and express your answer in the form $a + ib$.

Division	Show work here		Answer
21. $\dfrac{2i^2 + 1}{i^4 - i^3} =$	$\dfrac{-2 + 1}{1 + i} = \dfrac{-1}{1 + i} \cdot \dfrac{1 - i}{1 - i}$ $= \dfrac{-1 + i}{1 - i^2} = \dfrac{-1 + i}{2}$	21.	$-\dfrac{1}{2} + \dfrac{1}{2}i$

22. $\dfrac{2}{2 + i} =$ 22. _____

23. $\dfrac{1 - i}{1 + i} =$ 23. _____

24. $\dfrac{2 - i}{3 + 2i} =$ 24. _____

25. $(4 - i + 2i^3) \div (i^4 - 2i^3) =$ 25. _____

Chapter Summary

1. **Definition of nth root**   If $b = a^n$ then $a = b^{1/n}$.
2. **Radical form of the nth root** $\sqrt[n]{a}$ is also the nth root of a. The number n is a positive integer and is called the *index*. The number a is called the *radicand*.
3. **Square root**   When no index appears in a radical expression, the index is assumed to be 2. Thus, $\sqrt[2]{3}$ or $\sqrt{3}$ is the square root of three.

4. **Principal root** The principal n^{th} root of a positive number is positive. The principal n^{th} root of a negative number is negative if n is odd. The principal n^{th} root of a negative number, if n is even, is not defined in the reals. For the principal root

$$x^{1/n} = \sqrt[n]{x}$$

5. **Rule 9.1** $\quad x^{m/n} = \sqrt[n]{x^m}$
The denominator goes as the index
The numerator goes as the exponent in the radicand

6. **Simplifying radicals** Simplifying radicals means: (1) pull out of the radicand as many factors as possible, (2) reduce the index completely, (3) rationalize the denominator.

7. **Rule 9.2** $\quad \sqrt[n]{x \cdot y} = \sqrt[n]{x} \sqrt[n]{y}$

$$\sqrt[n]{\frac{x}{y}} = \frac{\sqrt[n]{x}}{\sqrt[n]{y}}$$

8. **To simplify** $\sqrt[n]{x^m}$ Reduce the fraction $\frac{m}{n}$ completely,

   1. If $\frac{m}{n}$ is a whole number, say p, $\sqrt[n]{x^m} = x^p$
   2. If $\frac{m}{n}$ is a proper fraction, say $\frac{p}{q}$, then $\sqrt[n]{x^m} = \sqrt[q]{x^p}$
   3. If $\frac{m}{n}$ is an improper fraction and equals say $p + \frac{q}{r}$ then $\sqrt[n]{x^m} = x^p \cdot \sqrt[r]{x^q}$

9. **To simplify** $\sqrt[c]{\frac{x^a}{y^b}}$
Case 1: If b is a multiple of c, say b = mc then

$$\sqrt[c]{\frac{x^a}{y^b}} = \sqrt[c]{\frac{x^a}{y^{mc}}} = \frac{\sqrt[c]{x^a}}{y^m}$$

Case 2: If b is not a multiple of c, we find another number, say b_1, so that $b + b_1 = mc$. Then we multiply and divide the radicand by y^{b_1}. That is,

$$\sqrt[c]{\frac{x^a}{y^b}} = \sqrt[c]{\frac{x^a y^{b_1}}{y^{b+b_1}}} = \frac{\sqrt[c]{x^a y^{b_1}}}{y^m}$$

10. **Adding radicals** We can add or subtract expressions involving radicals, just as we do for algebraic expressions by combining like terms. In radicals, like terms means terms with the same radicand and the same index. All the radicals in the expression must be simplified to identify which are the like terms.

11. **Multiplying radicals**
Case 1: If the index of the factors are the same we use rule 9.2

$$\sqrt[n]{x} \sqrt[n]{y} = \sqrt[n]{xy}$$

Case 2: If the index of the factors is not the same then,
   (a) find the LCM of the index of all the radical factors.
   (b) change each radical factor to an equivalent radical with index equal to the LCM.
   (c) use Rule 9.2
Simplify the resulting radical.

12. **Complex numbers** If the radicand is negative and the index is even, then the number is a complex number. For example,

$$\sqrt{-3}, \sqrt[4]{-5}.$$

We define i = $\sqrt{-1}$. Any complex number can be expressed in the form a + ib, where

a = Real part of a + bi
b = Imaginary part of a + bi

13. **Conjugate complex numbers** The conjugate of a complex number is obtained by changing the sign of i in a + ib.

a − bi is conjugate of a + bi, and a + bi is conjugate of a − bi.

14. **Sum and product of conjugate complex numbers** (a + bi) + (a − bi) = 2a
(a + bi)(a − bi) = a² + b²

15. **Equivalent forms of i^n** i^n is always equal to one of the numbers 1, −1, i, −i, depending on the value of n.

16. **Adding and multiplying complex numbers** We can add or multiply complex numbers the same way we perform these operations on algebraic fractions. The final result is always simplified to the form a + ib.

17. **Dividing complex numbers** When we divide two complex numbers, the problem reduces to simplifying a fraction

$$\frac{a + bi}{c + di}.$$

This fraction can be simplified by multiplying and dividing by the conjugate of the denominator.

Review Problems

In Problems 1–7, simplify. There should be no negative exponents in the answer.

1. $3x^0 x^{-2} =$

2. $3x^{-4} y^3 =$

3. $(-3)^{-3} =$

4. $(2x^2)^3 x^2 =$

5. $\left(\dfrac{x^2 y^{-3}}{z^{-2}}\right)^{-5} =$

6. $\dfrac{(-2x^0)(-2x)^4}{(-x)^3} =$

7. $(y^2 x^{-3})^{-2} (x^{-1})^{-3} =$

In Problems 8–13, find the indicated root.

8. $\sqrt{4x^2} =$

9. $\sqrt[3]{x^6} =$

10. $\sqrt[4]{16x^4} =$

11. $\sqrt[3]{-27} =$

12. $\sqrt[3]{-8x^3} =$

13. $\sqrt[5]{243x^{15}} =$

Find the value of the expressions in Problems 14–17.

14. $(-8)^{1/3} =$ _____ 15. $(x^{-3}y^6)^{-1/6} =$ _____

16. $\left(\dfrac{1}{27}\right)^{-1/3} =$ _____ 17. $\dfrac{1}{(-8)^{-3}} =$ _____

In Problems 18–22, simplify by use of the exponent rules. Express the answer with positive exponents.

18. $\dfrac{x^{-2/3}x^{4/3}}{x^{-1/3}} =$
18. _____

19. $(x^{-3}y^6)^{-1/3} =$
19. _____

20. $\left(\dfrac{4x^2y^{-4}}{z^{-10}}\right)^{-1/2} =$
20. _____

21. $\left(\dfrac{2x^0y^{-4}z^{-4}}{z^{-12}}\right)^{-3/4} =$
21. _____

22. $\left(\dfrac{3x^2y^{-2}}{z^{-3}}\right)^{1/6}$
22. _____

In Problems 23–30, rewrite the expression in exponent form.

23. $\sqrt[3]{x^4} =$ _____ 24. $\sqrt[7]{x^5} =$ _____

25. $\sqrt[3]{y^6} =$ _____ 26. $\sqrt[6]{y^3x^2} =$ _____

27. $x^3\sqrt{x} =$ _____ 28. $\sqrt{x^7} =$ _____

29. $\dfrac{1}{\sqrt[3]{x^2}} =$ _____ 30. $\sqrt[6]{x^6y^{21}} =$ _____

In Problems 31–44, simplify the radicals. Reduce the index whenever possible and there should be no radicals in the denominator of the answer.

31. $\sqrt[3]{x^9} =$ _____ 32. $\sqrt[5]{x^{10}} =$ _____

33. $\sqrt[4]{x^2} =$ _____ 34. $\sqrt[6]{x^4} =$ _____

35. $\sqrt[6]{x^9} =$ _____ 36. $\sqrt[5]{-x^{11}} =$ _____

37. $\sqrt[3]{\dfrac{x^3}{y^3}} =$ _____ 38. $\sqrt{\dfrac{1}{2}} =$ _____

39. $\sqrt{\dfrac{1}{y}} =$
39. _____

40. $\sqrt[3]{\dfrac{x^5}{y^3}} =$

41. $\sqrt[12]{x^{18}y^{42}} =$

42. $\sqrt[3]{-27x^4y^6z^{10}} =$

43. $\sqrt[3]{\dfrac{y}{z^4}} =$

44. $\sqrt[4]{\dfrac{y^4}{z^6}} =$

In Problems 45–50, simplify.

45. $2\sqrt{3} - 4\sqrt{3} - 2\sqrt{3} =$

46. $5\sqrt{12} - 2\sqrt{27} =$

47. $\sqrt[4]{x^2} - 3\sqrt{x} =$

48. $\sqrt[3]{x} \cdot \sqrt[3]{x^5} =$

49. $\sqrt[4]{4x^2} \cdot \sqrt[4]{4x^6} =$

50. $\sqrt[3]{x^2}\,\sqrt[4]{y} =$

In Problems 51–55, rationalize the denominators.

51. $\dfrac{1}{\sqrt{3}} =$

52. $\dfrac{3}{\sqrt{2}} =$

53. $\dfrac{1}{\sqrt{3} - 2} =$

54. $\dfrac{3}{\sqrt{5} - \sqrt{2}} =$

55. $\dfrac{\sqrt{2} + \sqrt{3}}{\sqrt{2} - \sqrt{3}} =$

In Problems 56–60, simplify and express the result in the form $a + ib$.

56. $2i^3 - 2i^2 + 4i - 5 =$

57. $(3 - i^2)(i^2 - i + 2) =$

58. $\dfrac{2}{i+1} =$

59. $\dfrac{2i-1}{3i-4} =$

60. $\dfrac{i^3 - 2i^2 + 5}{i^5 - 3i^3 + 4} =$

Chapter Test

Solve each of the following problems and match your answer with the responses.

1. Solve for x: $\dfrac{x-4}{3} = \dfrac{1}{x-2}$
   (a) 3,4 (b) 1,2 (c) 1,5 (d) None of these

2. Solve for x: $\dfrac{x^2}{6} - \dfrac{2x}{3} = 2$
   (a) $-2,6$ (b) $2,-6$ (c) $-2,-6$ (d) None of these

3. Simplify: $(x^{-1/2} y^{1/3})^{-6} =$
   (a) $x^3 y^2$ (b) $\dfrac{y^2}{x^3}$ (c) $\dfrac{x^3}{y^2}$ (d) None of these

4. Simplify and write your answer with positive exponent only: $\left(\dfrac{x^{-2/5}}{x^2 y^{-3/5}}\right)^{-5}$
   (a) $\dfrac{1}{x^8 y^2}$ (b) $\dfrac{x^8}{y^3}$ (c) $\dfrac{x^{12}}{y^3}$ (d) $x^{12} y^3$ (e) None of these

5. Simplify: $\sqrt[3]{54} =$
   (a) $27\sqrt[3]{2}$ (b) $3\sqrt[3]{2}$ (c) $9\sqrt[3]{2}$ (d) $3\sqrt[3]{17}$ (e) None of these

6. Simplify completely: $\sqrt{x^6 y^7} =$
   (a) $x^3 y^3$ (b) $x^3 y^3 \sqrt{y}$ (c) $x^2 y^3 \sqrt{y}$ (d) $xy \sqrt{x^4 y^3}$ (e) None of these

7. Consider the following statements P and Q:
   P: $x^{m/n} = \sqrt[n]{x^m}$.
   Q: $\sqrt[3]{16} = 2\sqrt[3]{2}$.
   Which of the above statements are true?
   (a) P only (b) Q only (c) P and Q (d) None

8. Consider the following statements P and Q:
   P: $\sqrt[4]{x^8} = x^4$.
   Q: $\sqrt[8]{x^4} = \sqrt{x}$.
   Which of the above statements are true?
   (a) P only (b) Q only (c) P and Q (d) None

9. Simplify: $\sqrt[3]{12}\ \sqrt[3]{18} =$
   (a) 24   (b) 36   (c) $\sqrt[6]{12 \times 18}$ (d) 6   (e) None of these

10. Simplify completely: $\sqrt[6]{64x^7 y^{15}} =$
    (a) $2x^2y\ \sqrt[6]{y^2}$ (b) $2x\ \sqrt[6]{x^2 y}$ (c) $2xy\ \sqrt[6]{xy^3}$ (d) $2xy^2\ \sqrt[6]{xy^3}$ (e) None of these

11. $\sqrt[3]{\dfrac{x^3}{y^2}} = \dfrac{x}{y}\ \sqrt{y}$.
    (a) True   (b) False

12. $\sqrt[3]{x^2y}\ \sqrt[3]{xy^2} = xy$.
    (a) True   (b) False

13. Simplify: $(\sqrt{5} - \sqrt{3})(\sqrt{5} + \sqrt{3}) = 2$.
    (a) True   (b) False

14. Rationalize: $\dfrac{-2}{\sqrt{3} - 2} =$
    (a) $-2\sqrt{3} + 4$ (b) $2\sqrt{3} - 4$ (c) $2\sqrt{3} + 4$ (d) $-2\sqrt{3} - 4$

Chapter

10

Graphing and Systems of Linear Equations

In previous chapters we have learned how to solve linear equations of the form ax + b = c. In these equations there was one unknown and we used algebraic techniques to find the solution.

Now we encounter a new situation. We will be interested in dealing with linear equations of the form ax + by = c, where x and y represent unknowns and a, b and c are given numbers. We will discover that an equation of the form ax + by = c has an infinite number of solutions. That is, for each value we assign to x there is a corresponding value for y. In order to obtain a single solution we must solve two linear equations in two unknowns.

We will also discover that we can represent the solution set for a linear equation ax + by = c as a straight line. We will call this straight line representation of the equation, the graph of the linear equation. We will learn that when two linear equations have a single solution, their graphs will intersect at a single point which also corresponds to the solution.

In this Chapter we will discuss

(a) the Cartesian coordinate system
(b) plotting points
(c) graphing straight lines and simple quadratics
(d) the solution of two linear equations by graphical means
(e) the solution of two linear equations by algebraic means

10.1 The Cartesian Coordinate System

In Chapter 1, we learned that we could represent real numbers geometrically by use of the number line. However, there are many situations in life which require two real numbers to describe them. For example, your position on the earth's surface requires knowledge of both longitude and latitude. The dimensions of a rectangular field requires knowledge of both length and width. To determine if you are overweight requires knowledge of your height as well as your weight.

We can represent pairs of real numbers geometrically by use of the *Cartesian coordinate system*. This system has much in common with our convention for locating ourselves in a city which has north-south and east-west streets.

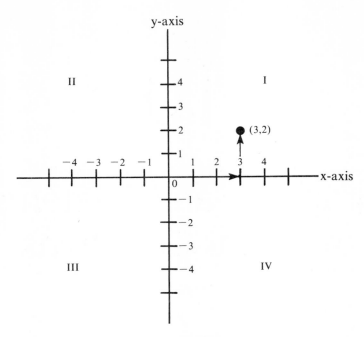

Figure 10.1 Cartesian coordinate system

To establish a Cartesian coordinate system we must draw two lines which intersect at a point called the *origin*. The usual convention is that these lines, called *axes*, be perpendicular to each other. See Figure 10.1.

The vertical line is often labelled as the y-axis and the horizontal line is often labelled as the x-axis. You should not feel bound by these conventions. If we were dealing with a problem involving time and distance it would probably make far more sense to label the x-axis as the t-axis (time) and the y-axis as the d-axis (distance). We label axes according to our needs. When we have no special application in mind we use the x-y convention described above.

Having established our axes, we now choose a unit of distance and mark off units of distance on each of the axes. That is, we make each axis into a number line. Again, there is no rule which requires us to use the same unit of length on the y-axis which we use on the x-axis. We allow the problem under discussion to lead us to the choice of units. However, in the absence of any modifying information we choose the same units on each axes. See Figure 10.1.

As a final remark, we note that by convention the points on the x-axis to the right of the origin are positive, points on the x-axis to the left of the origin are negative, points on the y-axis above the origin are positive and points on the y-axis below the origin are negative.

By convention we agree that when given a pair of numbers, such as (3,2), the first number in the pair is called the first or x coordinate (in an x-y coordinate system) and the second number in the pair is called the second or y coordinate. Thus, 3 is the first coordinate. It is also called the x coordinate. Two is the second coordinate or the y coordinate. When we plot the point (3,2) we move from the origin down the x-axis 3 units in the positive direction and then move vertically, on a line parallel to the y-axis, 2 units up. The point we reach is the point which represents the number pair (3,2). See Figure 10.1.

Example 1: *Plotting Points. Plot each of the following points in an x-y Cartesian coordinate system.*

(a) (1,2) (b) (4,1)
(c) (−1,1) (d) (−3,−4)
(e) (0,5) (f) (−2,0)

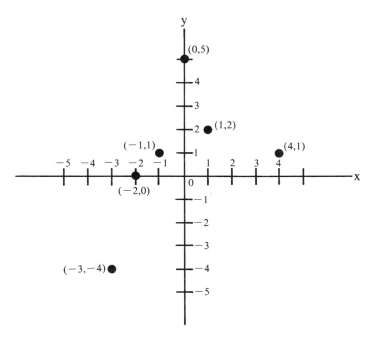

Figure 10.2 Plotting points

In order to plot the point (1,2) we first move one unit down the x-axis in the positive direction (i.e. to the right) and then two units up in the vertical direction (parallel to the y-axis). We use a similar strategy to locate each of the remaining points.

Example 2: *Name points. Give the Cartesian coordinates of each of the points labelled A, B, C, D, E, F, and G, and H in Figure 10.3.*

To reach the point labelled A, we must move one unit to the right on the x-axis and three units up parallel to the y-axis. Thus, the coordinates of this point are (1,3). To reach the point labelled B, we must move three units to the right on the x-axis and two units down parallel to the y-axis. Thus, the coordinates of this point are $(3,-2)$.

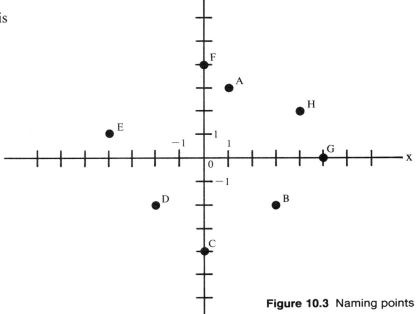

C   move no units right or left
      move four units down the y-axis
      The point is $(0,-4)$

D   two units left
      two units down
      The point is $(-2,-2)$

E   four units left
      one unit up
      The point is $(-4,1)$

F   no units left or right
      four units up
      The point is (0,4)

G   five units right
      no units up or down
      The point is (5,0)

Figure 10.3 Naming points

There is some further vocabulary which is useful in describing the general location of points. Notice that the coordinate axes separate the plane into four regions. These regions are called *QUADRANTS*. In Figure 10.1, we have labelled each of the quadrants with a Roman numeral I, II, III or IV. The points in Quadrant I have the property that their coordinates are both positive. The points in Quadrant II have a negative x-coordinate and a positive y-coordinate. The points in Quadrant III have both coordinates negative, and the points in Quadrant IV have a positive x-coordinate and a negative y-coordinate.

Example 3: *Locating Quadrants. Refer to Figure 10.2 and locate each point in its proper quadrant.*

(1,2)	is in Quadrant I since both coordinates are positive.
(4,1)	is in Quadrant I since both coordinates are positive.
(−1,1)	is in Quadrant II since the first coordinate is negative and the second coordinate is positive.
(−3,−4)	is in Quadrant III since both coordinates are negative.
(0,5)	is not in any quadrant. It is on the y-axis which separates Quadrants I and II.

Problem Set 10.1

Plot the points whose coordinates are given in Problems 1–6.

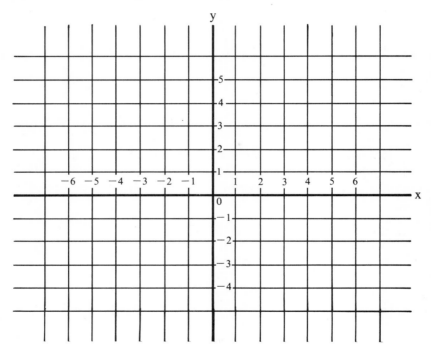

1. A(−2,3)
2. B(4,5)
3. C(4,−2)
4. D(−1,−3)
5. E(0,4)
6. F(−4,0)

In Problems 7–15, give the Cartesian coordinates of each of the points. Also, indicate in which quadrant or coordinate axis the point is.

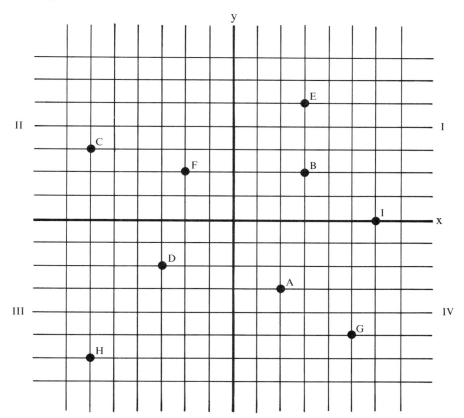

	Point	Coordinates	Lies in Quadrant or Axis
7.	A	(2,−3)	IV
8.	B	(,)	
9.	C	(,)	
10.	D	(,)	
11.	E	(,)	
12.	F	(,)	
13.	G	(,)	
14.	H	(,)	
15.	I	(,)	

Plot the points whose coordinates are given in 16–21.

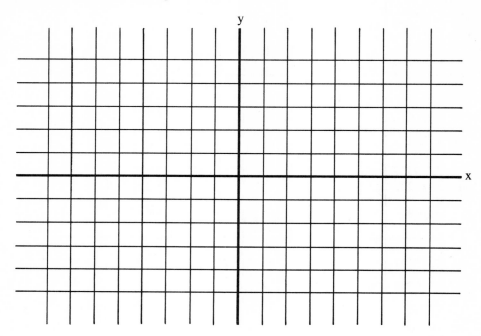

16. $A\left(\frac{3}{2}, -\frac{1}{3}\right)$
17. $B\left(-2, \frac{7}{2}\right)$
18. $C\left(-\frac{4}{3}, -\frac{7}{2}\right)$
19. $D\left(-1, -\frac{5}{2}\right)$
20. $E\left(0, -\frac{8}{3}\right)$
21. $F\left(-\frac{11}{2}, -\frac{5}{3}\right)$

10.2 Graphing Straight Lines and Simple Quadratics

As we mentioned in the previous section, a linear equation in two unknowns has the form

$$ax + by = c,$$

where a, b and c are numbers and x and y are unknowns. *Linear equations can be represented graphically in the Cartesian plane and their graphs are straight lines.*

But, what do we mean by a graph. Suppose we are given an equation involving two unknowns. For the purposes of discussion let us take this equation to be the linear equation $y - x = 2$. Suppose we can solve the equation for one of the unknowns in terms of the other. In the present case, this is easy to do and we obtain

$$y = x + 2.$$

Now, the set of points (x,y) in the plane which satisfy the equation is called the *graph* of the equation. Put another way, the graph of an equation is the pictoral representation of its solution set. It is a fact that the graph of every linear equation in two unknowns is a straight line. With this information, it is easy to graph linear equations. Strictly speaking, two points determine a straight line. So, it suffices to find two points which satisfy the given linear equation. However, in order to reduce the chance of making an error, we will always plot three points. That way, if they don't lie on a straight line, we will know that we have made an error.

Example 1: *Graph $y - x = 2$*

Step 1. Solve for y in terms of x.

$y = x + 2$

Step 2. Assign three values to x and find the corresponding value of y.

Let $x = -1, 0,$ and 1
When $x = -1$, then $y = (-1) + 2 = 1$
When $x = 0$, then $y = (0) + 2 = 2$
When $x = 1$, then $y = (1) + 2 = 3$

This can be done more simply in a tabular form

x	−1	0	1
y	1	2	3

▶ Note: It does not matter what values you choose for x. However, it is wise to make choices which will make your computations easy.

Step 3. Construct a Cartesian coordinate system and plot the three points.

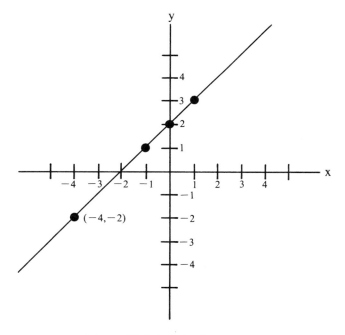

Figure 10.4 Graph of $y - x = 2$

Step 4. Draw a straight line through the three points plotted. This straight line represents the graph of the straight line $y - x = 2$.

You can verify that any point on the graph satisfies the equation. For example, $(-4, -2)$ is on the graph of

$y - x = 2$

because $(-2) - (-4) = 2$.

Example 2: *Graph $2y + 3x = 6$*

Step 1. Solve for y in terms of x
$$2y + 3x = 6$$
$$-3x -3x$$
$$2y = -3x + 6$$
$$y = -\frac{3}{2}x + \frac{6}{2}$$
$$y = -\frac{3}{2}x + 3$$

Step 2. Assign three values to x. Here we will choose -2, 0, and 2 to avoid having to deal with fractions.

When $x = -2$, then $y = -\left(\frac{3}{2}\right)(-2) + 3 = 6$

When $x = 0$, then $y = -\left(\frac{3}{2}\right)(0) + 3 = 3$

When $x = 2$, then $y = -\left(\frac{3}{2}\right)(2) + 3 = 0$

You should realize that any three values of x are alright to use. We have chosen values which seem convenient to us. In a more compact form we have

x	−2	0	2
y	6	3	0

Step 3. Construct a Cartesian coordinate system and plot the three points.

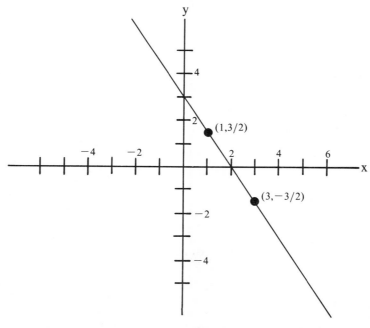

Figure 10.5 Graph of $2y + 3x = 6$

Example 6: *Graph* $y = \frac{1}{2}x^2$

x	−4	−3	−2	−1	0	1	2	3	4
y	8	$\frac{9}{2}$	2	$\frac{1}{2}$	0	$\frac{1}{2}$	2	$\frac{9}{2}$	8

See Figure 10.7 for the graph.

From Example 5 and 6 we can see that the graphs of $y = ax^2$ when $a > 0$ will always be a parabola opening up with bottom (*vertex*) at the origin. As "a" increases in value, the shape of the parabola will become narrower. We will leave it to you to discover in the problem set that when $a < 0$, the graph has the same shape but opens downward below the x-axis.

Problem Set 10.2

In Problems 1–10, solve for y in terms of x (if possible), complete the table, and then graph the straight lines.

1. $y - x = 3$ or $y = 3 + x$

x	y
0	3
1	4
−1	2

2. $y + 2x = 1$ or $y =$

x	y

3. $x + 2y = 2$ or $y =$

x	y

4. $y = 4$

x	y

5. $x = 5$

x	y

6. $2x + 3y = 12$

x	y

7. $3x - 4y + 12 = 0$

x	y

8. $6x + 9y + 12 = 0$

x	y

9. $5x + 3y = 10$

x			
y			

10. $2x - 5y - 10 = 0$

x			
y			

You can and should verify that any other point on the line graphed in Figure 10.5 is also a solution to $2y + 3x = 6$. For example, $\left(1, \frac{3}{2}\right)$ and $\left(3, \frac{-3}{2}\right)$ lie on the line, because

$$2\left(\frac{3}{2}\right) + 3(1) = 3 + 3 = 6$$

and

$$2\left(-\frac{3}{2}\right) + 3(3) = -3 + 9 = 6.$$

Example 3: *Graph $y = 4$*

Step 1. Solve for y in terms of x. In this example y is already solved in terms of x.

Step 2. Assign three values to x. Let us choose $x = -1, 0,$ and 1. Since y is always 4 we have

x	−1	0	1
y	4	4	4

Step 3. Construct a Cartesian coordinate system and plot the three points.

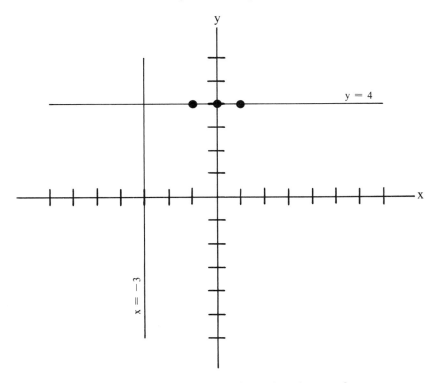

Figure 10.6 The graphs of $y = 4$ and $x = -3$

It is important to realize that if one of the variables is missing in a linear equation, then the graph of the solution set will be a straight line parallel to one of the coordinate axes.

Example 4: *Graph $x = -3$*

Step 1. y is missing as an unknown so we solve for the other unknown x.

Step 2. We cannot assign different values to x, but that is not necessary because x must always take on the value -3 and that will occur on the line parallel to the y-axis and three units to its left.

Step 3. See Figure 10.6.

We will now learn how to graph the solution set of equations of the form

$$y = ax^2.$$

This is a special case of the quadratic equation $y = ax^2 + bx + c$ whose graph is always a *parabola* of some form.

Example 5: *Graph $y = x^2$*

Step 1. Solve for y in terms of x. This is already done.

Step 2. We must now plot more than three points. We must plot enough points to be confident that we can sketch the graph.

x	−3	−2	−1	0	1	2	3
y	9	4	1	0	1	4	9

Step 3. Plot the points.

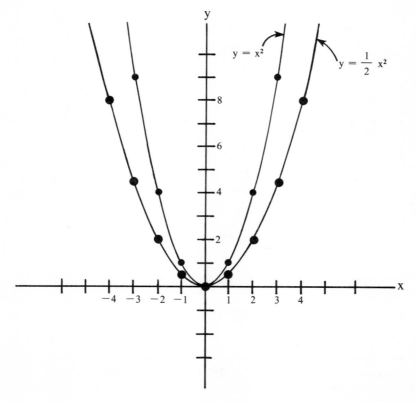

Figure 10.7 The graph of $y = x^2$ and $y = \frac{1}{2}x^2$

In Problems 11–18, complete the table and draw a neatly labelled graph.

11. $y = x^2$

x	y
0	0
1	1
−1	1
2	4
−2	4
1.5	2.25
−1.5	2.25
3	9
−3	9

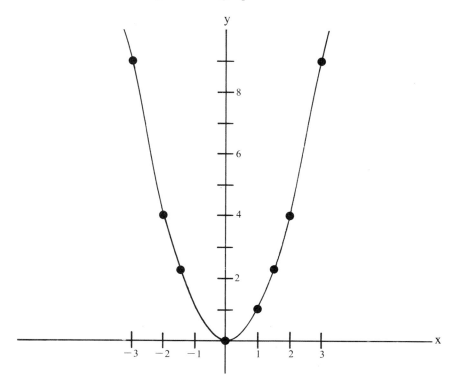

12. $y = 2x^2$

x	y

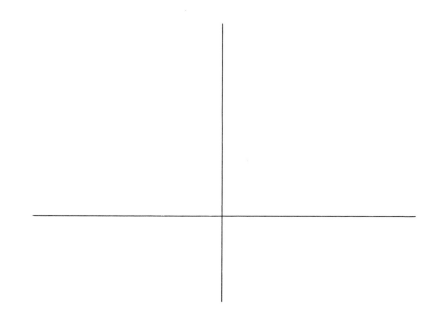

13. $y = \frac{1}{2}x^2$

x	y

14. $y = 4x^2$

x	y

In Problems 11–18, complete the table and draw a neatly labelled graph.

11. $y = x^2$

x	y
0	0
1	1
−1	1
2	4
−2	4
1.5	2.25
−1.5	2.25
3	9
−3	9

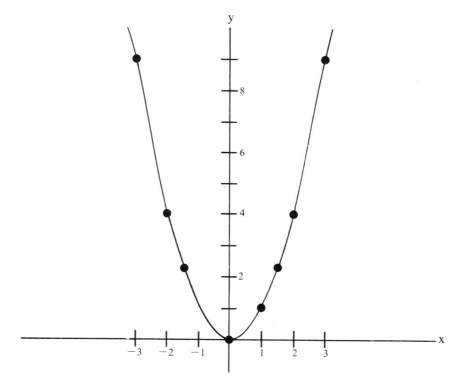

12. $y = 2x^2$

x	y

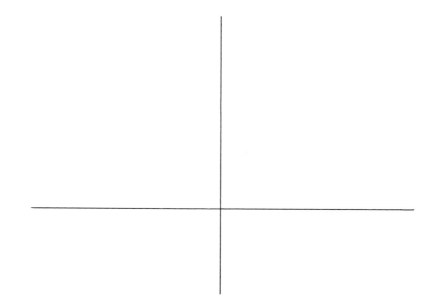

13. $y = \frac{1}{2}x^2$

x	y

14. $y = 4x^2$

x	y

15. $y = -x^2$

x	y
0	0
1	−1
−1	−1
−1.5	−2.25
1.5	−2.25
2	−4
−2	−4

16. $y = -2x^2$

x	y
0	0
1	−2
−1	−2
1.5	−4.5
−1.5	−4.5
2	−8
−2	−8

17. $y = -\frac{1}{2}x^2$

x	y

18. $y = -3x^2$

x	y

10.3 Graphical Solution of Two Linear Equations

We have learned in the previous section that an equation of the form $ax + by = c$ has an infinite number of solutions. That is, there is an infinite set of pairs of real numbers which makes the conditional equation true. We also learned that we could represent this set of solutions as a straight line in the Cartesian coordinate system.

Now, if you are given two linear equations, their solution sets will be straight lines as before, but if we plot them both on the same coordinate system, three possibilities arise:

(a) the two lines cross at exactly one point (see Figure 10.8(a))

> In this case the pair of equations in two unknowns has a unique solution. The x and y coordinates of the point of intersection of these lines represent the solution.

(b) the two lines are parallel and never meet (see Figure 10.8(b))

> In this case the pair of equations has no solution. That is, there are no values of x and y which satisfy both equations simultaneously.

(c) the two lines have the same graph and coincide (see Figure 10.8(c)).

> In this case the two equations are equivalent. Thus, every point which is on one line is also on the other. Hence, every point on the line represents a solution.

We will now illustrate these three situations in the following three examples.

Example 1: *Solve the following system graphically, if possible.*

$$x + 2y = 8$$
$$3x - 2y = 8$$

We first solve each equation for y and construct a table to plot the solutions.

$$\begin{aligned} x + 2y &= 8 \\ 2y &= -x + 8 \\ y &= -\frac{1}{2}x + 4 \end{aligned} \qquad \begin{aligned} 3x - 2y &= 8 \\ -2y &= -3x + 8 \\ y &= \frac{3}{2}x - 4 \end{aligned}$$

x	−2	0	2
y	5	4	3

x	−1	0	2
y	$-\frac{11}{2}$	−4	−1

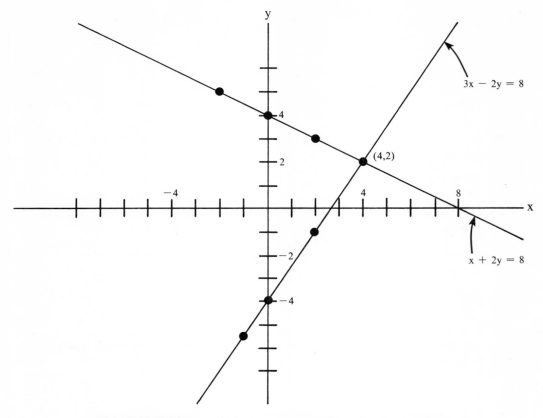

Figure 10.8(a) One solution to a system of two linear equations

In Figure 10.8(a), we can see that the two graphs cross at the point (4,2). This point is simultaneously on both graphs, hence must be a solution to both equations. This is easy to verify:

$$4 + 2(2) = 4 + 4 = 8$$
$$3(4) - 2(2) = 12 - 4 = 8.$$

The point (4,2) is called the solution to the system of equations.

Example 2: *Solve the following system graphically, if possible.*

$$x + 2y = 2$$
$$2x + 4y = 1$$

As before, we first solve each equation for y in terms of x, and construct a table to plot the solutions.

$$x + 2y = 2 \qquad\qquad 2x + 4y = 1$$
$$2y = -x + 2 \qquad\qquad 4y = -2x + 1$$
$$y = -\frac{1}{2}x + 1 \qquad\qquad y = -\frac{1}{2}x + \frac{1}{4}$$

x	−2	0	2
y	2	1	0

x	−2	0	2
y	$\frac{5}{4}$	$\frac{1}{4}$	$-\frac{3}{4}$

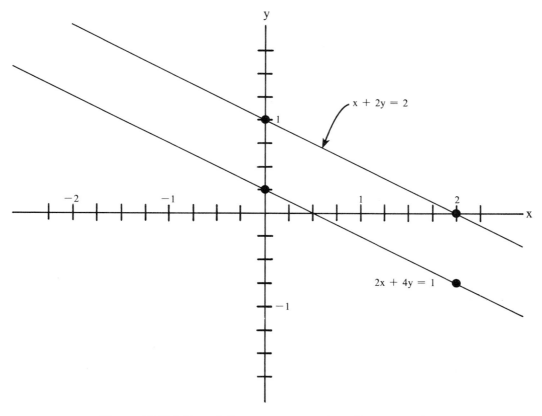

Figure 10.8(b) No solution to a system of two linear equations

In Figure 10.8(b) we can see that the two graphs are parallel. They have no points in common, hence they have no solutions in common. We must conlcude that the given system of equations has no solution.

Example 3: *Solve the following system graphically, if possible.*

$$2x - y = 3$$
$$4x - 2y = 6$$

Again, we solve each equation for y and construct tables for plotting.

$$\begin{aligned} 2x - y &= 3 \\ -y &= -2x + 3 \\ y &= 2x - 3 \end{aligned} \qquad \begin{aligned} 4x - 2y &= 6 \\ -2y &= -4x + 6 \\ y &= 2x - 3 \end{aligned}$$

We see that both equations reduce to exactly the same relationship between x and y. Thus, the graphs of both linear equations must coincide.

x	−1	0	1
y	−5	−3	−1

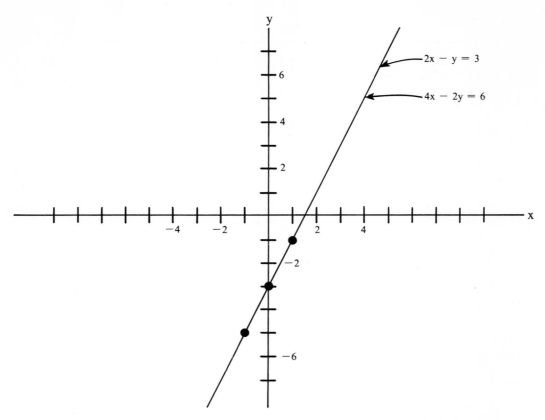

Figure 10.8(c) An infinite number of solutions to a system of linear equations

In Figure 10.8(c), we can see that the graphs of each straight line coincide, so they have an infinite number of points in common. That is, the system has an infinite number of solutions.

We now summarize the observations we have made above:

> A system of two linear equations in two unknowns can have
>
> (a) exactly one solution and, in that case the graphs of the linear equations cross at the solution point,
> (b) no solution, and in that case the graphs of the linear equations are parallel lines, and
> (c) infinitely many solutions, and in that case the graphs coincide.

Problem Set 10.3

In Problems 1–10, complete the tables by finding three solutions for each equation. Draw the graphs of these equations and find the solutions, if possible.

1. $2x + 3y = 6$ $x - 2y = 3$
   $y = -\dfrac{2x}{3} + 2$ $y = \dfrac{x}{2} - \dfrac{3}{2}$

x	y
0	2
3	0
−3	4

x	y
0	$-\dfrac{3}{2}$
2	$-\dfrac{1}{2}$
3	0

   Solution: (3,0)

2. $y = 2x - 1$ $y = x + 3$

x	y

x	y

   Solution: _____

3. $y = 4x - 2$ $y = 3 - x$

x	y

x	y

   Solution: _____

4. $y = 5x - 3$ $y = 2x - 6$

x	y

x	y

   Solution: _____

5. $y = 2 - 4x$ $y = 2(x - 5)$

x	y

x	y

Solution: _____

6. $x + y = 2$ $2x - y = 1$

x	y

x	y

Solution: _____

7. $2x + 5y = 20$ $3x - 4y = 7$

x	y

x	y

Solution: _____

8. $x + 2y = 4$ $2x + 4y = 7$

x	y

x	y

Solution: _____

9. $x - 2y = 1$ $2x - 4y = 2$

x	y

x	y

Solution: _____

10. $2x + 3y = 2$ $4x - 6y = 12$

x	y

x	y

Solution: _____

10.4 Algebraic Solution of Two Linear Equations—Substitution

It is important to have geometric insight into what can happen when we attempt to solve a system of linear equations. However, we can only obtain approximate solutions by graphical means because our graphs are often too crude to give precise answers. We will now learn how to solve a system of two linear equations by algebraic means.

There are two basic methods for solving linear equations: the method of substitution and the method of elimination. We will study the method of substitution in this section.

Example 1: *Solve the following system by the method of substitution.*

$$x + 2y = 8 \quad (1)$$
$$3x - 2y = 8 \quad (2)$$

In the method of substitution we solve one of the equations for one of the unknowns and substitute this expression into the second equation. In this problem it is obviously easy to solve (1) for x. We do this and obtain

$$x = 8 - 2y.$$

Now we replace x in (2) by $8 - 2y$ and obtain

$$\begin{aligned} 3x - 2y &= 8 & \text{original equation} \\ 3(8 - 2y) - 2y &= 8 & \text{substitution complete} \\ 24 - 6y - 2y &= 8 & \text{now solve for y} \\ 24 - 8y &= 8 \\ -8y &= -16 \\ y &= 2 \end{aligned}$$

From $x = 8 - 2y$ and $y = 2$ we obtain

$$x = 8 - 2(2) = 8 - 4 = 4.$$

The solution to this system is (4,2). Now refer back to Figure 10.8(a), where this same system was solved graphically and we obtained the same solution.

Let us summarize the steps we followed in the following rule.

Rule 10.1 The Method of Substitution

Step 1. Solve one of the equations for one variable in terms of the other.
Step 2. Substitute the relationship obtained in Step 1 into the other equation. This will produce one linear equation in one unknown.
Step 3. Solve the linear equation obtained in Step 2.
Step 4. Use the value obtained in Step 3 to solve for the other variable. Use the relationship obtained in Step 1.
Step 5. Check your answer.

Example 2: *Solve the system of equations.*

$$3x + 2y = 4$$
$$x - y = 7$$

Step 1. From the second equation, we have
$$x = y + 7$$

Step 2.
$$3x + 2y = 4$$
$$3(y + 7) + 2y = 4$$
$$3y + 21 + 2y = 4$$
$$5y + 21 = 4$$

Step 3.
$$5y + 21 = 4$$
$$5y = -17$$
$$y = -\frac{17}{5}$$

Step 4. $x = y + 7 = -\frac{17}{5} + 7 = -\frac{17}{5} + \frac{35}{5} = \frac{18}{5}$

The solution is $\left(\frac{18}{5}, -\frac{17}{5}\right)$

Step 5. **Check:** $3\left(\frac{18}{5}\right) + 2\left(-\frac{17}{5}\right) = \frac{54}{5} - \frac{34}{5} = \frac{20}{5} = 4$

$\left(\frac{18}{5}\right) - \left(-\frac{17}{5}\right) = \frac{18}{5} + \frac{17}{5} = \frac{35}{5} = 7$

Example 3: Solve

$$x = 4 - 2y$$
$$y = 2x - 2$$

Step 1. $x = 4 - 2y$

Step 2. $y = 2x - 2 = 2(4 - 2y) - 2$
$$y = 8 - 4y - 2$$
$$y = 6 - 4y$$

Step 3. $5y = 6$
$$y = \frac{6}{5}$$

Step 4. $x = 4 - 2y = 4 - 2\left(\frac{6}{5}\right) = 4 - \frac{12}{5} = \frac{20}{5} - \frac{12}{5} = \frac{8}{5}$

Thus, the solution is $\left(\frac{8}{5}, \frac{6}{5}\right)$

Step 5. **Check:** $\frac{8}{5} = 4 - 2\left(\frac{6}{5}\right) = \frac{20}{5} - \frac{12}{5} = \frac{8}{5}$

$\frac{6}{5} = 2\left(\frac{8}{5}\right) - 2 = \frac{16}{5} - \frac{10}{5} = \frac{6}{5}$

Example 4: Solve

$$x + 2y = 2$$
$$2x + 4y = 1$$

Step 1. $x = 2 - 2y$
Step 2. $2x + 4y = 1$
$2(2 - 2y) + 4y = 1$
$4 - 4y + 4y = 1$
$4 = 1$

At this point we have a false statement. We must conclude that there are no solutions to the given system. Refer to Figure 10.8(b) for a reminder of what happens geometrically in this situation.

Problem Set 10.4

In Problems 1–6, test to see if the indicated point is a solution to the given system of equations.

	Point	System of equation	Test	Answer Yes No
1.	$(2, -3)$	$x + 2y = -4$ $3x - y = 9$	$2 - 6 = -4$ $6 + 3 = 9$	yes
2.	$(0, -2)$	$2x - 3y = 6$ $x + 2y = -4$		
3.	$(-6, 0)$	$x - 3y = -6$ $\frac{1}{2}x + y = -3$		
4.	$(2, 1)$	$3x - y = 5$ $x + 2y = 3$		
5.	$(-1, 2)$	$2x + 4y = 6$ $x + y + 1 = 0$		
6.	$(-2, -4)$	$x - 3y = 10$ $4x - 5y = 12$		

In Problems 7–18, solve by substitution, and check your solution. Show work as in Problem 7.

	Equations	Show work here	Solution	Check
7.	$y = 2x + 1$ $y = x + 4$	$2x + 1 = x + 4;\ x = 3$ $y = 2 \cdot 3 + 1 = 7$	$(3, 7)$	$7 = 2 \cdot 3 + 1$ $7 = 3 + 4$
8.	$y = 4x - 7$ $y = 2x + 1$		(,)	
9.	$y = \frac{1}{2}x + 1$ $y = 2x - 3$		(,)	

10. $y - 2x = 3$
    $y + x = 4$ (,)

11. $2x + y = 4$
    $x - y = 3$ (,)

12. $2x + 3y = 8$
    $x - 2y = 5$ (,)

13. $2x + 4y - 5 = 0$
    $x - 2y + 3 = 0$ (,)

14. $x + 3y = 6$
    $2x + 6y - 5 = 0$ (,)

15. $2x - 4y = 1$
    $x + 2y = 3$ (,)

16. $2x - 4y = 3$
    $3x - 4y = 5$ (,)

17. $2(3 - x) + 2y = 5$
    $x - 3y = 2$ (,)

18. $\frac{2}{3}x - \frac{1}{5}y = 3$
    $x - \frac{4}{5}y = \frac{5}{3}$ (,)

10.5 Algebraic Solution of Two Linear Equations—Elimination

We will now re-examine each of the examples in the last section and solve the problems presented there by the method of elimination.

Example 1: *Solve the following system of equations by use of the method of elimination.*

$$x + 2y = 8 \quad (1)$$
$$3x - 2y = 8 \quad (2)$$

Recall that if we multiply or divide an equation by a number we obtain an equivalent equation, i.e. one with the same solutions. If we add or subtract two equations we still obtain an equation. We will use these operations to eliminate one of the variables in the two equations.

It is clear that if we add Equations (1) and (2) we obtain

$$4x = 16$$

The y variable has been eliminated. Now solve for x to obtain $x = 4$. This value can be substituted in either Equation (1) or Equation (2) to find y. From (1) we have

$$4 + 2y = 8$$
$$2y = 4$$
$$y = 2$$

The solution is (4,2) as we obtained in Section 10.4.

We would also have obtained $y = 2$ if we had substituted $x = 4$ in Equation (2). Try it!

Example 2: Solve

$$3x + 2y = 4 \quad (1)$$
$$x - y = 7 \quad (2)$$

In this problem, it should be clear that if we multiply Equation (2) by 2, we will be able to eliminate y by adding the two equations.

$$
\begin{aligned}
3x + 2y &= 4 \\
2x - 2y &= 14 \quad &&\text{multiply (2) by 2} \\
5x &= 18 \quad &&\text{add both equations} \\
x &= \frac{18}{5} \quad &&\text{solve for x} \\
3\left(\frac{18}{5}\right) + 2y &= 4 \quad &&\text{substitute into (1) for x} \\
\frac{54}{5} + 2y &= \frac{20}{5} \\
2y &= -\frac{34}{5}, \\
\text{or} \quad y &= -\frac{17}{5}
\end{aligned}
$$

The solution is $\left(\frac{18}{5}, -\frac{17}{5}\right)$ which is what we obtained earlier by the method of substitution in Example 2 of the last Section.

Example 3: Solve

$$x = 4 - 2y$$
$$y = 2x - 2$$

In this problem, we should first rewrite the two equations so that the like variables are on the left side of the equation under each other and the constant terms are on the right side of the equation.

$$
\begin{aligned}
x + 2y &= 4 \quad &&(1) \\
-2x + y &= -2 \quad &&(2)
\end{aligned}
$$

$$
\begin{aligned}
2x + 4y &= 8 \quad &&\text{multiply (1) by 2} \\
-2x + y &= -2 \quad &&\text{equation (2)} \\
5y &= 6 \quad &&\text{add both equations} \\
y &= \frac{6}{5} \\
-2x + \frac{6}{5} &= -2 \quad &&\text{substitute in (2) for y} \\
-2x &= -\frac{16}{5}, \\
\text{or} \quad x &= \frac{8}{5}
\end{aligned}
$$

Thus, the solution of the system of equations is $\left(\frac{8}{5}, \frac{6}{5}\right)$. Again, we have obtained the same solution as we obtained by the method of substitution in Example 3 of the last Section.

Example 4: Solve

$$x + 2y = 2 \quad (1)$$
$$2x + 4y = 1 \quad (2)$$

$$-2x - 4y = -4 \quad \text{multiply (1) by } -2$$
$$\underline{2x + 4y = 1} \quad \text{equation (2)}$$
$$0 = -3 \quad \text{add both equations}$$

Since this last equation is never true we know that the original system had no solution.

Example 5: Solve

$$2x + 3y = 4 \quad (1)$$
$$3x + 4y = 2 \quad (2)$$

As it stands, we cannot employ the strategy we used on the previous examples. However, a minor modification will work. We will produce two equivalent equations which when added together will cancel out one of the variables. We choose suitable multipliers for *both* equations to create the situation we want.

Multiply Equation (1) by -3 and Equation (2) by 2. When we add the equivalent equations, x will be eliminated.

$$-6x - 9y = -12 \quad \text{multiply (1) by } -3$$
$$\underline{6x + 8y = 4} \quad \text{multiply (2) by 2}$$
$$-y = -8 \quad \text{add both equations}$$
$$y = 8$$

$$2x + 3(8) = 4 \quad \text{substitute in (1) for y}$$
$$2x + 24 = 4$$
$$2x = -20$$
$$x = -10$$

Thus, the solution of the system is $(-10, 8)$.

Check: $2(-10) + 3(8) = -20 + 24 = 4$
$3(-10) + 4(8) = -30 + 32 = 2$

You should now be in a position to solve any pair of linear equations by the method of elimination. It is important that you keep in mind the geometric meaning of the solution to a pair of linear equations. That is, the graphs of the two linear equations intersect at the solution point. Remember, if the two equations have no solution, it is because their graphs are parallel lines and if the two equations have an infinite number of solutions it is because the two graphs are identical.

Problem Set 10.5

In Problems 1–20, solve the system of equations by the elimination method and check your solutions. Show work as in Problem 1.

	Equations	Show work here	Solution	Check
1.	$2x + y = 4$ $3x + 2y = 3$	$-4x - 2y = -8$ $\underline{3x + 2y = 3}$ $-x = -5, x = 5$ $2 \cdot 5 + y = 4, y = -6$	$(5, -6)$	$10 - 6 = 4$ $15 - 12 = 3$
2.	$x - y + 3 = 0$ $2x - 5y = -21$		(,)	
3.	$7x - y = 4$ $3x - 2y = -3$		(,)	
4.	$2x + y = 1$ $-3x + y = -4$		(,)	
5.	$4x - y = 3$ $x - 3y + 4 = 0$		(,)	
6.	$3x - 2y = 6$ $x + y = 3$		(,)	
7.	$2y - 3x + 4 = -3$ $2x + 4y = 5$		(,)	
8.	$4x - 5y - 7 = 0$ $2x = 3y - 1$		(,)	
9.	$5x + 3y - 2 = 0$ $2x - 4y + 5 = 0$		(,)	
10.	$-x = -2y + 7$ $\frac{x}{2} + 3y = -\frac{3}{2}$		(,)	
11.	$x - \frac{y}{2} = 3$ $2x - y = 1$		(,)	
12.	$2x - 4y = 5$ $3x + 5y = 3$		(,)	
13.	$\frac{x}{3} - \frac{y}{2} = \frac{1}{6}$ $\frac{2x}{3} - \frac{5y}{2} = 3$		(,)	

14. $3x - \frac{y}{2} = 5$ (,)
    $\frac{x}{5} - 3y = 4$

15. $5x - \frac{y}{2} = 3$ (,)
    $\frac{x}{5} = 5 - \frac{2y}{5}$

16. $4s - 7 = \frac{3t}{2}$ (,)
    $5 - 3t = \frac{2s}{3}$

17. $2x - \frac{y}{3} = 7$ (,)
    $6x - y = 4$

18. $1 - 2x = 3y$ (,)
    $y - 3 = 2x$

19. $\frac{x}{2} - \frac{4}{3}y = 1$ (,)
    $-x + \frac{8}{3}y = -2$

20. $10u + 7v = 4$ (,)
    $5u - 2v = 7$

Chapter Summary

1. **Coordinate axis** Two lines, one horizontal and the other vertical, intersecting each other at right angles. The horizontal line is called the x-axis, and the vertical line is called the y-axis. The point where these two lines intersect is called the *origin*.

2. **Coordinates** A pair of numbers that describes the position of a point. If P(x,y) is a point then, P is the name of that point, and (x,y) are the coordinates.

3. **Linear equation in two unknowns** A linear equation in two unknowns has the form
   $$ax + by = c$$
   where a, b, and c are some constants.

4. **Graph of linear equations** For every value of x, there is a unique value of y which can be obtained from the linear equation. Each pair (x,y) represent some point in the coordinate plane. For a linear equation, all these points will lie on a straight line. We call this line the graph of the linear equation. For graphing straight lines, two points are enough, but we usually take three points to make sure that the graph is ok.

5. **Graphs of simple quadratics** The relation $y = ax^2$ is a simple quadratic. To find the graph of such a relation, we take several values of x (at least 7) and obtain the corresponding values of y. This gives us several points. If we join the plots of these points, we get the graph. The graph of $y = ax^2$ opens upward if $a > 0$, and downward if $a < 0$.

6. **Solution of a linear equation** The coordinates of a point which satisfy the equation. The graph of a linear equation is a straight line. There are infinitely many points on a line. The coordinates of all these points satisfy the linear equation. Thus a linear equation has infinitely many solutions.

7. **Graphical solutions of a system of equations** If we have two linear equations in two unknowns then their graphs are two straight lines.
   (a) If these lines intersect at a point, then the coordinates of that point is the unique solution of the system,
   (b) If these lines are parallel, then there is no solution of the system,
   (c) If these lines coincide, then there are infinitely many solutions of the system.

8. **Algebraic solution of a system of equations** There are two methods.
   1. Method of substitution. Find the value of one unknown in terms of the other from one of the equations and substitute this value in the second equation.
   2. Method of elimination. Add or subtract multiples of the given equations so that one of the unknowns is eliminated.

Review Problems

In Problems 1–2, use the following coordinate system.

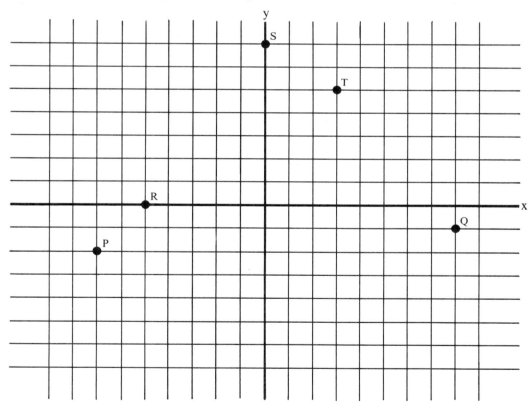

1. Plot the following points
   A(2,5)   B(−3,4)   C(−2.5,0)   D(−3,−4)   E(0,−5)   F(1.5,−1.2)

2. Give the Cartesian coordinates of the following points.
   (a) P   (b) Q   (c) R   (d) S   (e) T

In Problems 3-7, draw the graphs of the equations.

3. $y = 2x - 4$

x	y

4. $2x + 3y = -1$

x	y

5. $3x - 4y - 6 = 0$

x	y

6. $y = 2x^2$

x	y

7. $y = -x^2$

x	y

In Problems 8–11, find graphically the approximate solution of the system of equations.

8. $y = x + 1$ $\qquad$ $y = -x + 2$

x	y

x	y

The solution: _____

9. $y = 2x - 5$ $\qquad$ $y = 4 - x$

x	y

x	y

The solution: _____

10. $2x - 3y = 1$ $\qquad$ $x + 2y = 4$

x	y

x	y

The solution: _____

11. $2x - y = 3$ $y = x + 4$

x	y

x	y

The solution: _____

In Problems 12–25, solve algebraically and check your solutions.

	Equations	Show work here	Solution	Check
12.	$2x - 3y = 4$ $x + 3y = 5$		(,)	
13.	$x - 3y = 2$ $x + 2y = 8$		(,)	
14.	$2x - 3y = 1$ $x + 2y = 4$ (compare the solution with problem 10)		(,)	
15.	$5x - 4y + 4 = 0$ $2x - 3y = 8$		(,)	
16.	$2x - 7y = -2$ $3x + 5y + 4 = 0$		(,)	
17.	$a - 3b = 4$ $b - 3a = 4$		(,)	
18.	$a + 3b = 1$ $b + 3a = 0$		(,)	
19.	$\frac{1}{2}a + \frac{1}{3}b = 2$ $\frac{1}{3}a - \frac{1}{4}b = 1$		(,)	
20.	$r + s = 7$ $\frac{r}{3} - \frac{s}{7} = 1$		(,)	
21.	$2r + 3s = 0$ $3r + \frac{9}{2}s = 1$		(,)	
22.	$r - \frac{1}{2}s = 1$ $2r - s = 1$		(,)	

23. $2x + 3y = 1$
    $3x + \frac{9}{2}y = \frac{3}{2}$
    (,)

24. $x = 2(y - 1) + 1$
    $y = 3(x + 1) + 2$
    (,)

25. $x - y = 3$
    $3x = 4 - 2y$
    (,)

Chapter Test

1. One number is three more than another number. A third number is two less than the smaller of the first two numbers. If the sum of the three numbers is 31, what are the numbers?
   (a) 5,7,19 (b) 10,14,9 (c) 12,14,5 (d) 13,10,8 (e) None of these

2. Perform the long division $(x^3 + 1) \div (x + 1)$
   The quotient Q and remainder R are:
   (a) $Q = x^2 + 2x - 1, R = 1$ (b) $Q = x^2 + 2x + 1, R = 1$
   (c) $Q = x^2 - x + 1, R = 0$ (d) $Q = x^2 - x + 1, R = 2$
   (e) None of these

3. One of the factors common in the two expressions $x + x^2$ and $1 - x^2$ is
   (a) $x + 1$ (b) $x - 1$ (c) x (d) None

4. One of the solutions of $4x^2 - 5x - 6 = 0$ is:
   (a) $\frac{3}{4}$ (b) -2 (c) $\frac{5 - \sqrt{71}}{8}$ (d) $\frac{-5 + \sqrt{71}}{8}$ (e) None

5. One of the solutions of $y^2 + 2y = 24$ is:
   (a) 6 (b) $-1 - \frac{2i\sqrt{23}}{3}$ (c) 4 (d) $-1 - i\sqrt{23}$

6. The expression $\frac{2x^2 + 9x + 9}{x^2 - 9}$ reduced completely equals:
   (a) $\frac{2x^2 + 9x + 9}{x^2 - 9}$ (b) $\frac{2x + 3}{x - 3}$ (c) $\frac{2x - 3}{x - 3}$ (d) $\frac{2x + 3}{x + 3}$ (e) None

7. Simplify: $\frac{x - 2}{2x^2 - x} - \frac{2}{x} =$
   (a) $\frac{-3}{2x - 1}$ (b) $\frac{3}{2x - 1}$ (c) $\frac{1}{2x - 1}$ (d) None

8. Which of the following pairs is a solution of $\frac{2}{x - 1} - \frac{6}{2x + 1} = 5$.
   (a) $1, \frac{13}{10}$ (b) $-1, \frac{13}{10}$ (c) $1, -\frac{13}{10}$ (d) $-\frac{13}{10}, -1$ (e) None

9. Simplify: $(x^{-1/3} y^{-1/2})^{-6} =$
   (a) $\frac{y^3}{x^2}$ (b) $\frac{x^2}{y^3}$ (c) x^2y^3 (d) $x^{18}y^{12}$ (e) None of these

10. Consider the following statements P and Q:
    P: $\dfrac{3}{\sqrt{2}+1} = 3(\sqrt{2}-1)$.
    Q: $\dfrac{3}{\sqrt{3}+1} = \dfrac{3\sqrt{3}-3}{2}$
    Which of the above statements are true?
    (a) P and Q   (b) P only   (c) Q only   (d) None

11. Consider the following statements P and Q:
    P: The graph of $y = 2$ is a line parallel to the y − axis.
    Q: The graph of $2x = 9$ is a line parallel to the x − axis.
    Which of the above statements are true?
    (a) P only   (b) Q only   (c) P and Q   (d) None

12. The two straight lines given by the equations $3x - y = 4$ and $5x - y = 6$ intersect at a point whose coordinates are:
    (a) $(1,1)$ (b) $\left(-\dfrac{4}{3},1\right)$ (c) $(1,-1)$ (d) $\left(1,-\dfrac{4}{3}\right)$ (e) None

13. The two straight lines given by the equations $4x - \dfrac{3}{2}y = -77$ and $3x - 2y = -77$ intersect at a point whose coordinates are:
    (a) $\left(1,-\dfrac{1}{2}\right)$ (b) $\left(-\dfrac{1}{2},1\right)$ (c) $(11,22)$ (d) $(-11,22)$ (e) None

14. Solve for the values of x and y.
    $8x + 5y = 15$.
    $3x + 15y = 13$.
    (a) $\left(\dfrac{32}{105},\dfrac{59}{21}\right)$ (b) $\left(\dfrac{32}{21},\dfrac{59}{105}\right)$ (c) $\left(\dfrac{63}{41},\dfrac{31}{21}\right)$ (d) $\left(\dfrac{21}{31},\dfrac{31}{21}\right)$ (e) None

15. Solve for the values of x and y.
    $4x + 3y = 8$.
    $x - 2y = 9$.
    (a) $\left(\dfrac{43}{11},\dfrac{28}{11}\right)$ (b) $\left(\dfrac{24}{11},-\dfrac{12}{11}\right)$ (c) $(1,1)$ (d) $\left(-\dfrac{28}{11},\dfrac{43}{11}\right)$ (e) None

Chapter 11

Logarithms

The concept of logarithms is very closely related to the concept of exponents. Exponents were discussed in Chapter 2 and again in Chapter 9. We will use that information to introduce logarithms. In fact, all the rules of logarithms will be deduced from our knowledge of exponent rules. After dealing with the different properties of logarithms, we will discuss how to estimate the value of a logarithm. We will then see how to compute the value of a logarithm and use the estimated value as a validity check.

In this Chapter we will discuss

(a) More on exponents
(b) An introduction to logarithms
(c) Properties of logarithms
(d) Logarithmic equations
(e) Computing logarithms on a calculator
(f) The base changing formula

11.1 More on Exponents

In Chapter 2 you have seen that x^n is always defined for all values of x if n is an integer.

$$2^3 = 8, \qquad (-2)^3 = -8$$
$$2^{-3} = \frac{1}{8}, \qquad (-2)^{-3} = -\frac{1}{8}$$

However, as seen in Chapter 9, if n is a rational number other than an integer (fraction) and x is negative, then the expression x^n may not exist as a real number.

$(-2)^{1/2}$ is not a real number

In order to exclude expressions of the type $(-2)^{1/2}$, we restricted the base in x^n to be always positive. With the restriction that $x > 0$ we defined

$$x^{p/q} = \sqrt[q]{x^p} \qquad \text{p,q are any two integers and } q > 1$$

That is,
$$9^{1/2} = \sqrt{9} = 3 \qquad 27^{1/3} = \sqrt[3]{27} = 3$$
$$(25)^{-1/2} = \frac{1}{25^{1/2}} = \frac{1}{\sqrt{25}} = \frac{1}{5}$$

We want to extend the definition of x^n to include the cases when n is any real number—rational or irrational. For example, we want to give meaning to expressions such as $2^{\sqrt{3}}$.

At this stage we can only state that numbers of the type $2^{\sqrt{3}}$ are defined. The justification for this statement is beyond the scope of this book. You can convince yourself that you can approximate $2^{\sqrt{3}}$ on your calculator.

$$2^{\sqrt{3}} = 2 \boxed{y^x} \; 3 \boxed{\sqrt{}} \boxed{=} \; 3.3219971$$

Also note that

$$4^{\sqrt{5}} = 4 \boxed{y^x} \; 5 \boxed{\sqrt{}} \boxed{=} \; 22.194587$$

Finally, we state

> 1. x^n is defined for $x > 0$ and for any real number n.
> 2. All the exponent rules stated in Chapter 2 and Chapter 9 are applicable for real number exponents.

Another property of exponents which will be used in the next section and which is also a handy tool for solving equations involving exponents is stated below.

> **Rule 11.1**
>
> If $x > 0$, $x \neq 1$, and m, n are real numbers,
>
> then $x^m = x^n$ implies $m = n$

Notice that we imposed a restriction, $x \neq 1$, because for $x = 1$ the statement $x^m = x^n$ is true for any m and n.

Example 1: *Solving equations involving exponents by use of Rule 11.1*

(a) $2^x = 2^3$ implies, by Rule 11.1, that $x = 3$.

(b) $2^x = \frac{1}{2^2}$ means $2^x = 2^{-2}$ which implies, by Rule 11.1, that $x = -2$.

(c) $5^{4x} = 5^2$ implies, by Rule 11.1, that $4x = 2$ or $x = \frac{1}{2}$.

(d) $3^{4x} = 3^0$ implies, by Rule 11.1, that $4x = 0$ or $x = 0$.

(e) $(2^{-1})^{3x-1} = 2^3$ means that $2^{-3x+1} = 2^3$. By Rule 11.1, we have

$$-3x + 1 = 3$$
$$-3x = 2$$
$$x = -\frac{2}{3}.$$

Problem Set 11.1

In Problems 1–20, complete the given statement.

1. The expression x^n is always real when x is _____ and n is _____ .

2. $4^{1/2} =$ _____

3. $8^{1/3} =$ _____

4. $(27)^{1/3} =$ _____

5. $(25)^{1/2} =$ _____

6. $(16)^{1/4} =$ _____

7. $(216)^{1/3} =$ _____

8. $(125)^{1/3} =$ _____

9. $\left(\dfrac{1}{8}\right)^{1/3} =$ _____

10. $\left(\dfrac{1}{64}\right)^{1/3} =$ _____

11. $(8)^{-1/3} =$ _____

12. $(64)^{-1/3} =$ _____

13. $(8)^{-2/3} =$ _____

14. $(27)^{-2/3} =$ _____

15. $(32)^{-2/5} =$ _____

16. $(16)^{3/4} =$ _____

17. $(16)^{-3/4} =$ _____

18. $\left(\dfrac{1}{16}\right)^{-1/4} =$ _____

19. $(243)^{-2/5} =$ _____

20. $\left(\dfrac{1}{243}\right)^{2/5} =$ _____

In Problems 21–30, evaluate the given expression with the aid of your calculator.

21. $2^{\sqrt{5}} =$ _____

22. $3^{\sqrt{2}} =$ _____

23. $3^{\sqrt{7}} =$ _____

24. $2^{\sqrt[3]{2}} =$ _____

25. $5^{\sqrt[3]{5}} =$ _____

26. $(281)^{\sqrt[4]{3}} =$ _____

27. $8^{\sqrt[3]{5}} =$ _____

28. $8 \cdot 2^{\sqrt{3}} =$ _____

29. $2^{\sqrt{2}} - 5 \cdot 4^{\sqrt[3]{2}} =$ _____

30. $3^{\sqrt{2}} \cdot 4^{\sqrt[3]{2}} =$ _____

In Problems 31–40, solve for x.

Equation	Show work here	Solution
31. $9^x = 27$	$(3^2)^x = 3^3$ $2x = 3$	31. $x = \dfrac{3}{2}$
32. $3^x = 27$		32. _____
33. $2^{2x} = 16$		33. _____

34. $2^{2x} = \dfrac{1}{8}$ 34. _____

35. $\left(\dfrac{1}{4}\right)^x = \dfrac{1}{16}$ 35. _____

36. $\left(\dfrac{1}{9}\right)^{2x-1} = 27$ 36. _____

37. $5^{2x-3} = 125$ 37. _____

38. $3^{2x} - 1 = 26$ 38. _____

39. $3^{4x+1} - 3^{4x} = 18$ 39. _____

40. $13 + 2^{2x-7} = 1 + 2^{2x-5}$ 40. _____

11.2 Introduction to Logarithms

Consider the numbers 5, 6 and 30. We can express each of these three numbers in terms of the other as follows.

$$30 = 5 \cdot 6$$
$$5 = \dfrac{30}{6}$$
$$6 = \dfrac{30}{5}$$

Now, consider another set of three numbers, 3, 5 and 125. Let us try to express each of these three numbers in terms of the other two.

$$125 = 5^3$$
$$5 = \sqrt[3]{125} \text{ or } (125)^{1/3}$$
$$3 = ?$$

We know that *"3 is a number which when used as a power to the base 5 produces 125"*.

Another way to express the statement above is to say that "3 is the logarithm of 125 to the base 5". In symbols we express this statement as

$$\boxed{3 = \log_5 125}$$

Illustration 1: Evaluating Logarithms

    (a) $\log_2 8$

        $\log_2 8$ is a number which when used as power to the base 2 produces 8.
        That is $2^{(\log_2 8)} = 8$
        We also know $2^3 = 8$
        Thus, $2^{\log_2 8} = 2^3$
        or $\log_2 8 = 3$ Rule 11.1

➥   Notice that $2^{\log_2 8} = 8$, and $5^{\log_5 125} = 125$
       In general, if $a > 0$, then $a^{\log_a b} = b$

    (b) $\log_3 27$

        $\log_3 27$ is a number which when used as a power to the base 3 produces 27
        $3^{\log_3 27} = 27 = 3^3$
        That is, $3^{\log_3 27} = 3^3$.
        Hence, $\log_3 27 = 3$.   Rule 11.1
        We can also say that $\log_3 27 = 3$ since $3^3 = 27$

Example 1: *To find the value of a logarithm*

    (a) $\log_5 25 = 2$ since $5^2 = 25$
    (b) $\log_{10} 100 = 2$ since $10^2 = 100$
    (c) $\log_{10} 10 = 1$ since $10^1 = 10$
    (d) $\log_{36} 6 = \dfrac{1}{2}$ since $36^{1/2} = 6$
    (e) $\log_{125}\left(\dfrac{1}{5}\right) = -\dfrac{1}{3}$ since $(125)^{-1/3} = \dfrac{1}{5}$

The statement $\log_2 8 = 3$ implies $2^3 = 8$ and $2^3 = 8$ implies $\log_2 8 = 3$

Example 2: *To convert a statement from exponent form to logarithmic form*

    (a) $2^5 = 32$,   so $\log_2 32 = 5$
    (b) $3^4 = 81$,   so $\log_3 81 = 4$
    (c) $2^{-2} = \dfrac{1}{4}$,   so $\log_2\left(\dfrac{1}{4}\right) = -2$
    (d) $2^0 = 1$,   so $\log_2 1 = 0$
    (e) $25^{1/2} = 5$,   so $\log_{25} 5 = \dfrac{1}{2}$
    (f) $8^{-1/3} = \dfrac{1}{2}$,   so $\log_8\left(\dfrac{1}{2}\right) = -\dfrac{1}{3}$

We make the following observations about a logarithm.

    1.   The base in both the exponent form and the logarithm form, is the same.
    2.   The value of the logarithm is the exponent, in the exponent form.

> In general,
> $$\log_b a = c \text{ implies } b^c = a \text{ and } b^c = a \text{ implies } \log_b a = c$$
> where b, c and a are real number, $b > 0$ and $b \neq 1$.

We exclude 1 being the base since $1^n = 1$ for all n and is therefore of no importance to us. The statement $\log_b a = c$ involves three numbers a, b and c. If you are given any two of these numbers, then the third number can be obtained by converting the logarithmic statement to exponent form.

3. Since we have agreed to take the base in b^x to be always positive, the value of this expression is always positive.

$$4^2 = 16, \; 4^{1/2} = 2, \; 4^{-1/2} = \frac{1}{2}$$

$$4^{\sqrt{3}} = 4 \; \boxed{y^x} \; 3 \; \boxed{\sqrt{}} \; \boxed{=} \; 11.035665$$

Thus, we observe that $4^2, 4^{1/2}, 4^{-1/2}, 4^{\sqrt{3}}$ are all positive quantities. That is if $b^x = c$, then since $b > 0$ we have $c > 0$.

The observations made above in 1, 2, and 3 are summarized in the following two boxes.

For $b^x = c$ to be defined	For $\log_b c = x$ to make sense;
(i) $b > 0$, and $b \neq 1$	(i) $b > 0$ and $b \neq 1$
(ii) $c > 0$	(ii) $c > 0$
(iii) x is any real number.	(iii) x is any real number.

Example 3: *To solve equations containing logarithms*

(a) $\log_2 x = 1$ implies $2^1 = x$
Thus, $x = 2$ (Note: $\log_2 2 = 1$.)

(b) $\log_5 x = 0$ implies $5^0 = x$
Thus, $x = 1$, and $\log_5 1 = 0$ (Note: $\log_5 1 = 0$.)

➡ Notice that, in general,

1. $\log_b b = 1$ for all $b > 0$
2. $\log_b 1 = 0$ for all $b > 0$

(c) $\log_5 x = 2$ implies $5^2 = x$
Thus, $x = 25$

(d) $\log_x 4 = 2$ implies $x^2 = 4$
Thus, $x = 2$ or -2. but since the base cannot be negative, the only solution is 2.

(e) $\log_5 25 = x$ implies $5^x = 25 = 5^2$
Thus, $x = 2$ Rule 11.1

(f) $\log_8 1 = x$ implies $8^x = 1 = 8^0$
Thus, $x = 0$ Rule 11.2
(Also see the note above)

(g) $\log_{27}\left(\frac{1}{3}\right) = x$ implies $27^x = \frac{1}{3}$
or $(3^3)^x = 3^{-1}$
or $3^{3x} = 3^{-1}$
Thus, $3x = -1$ or $x = -\frac{1}{3}$

(h) $\log_4(-4) = x$ implies $4^x = -4$
$4^x = -4$ has no solution because 4^x for all real numbers x is a positive quantity and therefore can never be equal to -4.

▶ The logarithm of a negative number is not defined.

A logarithm is called a *common logarithm* if its base is 10. Since logarithms to the base 10 are frequently used it is customary to write $\log_{10}x$ as log x. That is, if no base is indicated in a logarithm then it is understood that the base is 10.

$\log 100 = \log_{10} 100 = 2$

Another base used in logarithms is the irrational number e, approximately equal to 2.71828. . . . A logarithm to the base e is called a *natural* logarithm and expressed by the symbol "Ln". That is,

$\log_e x = \text{Ln } x$.

Example 4: *Logarithm with the base 10 or with the base e*

(a) $\log 1000 = 3$, since $10^3 = 1000$

(b) $\log .1 = -1$, since $10^{-1} = \frac{1}{10} = .1$

(c) $\text{Ln } e = 1$, since $e^1 = e$

Problem Set 11.2

In Problems 1–4, complete the statements.

1. $\log_2 16$ is a number which when used as an exponent to the base _____ produces _____ .

2. $\log_b x$ is defined only when b is _____ and x is _____ .

3. The logarithm of 25 to the base _____ is 1.

4. The logarithm of 1 to the base any number is _____ .

In Problems 5–10, mark the correct response.

5. If $\log_x 27 = 3$, then $x =$
   (a) 9 (b) 3 (c) 3^{27} (d) 27^3 (e) None of these

6. If $\log_5\left(\frac{1}{5}\right) = x$, then $x =$
   (a) $\frac{1}{25}$ (b) -1 (c) $\frac{1}{125}$ (d) 5^5 (e) None of these

7. If $\log_{10}x = -2$, then $x =$
   (a) -20 (b) -5 (c) $-\frac{1}{5}$ (d) 100 (e) $\frac{1}{100}$

8. Consider the following statements P and Q
   P: $\log_2 x = -1$ implies $2^x = -1$
   Q: $\log_x 5$ is defined only when $x > 0$ and $x \neq 1$
   Which of the above statements is true?
   (a) P only (b) Q only (c) P and Q (d) None

9. Consider the following statements P and Q
   P: $\log_x 1 = 0$ for all positive x
   Q: $\log_x x = 1$ for all positive x
   Which of the above statements is true?
   (a) P only (b) Q only (c) P and Q (d) None

10. Consider the following statements P and Q
    P: $\log_5 x$ is not defined when x is negative
    Q: $2^{\log_2 5} = 5$
    Which of the above statements is true?
    (a) P only (b) Q only (c) P and Q (d) None

In Problems 11–20, convert the exponential statements to logarithmic form.

11. $2^5 = 32$ implies $\log_2 32 = 5$

12. $5^3 = 125$ implies   _____

13. $8^1 = 8$ implies   _____

14. $5^0 = 1$ implies   _____

15. $4^{-2} = \frac{1}{16}$ implies   _____

16. $8^{1/3} = 2$ implies   _____

17. $4^{-1/2} = \frac{1}{2}$ implies   _____

18. $\left(\frac{4}{5}\right)^0 = 1$ implies   _____

19. $(1000)^{-1/3} = \frac{1}{10}$ implies   _____

20. $64^{5/6} = 32$ implies   _____

In Problems 21–27, find the value.

21. $\log_4\left(\frac{1}{16}\right) =$ -2 22. $\log_2 4 =$ _____

23. $\log_3 27 =$ _____   24. $\log_2 2 =$ _____

25. log .01 = _____   26. Ln e⁴ = _____

27. $\log_5\left(\frac{1}{5}\right) =$ _____

In Problems 28–38, convert the logarithmic statements to exponential form.

28. $\log_5 25 = 2$ implies $\underline{\quad 5^2 = 25 \quad}$

29. $\log_2 4 = 2$ implies   _____

30. $\log_3 x = 4$ implies   _____

31. $\log_{25} 5 = \frac{1}{2}$ implies   _____

32. $\log_2\left(\frac{1}{8}\right) = -3$ implies   _____

33. $\log 100 = 2$ implies   _____

34. $\log_x x^2 = 2$ implies   _____

35. $\log_{36} 6 = \frac{1}{2}$ implies   _____

36. $\log_5\left(\frac{1}{5}\right) = -1$ implies   _____

37. $\log_{29} 1 = 0$ implies   _____

38. $\log_{64}\left(\frac{1}{8}\right) = -\frac{1}{2}$ implies   _____

In Problems 39–50, convert to exponent form and solve.

	Show work here		Solution

39. $\log_2(2x - 1) = 4$ $2^4 = 2x - 1$ 39. $x = \frac{17}{2}$
     $2x - 1 = 16$
     $2x = 17$

40. $\log_2 x = -2$ 40. _____

41. $\log_x 25 = 2$ 41. _____

42. $\log_x 64 = 3$ 42. _____

43. $\log_{10} x = 3$ 43. _____

44. $\log_x\left(\frac{1}{2}\right) = \frac{1}{2}$ 44. _____

45. $\log_x\left(\frac{1}{2}\right) = -\frac{1}{2}$ 45. _____

46. $\log_8 1 = x$ 46. _____

47. $\log_x(.1) = -1$ 47. _____

48. $\log_{2.5} 2.5 = x$ 48. _____

49. $\log_{1/2} 4 = x$ 49. _____

50. $\log_2 x = -3$ 50. _____

11.3 Properties of Logarithms

We have seen in the last section that any logarithmic statement may be written as an exponential statement. We shall use this information and the exponent laws to prove three special properties of logarithms.

Property 1:

$$\log_b(x \cdot y) = \log_b x + \log_b y$$
$$\log_b x + \log_b y = \log_b(x \cdot y)$$

We may state this property in words as "the log of the product of factors is the sum of the logs of the factors".

Proof of Property 1

$$\text{Let } \log_b x = A \text{ and } \log_b y = B \quad \text{(I)}$$
$$\log_b x = A \text{ implies } x = b^A$$
$$\log_b y = B \text{ implies } y = b^B$$
$$\text{then } x \cdot y = b^A \cdot b^B$$
$$= b^{A+B}$$

But we know that

$$(x \cdot y) = b^{A+B} \text{ or } b^{A+B} = xy \text{ implies } \log_b xy = A + B.$$

Thus,

$$\log_b(x \cdot y) = A + B$$
$$= \log_b x + \log_b y \qquad \text{from (I)}$$

which is the same thing as

$$\log_b x + \log_b y = \log_b(xy)$$

➡ Notice that $\log_b x + \log_b y \neq \log_b(x + y)$
For example, $\log_2 8 + \log_2 8 = 3 + 3 = 6$
but $\log_2(8 + 8) = \log_2 16 = 4 \neq 6$

Example 1: *Given that $\log_2 4 = 2$, $\log_2 8 = 3$, and $\log_2 2 = 1$, find the value of the logarithm.*

(a) $\log_2 32 = \log_2 8 \cdot 4 = \log_2 8 + \log_2 4$
$\qquad\qquad\qquad\qquad\quad = \;\; 3 \;\; + \;\; 2$
$\qquad\qquad\qquad\qquad\quad = 5$

(b) $\log_2 64 = \log_2 8 \cdot 8 = \log_2 8 + \log_2 8$
$\qquad\qquad\qquad\qquad\quad = \;\; 3 \;\; + \;\; 3 = 6$

(c) $\log_2 256 = \log_2 4 \cdot 64 = \log_2 4 + \log_2 64$
$\qquad\qquad\qquad\qquad\quad = \;\; 2 \;\; + \;\; 6 = 8$

Example 2. *Simplify each of the following expressions as a single logarithm.*

(a) $\log_b 2x + \log_b x = \log_b(2x \cdot x) = \log_b 2x^2$
(b) $\log_b(3x - 1) + \log_b 2x = \log_b(3x - 1)2x = \log_b(6x^2 - 2x)$
(c) $\log 25 + \log 4 = \log 25 \cdot 4 = \log 100 = 2$. Why?
(d) $\log 5 + \log 5 + \log 4 = \log(5 \cdot 5) + \log 4$
$\qquad\qquad\qquad\qquad\quad = \log(5 \cdot 5 \cdot 4) = \log 100 = 2$

Property 2:

$$\boxed{\begin{array}{c} \log_b \dfrac{x}{y} = \log_b x - \log_b y \\ \log_b x - \log_b y = \log_b \dfrac{x}{y} \end{array}}$$

We may state this property in words as "the log of the ratio of two terms is the difference of the logs of the terms."

Property 2 can be proved exactly in the same way as Property 1. Try it yourself.

➡️
$$\text{Notice that } \log_b x - \log_b y \neq \log_b(x - y)$$
$$\text{For example, } \log_2 16 - \log_2 8 = 4 - 3 = 1$$
$$\text{but } \log_2(16 - 8) = \log_2 8 = 3 \neq 1$$

$$\text{Notice that } \log_b x - \log_b y \neq \frac{\log_b x}{\log_b y}$$
$$\text{For example, } \log_2 16 - \log_2 8 = 4 - 3 = 1$$
$$\text{but } \frac{\log_2 16}{\log_2 8} = \frac{4}{3} \neq 1$$

Example 3: *Given that $\log_3 9 = 2$ and $\log_3 27 = 3$, find the value of the logarithms.*

(a) $\log_3 3 = \log_3 \frac{27}{9} = \log_3 27 - \log_3 9 = 3 - 2 = 1.$

(b) $\log_3 \frac{1}{3} = \log_3 \frac{9}{27} = \log_3 9 - \log_3 27 = 2 - 3 = -1$

Example 4: *Simplify each of the following expressions as a single logarithm.*

(a) $\log_a x^2 - \log_a y = \log_a \frac{x^2}{y}$

(b) $\log 1000 - \log 10 = \log \frac{1000}{10} = \log 100 = 2$
$$\quad\;\; \downarrow \qquad\quad\;\; \downarrow \qquad\qquad\qquad\qquad\qquad\quad \downarrow$$
$$\quad\;\; 3 \quad - \quad 1 \qquad\qquad = \qquad\qquad\qquad 2$$

(c) $\log_a x + \log_a 2y - \log_a 2x = \log_a x(2y) - \log_a 2x$
$$= \log_a \frac{2xy}{2x} = \log_a y$$

Property 3:

$$\boxed{\log_b(x^y) = y \log_b x}$$

We may state this property in words as "the log of the exponential of a number is the exponent times the log of the number."

Proof of Property 3

Let $\log_b x = A$.
$\log_b x = A$ implies $b^A = x$
or $x = b^A$

Then,
$$x^y = (b^A)^y = b^{yA}$$

But, we know that
$$x^y = b^{yA}$$
or $b^{yA} = (x^y)$ implies $\log_b(x^y) = yA$
$$= y \log_b x$$

Thus, $\log_b x^y = y \cdot \log_b x$

Example 5: *Evaluate each of the following logarithms.*

(a) $\log_3 9 = \log_3 3^2 = 2 \log_3 3 = 2 \cdot 1 = 2$

(b) $\log_3 81 = \log_3 3^4 = 4 \log_3 3 = 4 \cdot 1 = 4$

(c) $\log_5 \frac{1}{5} = \log_5 5^{-1} = -1 \log_5 5 = -1 \cdot 1 = -1$

You can also evaluate this as

$\log_5 \frac{1}{5} = \log_5 1 - \log_5 5$ Property 2

$\qquad\quad = -\log_5 5 = -1$ $\log_5 1 = 0$
     $\log_5 5 = 1$

Example 6: *Simplify each of the following expressions.*

(a) $\log \sqrt[3]{x^2}$

$\log \sqrt[3]{x^2} = \log x^{2/3} = \frac{2}{3} \log x$ Property 3

(b) $\log \frac{1}{x^2} = \log 1 - \log x^2$ Property 2

$\qquad = -\log x^2$ Since $\log 1 = 0$

$\qquad = -2 \log x$ Property 3

You may simplify $\log \left(\frac{1}{x^2}\right)$ even without using property 2.

$\log \frac{1}{x^2} = \log x^{-2} = -2 \log x$ Property 3

(c) $\log_a(\sqrt[3]{a^4} \cdot a^{-4})$

$\log_a(\sqrt[3]{a^4} \cdot a^{-4}) = \log_a \sqrt[3]{a^4} + \log_a a^{-4}$ Property 1

$\qquad\qquad\qquad\quad = \log_a a^{4/3} + \log_a a^{-4}$

$\qquad\qquad\qquad\quad = \frac{4}{3} \log_a a + (-4)\log_a a$ Property 3

$\qquad\qquad\qquad\quad = \frac{4}{3} - 4$ $\log_a a = 1$

$\qquad\qquad\qquad\quad = -\frac{8}{3}$

Example 7: *Simplify each of the expressions as indicated.*

(a) Write $\log 2x + \log x^2$ as a single logarithm

$\log 2x + \log x^2 = \log (2x \cdot x^2)$ Property 2

$\qquad\qquad\qquad = \log(2x^3)$

(b) Write $\log_a \frac{x^2 y^3}{\sqrt{z}}$ in terms of $\log_a x$, $\log_a y$ and $\log_a z$

$\log_a \frac{x^2 y^3}{\sqrt{z}} = \log_a \frac{x^2 y^3}{z^{1/2}}$

$\qquad\quad = \log_a x^2 + \log_a y^3 - \log_a z^{1/2}$ Property 1 & 2

$\qquad\quad = 2 \log_a x + 3 \log_a y - \frac{1}{2} \log_a z$ Property 3

(c) Write $2 \log x + \frac{3}{2} \log y - 3 \log z$ as a single logarithm

$2 \log x + \frac{3}{2} \log y - 3 \log z = \log x^2 + \log y^{3/2} - \log z^3$ Property 3

$\qquad\qquad\qquad\qquad\qquad\quad = \log x^2 \cdot y^{3/2} - \log z^3$

$\qquad\qquad\qquad\qquad\qquad\quad = \log \frac{x^2 \cdot y^{3/2}}{z^3} = \log \frac{x^2 \sqrt{y^3}}{z^3}$ Property 2

(d) Write $\log_b \sqrt{\dfrac{x^5 - 4}{8x}}$ in terms of $\log_b 2$, $\log_b x$, and $\log_b(x^5 - 4)$

$$\log_b \sqrt{\dfrac{x^5 - 4}{8x}} = \log_b \left(\dfrac{x^5 - 4}{8x}\right)^{1/2}$$

$$= \dfrac{1}{2} \log_b \left(\dfrac{x^5 - 4}{8x}\right) \qquad \text{Property 3}$$

$$= \dfrac{1}{2} [\log_b(x^5 - 4) - \log_b 8x] \qquad \text{Property 2}$$

$$= \dfrac{1}{2} [\log_b(x^5 - 4) - (\log_b 8 + \log_b x)] \qquad \text{Property 1}$$

$$= \dfrac{1}{2} [\log_b(x^5 - 4) - \log_b 8 - \log_b x]$$

Finally, since $\log_b 8 = \log_b 2^3 = 3\log_b 2$ we may write the last line as

$$\dfrac{1}{2}[\log_b(x^5 - 4) - 3\log_b 2 - \log_b x]$$

which is the desired form.

Problem Set 11.3

In Problems 1–5, mark true or false.

1. $\log_a 5 + \log_a 3 = \log_a 8$
   (a) True   (b) False

2. $\log_b\left(\dfrac{5}{3}\right) = \log_b 2$
   (a) True   (b) False

3. $\log_2 \sqrt{2} = \dfrac{1}{2}$
   (a) True   (b) False

4. $\log x = \log_{10} x$
   (a) True   (b) False

5. The base in $\text{Ln } x$ is e
   (a) True   (b) False

In Problems 6–10, complete the statement.

6. $\log_x x = $ _____

7. $\log_x 1 = $ _____

8. $\log_b x^4 = $ _____ $\cdot \log_b x$

9. $\log_b 9 - \log_b 6 = \log_b (\quad)$

10. $\log_b 20 + \log_b 5 = $ _____ $\cdot \log_b 10$

In Problems 11–26, find the value of the logarithmic expressions.

11. $\log_a a^3 = $ _____ $3 \cdot \log_a a = 3$ _____ 12. $\log_x \sqrt{x} = $ _____

13. $\log_{10} 10 = $ _____ 14. $\log \sqrt{10} = $ _____

15. $\log_5 \sqrt[5]{5} = $ _____ 16. $\log .01 = $ _____

17. $\log_e \sqrt[3]{e^2} = $ _____ 18. $\log_3(3^{-2} \cdot 3^4)^{-4} = $ _____

19. $\log_3(27 \sqrt[3]{3}) = $ _____

20. $\operatorname{Ln} \dfrac{1}{e^2} = $ _____ 21. $\operatorname{Ln} e^5 \sqrt{e} = $ _____

22. $a^{\log_a 10} = 10$ [Recall $\log_a 10$ is a number which when used as an exponent on the base "a" produces 10.]

23. $10^{\log 100} = $ _____ 24. $e^{\operatorname{Ln}(9)} = $ _____

25. $3^{\log_3 10} = $ _____ 26. $\log_a a^x = $ _____

In Problems 27–35, write the expression as a single logarithm.

27. $\log_5 20 - \log_5 9 = $

28. $\log 22 - \log 7 = $

29. $3 \log_2 x^2 - 5 \log_2 x = \log_2(x^2)^3 - \log_2 x^5 = \log_2 x^6 - \log_2 x^5$
$= \log_2 \dfrac{x^6}{x^5} = \log_2 x$

30. $\log x + 2 \cdot \log y = $

31. $4 \cdot \log_3 x + 2 \cdot \log_3 y - \dfrac{1}{2} \log_3 z = $

32. $\dfrac{1}{3} \log_5 x + 2 \cdot \log_5 y + 4 \cdot \log_5 x = $

33. $2 \cdot \log_{15} x - \dfrac{1}{3} \log_{15} x^4 = $

34. $x \operatorname{Ln} y + y \operatorname{Ln} x = $

35. $3 \cdot \log 5 + 4 \cdot \log 2 - \log 2 = $

In Problems 36–44, write the expressions in terms of log x, log y, and log (2x − 1).

36. $\log(x^3 y^2) = $

37. $\log y^5 \sqrt{x} = $

38. $\log \dfrac{x^4}{2x-1} =$

39. $\log x^2 y^3 (2x-1)^4 =$

40. $\log\left(\dfrac{\sqrt[3]{x}\, y^5}{(2x-1)^2}\right) =$

41. $\log\left(\dfrac{x^2 y}{2x-1}\right)^4 =$

42. $\log[xy^2(2x-1)^2]^{-4} =$

43. $\log \sqrt{y(2x-1)} =$

44. $\log\left(\dfrac{x^3 \sqrt{y}}{\sqrt{2x-1}}\right) =$

11.4 Logarithmic Equations

If x and y are two *positive real numbers* and if $x = y$ then $\log_b x = \log_b y$. ($b > 0$, $b \neq 1$) Conversely, if $\log_b x = \log_b y$ then $x = y$. We use this fact for solving equations involving logarithmic terms.

Example 1: *To solve equations with logarithmic terms*

(a) $\log_4 5x = \log_4 15$
　　　$5x = 15$
　　　$x = 3$

(b) $\log_5 4 = \log_5 3x - \log_5 6$
　　$\log_5 4 + \log_5 6 = \log_5 3x$
　　　　$\log_5 4 \cdot 6 = \log_5 3x$　　　　　　　　　　　Property 1
　　　　　　$24 = 3x$
　　　　　　　$x = 8$

(c) $\log(x^2 - 1) - \log(x - 1) = \log 6$
　　$\log \dfrac{x^2 - 1}{x - 1} = \log 6$　　　　　　　　　Property 2
　　$\log(x + 1) = \log 6$
　　　　$x + 1 = 6$
　　　　　$x = 5$

It may happen, sometimes, that the solutions we obtain make no sense when we check for the solutions in the given equation. We shall call such a solution an *extraneous solution* and discard it from the set of solutions.

Example 2: *Extraneous solutions of an equation*

(a) $\log_5 x + \log_5(x - 2) = \log_5 8$
　　$\log_5[(x)(x - 2)] = \log_5 8$　　　　　　　　　Property 1
　　　　$x(x - 2) = 8$
　　$x^2 - 2x - 8 = 0$, $(x - 4)(x + 2) = 0$
　　　　$x = 4$, $x = -2$

$x = -2$ is an extraneous solution, because the logarithm of a negative number is not defined as a real number. Thus, the only solution is $x = 4$.

(b) $\log(x + 4) = \log 32 - \log x$
$\log(x + 4) + \log x = \log 32$
$\log[x(x + 4)] = \log 32$
$x(x + 4) = 32$
$x^2 + 4x - 32 = 0, (x - 4)(x + 8) = 0$
$x = 4, x = -8$

$x = -8$ is an extraneous solution because the logarithm of a negative number is not defined as a real number. The only solution is $x = 4$.

If the given equation contains some terms without logarithms then by using property 3 we can change these terms to the logarithm form with the same base as in other terms. For example

$2 = \log_b b^2$, since $\log_b b^2 = 2 \log_b b = 2$

In general

$$\boxed{n = \log_b b^n.}$$

Example 3: *To solve logarithmic equations involving non-logarithmic terms*

(a) $\log_5 x = 2$
$\log_5 x = \log_5 5^2$ \hfill $(2 = \log_5 5^2)$
$x = 5^2$ or $x = 25$

You may also solve this equation by converting the logarithmic form to exponential form, e.g.

$\log_5 x = 2$ implies $x = 5^2 = 25$

(b) $\log_3(2x + 1) = 4$
$\log_3(2x + 1) = \log_3 3^4$ \hfill $(4 = \log_3 3^4)$
$2x + 1 = 3^4 = 81$
$2x = 80, x = 40$

(c) $\log_4(3x + 1) = \log_4 5 + 2$
$\log_4(3x + 1) - \log_4 5 = 2$
$\log_4 \dfrac{3x + 1}{5} = \log_4 4^2$ \hfill Property 2, and
$\dfrac{3x + 1}{5} = 4^2, 3x + 1 = 80, x = \dfrac{79}{3}$ \hfill $2 = \log_4 4^2$

(d) $\log(x^2 - x - 6) - 3 = \log(x - 3) + \log 5$
$\log(x^2 - x - 6) - \log(x - 3) = \log 5 + 3$
$\log \dfrac{x^2 - x - 6}{x - 3} = \log 5 + \log 10^3$ \hfill Property 2, and
$\log \dfrac{(x - 3)(x + 2)}{(x - 3)} = \log(5 \cdot 10^3)$
$\log(x + 2) = \log(5000)$ \hfill $3 = \log_{10} 10^3$
$x + 2 = 5000, x = 4998$ \hfill $= \log 10^3$

Problem 11.4

In Problems 1–10, solve for the unknown.

	Equation	Show work here	Solution
1.	$\log_4 4x = \log_4 72 - \log_4 16$	$\log_4 4x = \log_4 \frac{72}{16}$ $\quad$ $4x = \frac{9}{2}, x = \frac{9}{8}$	$x = \frac{9}{8}$
2.	$\log_2 3x = \log_2 6$		$x =$
3.	$\log 5x = \log 8 - \log 2$		$x =$
4.	$\log(x + 1) = \log 4$		$x =$
5.	$\log_4(3x + 4) = \log_4(x + 7)$		$x =$
6.	$\log(x - 1) = \log 2 - \log 8$		$x =$
7.	$\log_4 2x - \log_4 5 = \log_4 8$		$x =$
8.	$\log x^2 = \log 2x + \log 5$		$x =$
9.	$\log(x^2 - 3x - 10) = \log(x + 2) + \log 7$		$x =$
10.	$\log_2(x^2 - 1) = \log_2(x + 1) - \log_2 3$ [Hint: $-\log_2 3 = \log_2 3^{-1} = \log_2\left(\frac{1}{3}\right)$]		$x =$

In Problems 11–15, change all the non-logarithmic terms to logarithmic form and solve.

Equation | Show work here | Solution

11. $\log_2 5x^2 = \log_2 x + 4$

$\log_2 5x^2 - \log_2 x = 4$

$\log_2 \dfrac{5x^2}{x} = \log_2 2^4$

$5x = 16$

$x = \dfrac{16}{5}$

12. $\log_3 5x = 2$ x =

13. $\log_5 x = 3 - \log_5 2$ x =

14. $\log_2(2x + 3) = \log_2 x + 3$ x =

15. $\log 4x^2 = \log x - \log 3$ x =

In Problems 16–20, solve the equations. Indicate the extraneous solution, if any.

Equation | | Solution / Ext. solution

16. $\log x + \log(x - 1) = \log(12)$

$\log x(x - 1) = \log(12)$
$x^2 - x = 12,\ x^2 - x - 12 = 0$
$(x - 4)(x + 3) = 0$

Sol. x = __4__
E.S. x = −3

17. $\log(x - 4) = \log 2x$

Sol. x = _____
E.S. x =

18. $\log(3x + 1) = \log 2x$

Sol. x = _____
E.S. x =

19. $\log_5 x = \log_5 2 - \log_5(2x - 3)$

Sol. x = _____
E.S. x =

20. $\operatorname{Ln} x = -\operatorname{Ln} 4x + 3$

Sol. x = _____
E.S. x =

11.5 Computing Logarithms on the Calculator

The most frequently used logarithms in scientific calculations have the base 10 or e. As discussed in Section 2, a logarithm to the base 10 is called a **common logarithm** and is written as log x, without mentioning the base 10, and a logarithm to the base e is called a **natural logarithm** and is written as Ln x. These two special types of logarithms are usually computed by use of a calculator.

When using a calculator, sometimes it may happen that,

(a) you punch a wrong key, or

(b) the battery is wearing out and you start getting numbers on the display that make no sense.

In order to be sure that there is nothing wrong and the answer on the calculator display is a reasonable answer, you must learn to make a rough estimate of the answer before you start computing a logarithm on the calculator, so that you have a check on the reasonableness of your answer.

You can easily see that

$$10 = 10^1 \qquad 100 = 10^2 \qquad 1000 = 10^3$$
$$.1 = 10^{-1} \qquad .01 = 10^{-2} \qquad .001 = 10^{-3}$$

Now consider the following common logarithms

$$\log 10 = \log_{10} 10^1 = 1$$
$$\log 100 = \log_{10} 10^2 = 2 \cdot \log_{10} 10 = 2$$
$$\log 1000 = \log_{10} 10^3 = 3 \cdot \log_{10} 10 = 3$$
$$\log .1 = \log_{10} 10^{-1} = -1 \cdot \log_{10} 10 = -1$$
$$\log .01 = \log_{10} 10^{-2} = -2 \cdot \log_{10} 10 = -2$$
$$\log .0001 = \log_{10} 10^{-4} = -4 \cdot \log_{10} 10 = -4$$

Thus, we see that if $n = 10^a$, then log n = a.

But, what about those numbers that cannot be expressed as 10 to an integer exponent, such as 2, 3, 5, 20, 129, etc? For estimating the logarithm of such a number we use the following results.

> If a and b are any two positive real numbers, then a < b implies log a < log b.

Example 1: *To estimate the common logarithm of numbers greater than 1*

(a) log 2

We know that log 1 = 0 and log 10 = 1.
Since 1 < 2 < 10,
we have log 1 < log 2 < log 10
$$0 < \log 2 < 1$$
Thus, log 2 is some number between 0 and 1.

(b) log 149

We know that log 100 = 2 and log 1000 = 3.
Since 100 < 149 < 1000,
we have log 100 < log 149 < log 1000
$$2 < \log 149 < 3$$
Thus, log 149 is some number between 2 and 3.

Example 2: *To estimate the common logarithms of numbers less than 1*

(a) log(.4)
since $.1 < .4 < 1$,
we have $\log(.1) < \log(.4) < \log 1$
$-1 < \log(.4) < 0$
Thus, log(.4) is a negative number with absolute value less than 1.

(b) log(.0079)
Since $.001 < .0079 < .01$
we have $\log .001 < \log(.0079) < \log .01$
$-3 < \log .0079 < -2$
Thus, log(.0079) lies between -3 and -2.

Once you have an estimate of the logarithm in mind, then computing the logarithm on your calculator is simple.

Example 3: *To compute the value of a common logarithm on a calculator*

(a) log 2
We know from Example 1(a) that log 2 lies between 0 and 1
log 2 = 2 $\boxed{\text{LOG}}$ $\boxed{=}$ .30103

(b) log(149)
We know from Example 1(b) that log(149) lies between 2 and 3
log(149) = 149 $\boxed{\text{LOG}}$ $\boxed{=}$ 2.1731863

(c) log(.4) = .4 $\boxed{\text{LOG}}$ $\boxed{=}$ $-$ 0.39794
The answer is reasonable since
$-1 < -0.39794 < 0$ \hfill (Example 2(a))

(d) log(.0079) = .0079 $\boxed{\text{LOG}}$ $\boxed{=}$ -2.1023729
The answer is reasonable since
$-3 < -2.1023729 < -2$ \hfill (Example 2(b))

Now consider the situation where you have to find n when log n is given. The value of n is also called the *anti-logarithm* of log n. For example,

if log n = 2.3579,
then n = anti-log (2.3579)

Anti-logarithms like logarithms can be computed very easily on your calculator. Again as before you should have an estimate of the answer in your mind before you actually compute the anti-log on your calculator. Consider, for example,

log n = 2.3579

We know that log 100 = 2 and log 1000 = 3 and $2 < 2.3579 < 3$. That is, log 100 < log n < log 1000. Therefore, $100 < n < 1000$. Thus, n = anti-log 2.3579 must lie between 100 and 1000.

We compute the anti-log of 2.3579 on our calculator as follows.

If log n = 2.3579,
then n = anti-log 2.3579
 = 2.3579 $\underbrace{\boxed{\text{INV}} \boxed{\text{LOG}}}_{\text{anti-log}}$ $\boxed{=}$ 227.98171

Example 4: *To find n when log n is given*

(a)   log n = 3.7983   log 1000 = 3, log 10000 = 4

         3 < 3.7983 < 4

     log 1000 < log n < log 10000

         1000 < n < 10000

     n = anti-log 3.7983

        = 3.7983 $\boxed{\text{INV}}$ $\boxed{\text{LOG}}$ $\boxed{=}$ 6284.9236

6284.9236 is a number between 1000 and 10000 as estimated earlier.

(b)   log n = −1.489

     log .1 = −1   log .01 = −2

       −2 < −1.489 < −1

     log .01 < log n < log .1

        .01 < n < .1

     n = anti-log (−1.489)

        = 1.489 $\boxed{+/-}$ $\boxed{\text{INV}}$ $\boxed{\text{LOG}}$ $\boxed{=}$ .032434

.032434 lies between .01 and .1 as estimated earlier.

So far in this Section, we computed the value of common logarithms. We follow exactly the same procedure for computing the value of natural logarithms. But, again as before we need to estimate the answer before we compute the natural logarithm on a calculator.

     $\log_e e = 1$

Therefore,

     e = anti-log$_e$ 1

       = anti-Ln 1

       = 1 $\boxed{\text{INV}}$ $\boxed{\text{LNx}}$ $\boxed{=}$ 2.7182818

For estimation consider e to be approximately equal to 3 and look for an integer "a" so that 3^a is close to N. Then "a" should be close to Ln N. For example,

(a)   Consider Ln 25. $3^2 = 9$, $3^3 = 27$

      Thus, 25 is close to 3^3. Therefore, the value of Ln 25 is close to 3.

(b)   Consider Ln 12. $3^1 = 3$, $3^2 = 9$, $3^3 = 27$

      Thus, 12 is close to 3^2. Therefore, the value of Ln(12) is close to 2.

Example 4: *To compute natural logarithms on a calculator*

(a)   Ln 25 = 25 $\boxed{\text{LNx}}$ $\boxed{=}$ 3.2188758

      As seen above it is close to 3.

(b)   Ln 12 = 12 $\boxed{\text{LNx}}$ $\boxed{=}$ 2.4849066

      As seen above it is close to 2.

(c)   Ln .1 = .1 $\boxed{\text{LNx}}$ $\boxed{=}$ − 2.3025851

      The answer seems reasonable because

      $.1 = \dfrac{1}{10} \simeq \dfrac{1}{9} = 3^{-2}$

(d) Ln 348 = 348 [LNx] [=] 5.8522025
The answer seems reasonable because
$3^5 = 243$, $3^6 = 729$
and $243 < 348 < 729$
Therefore $5 < $ Ln $348 < 6$.

Example 5: *To find n when Ln n is given*

(a) Ln(n) = 2.45
n = anti-Ln (2.45)
= 2.45 [INV] [LNx] [=] 11.588347

(b) Ln n = −3.97
n = anti-Ln (−3.97)
= 3.97 [+/−] [INV] [LNx] [=] .0188734

Example 6: *To solve exponential equations*

(a) $5^x = 32$
Take the common logarithm of both sides
log 5^x = log 32
x log 5 = log 32, x = $\dfrac{\log 32}{\log 5}$
= 32 [LOG] [÷] 5 [LOG] [=] 2.1533828

(b) $13^{2x+5} = 3782$
Take the common logarithm of both sides
log $13^{(2x+5)}$ = log 3782
(2x + 5)log 13 = log 3782
2x + 5 = $\dfrac{\log 3782}{\log 13}$
= 3782 [LOG] [÷] 13 [LOG] [=] 3.2117625
x = $\dfrac{3.2117625 - 5}{2}$
x = 3.2117625 [−] 5 [=] − 1.7882375 [÷] 2
[=] − 0.8941187

Example 7: *To compute logarithms*

(a) log(7.849 × 10^8)
= log 7.849 + log 10^8
= log 7.849 + 8
Since log 7.849 lies between 0 and 1,
log(7.849 × 10^8) lies between 8 and 9.
log(7.849 × 10^8) = 7.849 [EE] 8 [LOG] [=] 8.8948

➤ The number 7.849 × 10^8 could also be computed as 7.849 [x] 10 [y^x] 8, but key [EE] can do the same job in a simple way as 7.849 [EE] 8. Thus 7.849 [x] 10 [y^x] 8 and 7.849 [EE] gives the same result.

(b) $\log(2.8734 \times 10^{-7})$
$= \log 2.8734 + \log 10^{-7}$
$= \log 2.8734 - 7$
Since $\log 2.8734$ lies between 0 and 1,
$\log(2.8734 \times 10^{-7})$ lies between -7 and -6.
$\log(2.8734 \times 10^{-7}) = 2.8734$ EE 7 +/− LOG
$= -6.5416$.

Problem Set 11.5

In Problems 2–10, the given logarithm lies between what two integers?

	Logarithm	Show work here	Lies between
1.	log 325	$100 < 325 < 1000$ $2 < \log 325 < 3$	__2__ and __3__
2.	log 405		____ and ____
3.	log 3485		____ and ____
4.	log 9		____ and ____
5.	log .5		____ and ____
6.	log 18.73		____ and ____
7.	log 1893.457		____ and ____
8.	log .08		____ and ____
9.	log .00097		____ and ____
10.	log 2.08973		____ and ____

In Problems 11–15, the given logarithm is close to what integer?

	Logarithm	Show work here	Close to the integer
11.	Ln 347	$3^5 = 243,\ 3^6 = 729$ $3^5 < 347 < 3^6$	5 or 6
12.	Ln 8		____

13. Ln 32 _____

14. Ln 823 _____

15. Ln .01 _____

In Problems 16–26, compute the given logarithms after estimating the value.

	Logarithm	Lies between	The value
16.	log 273 $100 < 273 < 1000$	__2__ and __3__	__2.4361626__
17.	log 4773	___ and ___	_____
18.	log 2.79	___ and ___	_____
19.	log .479	___ and ___	_____
20.	log 1.3748	___ and ___	_____
21.	$\log(2.43 \times 10^{-8})$ $\log 2.43 + \log 10^{-8}$	__0−8__ and __1−8__ __−8__ and __−7__	__−7.6143937__
22.	$\log(7.278 \times 10^6)$	___ and ___ ___ and ___	_____
23.	Ln 74 $3^3 < 74 < 3^4$	close to 4	__4.3040651__
24.	Ln 283		_____
25.	Ln .073		_____
26.	Ln 2.7893		_____

In Problems 27–32, find n for a given log n.

27. log n = 1.345, n = 1.345 [INV] [LOG] [=]   __22.130947__

28. log n = 2.457, n =   _____

29. log n = −1.29, n = _____

30. log n = 4.754, n = _____

31. log n = −3.479, n = _____

32. log n = −8.459, n = _____

In Problems 33–37, find N for a given Ln N.

33. Ln N = 2.431, N = 2.431 [INV] [LNx] [=]   1.1371384

34. Ln N = 1, N = _____

35. Ln N = 8.3472, N = _____

36. Ln N = −2.743, N = _____

37. Ln N = 12.071, N = _____

In Problems 38–42, solve for x.

Equation | Show work here | Solution

38. $3^{(2x-7)} = 7892$
    $\log 3^{(2x-7)} = \log 7892$

    $(2x - 7)\log 3 = \log 7892$

    $x = \frac{1}{2}\left(\frac{\log 7892}{\log 3} + 7\right)$

    7.5840636

39. $2^x = 31$ _____

40. $5^{3x} = 239$ _____

41. $9^{(2x-5)} = .032$ _____

42. $23^{(.5 - 2x)} = 2.475$ _____

11.6 Base Changing Formula

In the last Section, we computed common and natural logarithms, that is, logarithms to the base 10 and logarithms to the base e by use of a calculator. To compute the value of a logarithm to any other base we use the following property.

Property 4: *(Base Changing Formula)*

> If a, b and c are any three positive real numbers where $b \neq 1$ and $c \neq 1$ then
> $$\log_b a = \frac{\log_c a}{\log_c b}.$$
> For the special cases $c = 10$ or $c = e$ (i.e., base 10 or base e logarithms)
> $$\log_b a = \frac{\log a}{\log b} = \frac{\text{Ln} a}{\text{Ln} b}.$$

Proof of Property 4

Let $\log_c a = x$, then $c^x = a$ (I)
$\log_c b = y$, then $c^y = b$ (II)
$\log_b a = z$, then $b^z = a$

Now, $b^z = a$.
Therefore, $(c^y)^z = c^x$ From (I) and (II)
$c^{yz} = c^x$,
That is, $yz = x$
or $z = \dfrac{x}{y}$.
Hence, $\log_b a = \dfrac{\log_c a}{\log_c b}$.

In the special case that $c = 10$ we have

$$\log_b a = \frac{\log a}{\log b}$$

and when $c = e$ we have

$$\log_b a = \frac{\text{Ln} a}{\text{Ln} b}.$$

Example 1: To find the value of a logarithm of a number to any base.

(a) $\log_5 20 = \dfrac{\log_{10} 20}{\log_{10} 5} = \dfrac{\log 20}{\log 5}$

$\log_5 20 = \dfrac{\log 20}{\log 5}$

$= 20 \boxed{\text{LOG}} \div 5 \boxed{\text{LOG}} = 1.8613531$

Example 2: To find n or b for a given $\log_b n$, when the base "b" is different from 10 or e.

(a) $\log_5 n = 3.477$

That is, $\frac{\log_{10} n}{\log_{10} 5} = 3.477$

$\log n = (3.477) \cdot \log 5$

$n = \text{anti-log}[(3.477) \log 5]$

$= 3.477 \;\boxed{\times}\; 5 \;\boxed{\text{LOG}}\; \boxed{=}\; 2.4303187$

$\boxed{\text{INV}}\; \boxed{\text{LOG}}\; \boxed{=}\; 269.35107$

(b) $\log_b 45 = 2.478$

$\frac{\log 45}{\log b} = 2.478,$

$\log b = \frac{\log 45}{2.478}$

$b = \text{anti-log}\left(\frac{\log 45}{2.478}\right)$

$= 45 \;\boxed{\text{LOG}}\; \boxed{\div}\; 2.478 \;\boxed{=}\; 0.667156$

$\boxed{\text{INV}}\; \boxed{\text{LOG}}\; \boxed{=}\; 4.6468214$

Problem Set 11.6

In Problems 1–5, use your calculator to find the value of the given logarithm.

Logarithm	Show work here	The value
1. $\log_4 22 =$	$\frac{\log 22}{\log 4} = 22 \;\boxed{\text{LOG}}\; \boxed{\div}\; 4 \;\boxed{\text{LOG}}\; \boxed{=}$	2.2297158

2. $\log_{31} 478 =$ _____

3. $\log_{19} 3.49 =$ _____

4. $\log_{1.7}(.01792) =$ _____

5. $\log_{20}(2.379 \times 10^{-9})$ _____

In Problems 6–10, find the value of the unknown in the given equation.

	Equation	Show work here	The value
6.	$\log_7 n = 2.387$	$\dfrac{\log n}{\log 7} = 2.387$ $n = \text{anti-log}[2.387 \times \log 7] =$	104.05166
7.	$\log_{22} n = 3.783$		_____
8.	$\log_{2.5} n = -1.457$		_____
9.	$\log_b 39 = 4.7389$		_____
10.	$\log_b 4893 = 2.879$		_____

Chapter Summary

1. **The expression x^n** x^n is defined for $x > 0$ and for any real number, n.
2. **Exponent rules** All exponent rules stated in Chapter 2 and Chapter 9 are applicable for real number exponents.
3. **Rule 11.1** If $x > 0$, and $x \neq 1$, and m, n are any two real numbers then $x^m = x^n$ implies m = n.
4. **Definition of logarithm** $\log_b x$ is a number which when used as an exponent for the base "b" produces "x".

    $\log_b x = a$ implies $b^a = x$, and $b^a = x$ implies $\log_b x = a$.

5. **Restrictions on the numbers involved in the definition of logarithms** $\log_b x$ is defined if $b > 0$, $b \neq 1$, and x is any positive real number.
6. **Special logarithms** $\log_b b = 1$ $(b > 0, b \neq 1)$
    $\log_b 1 = 0$
    $\log_{10} x = \log x$ common logarithms
    $\log_e x = \text{Ln } x$ natural logarithms
7. **Properties of logarithms**
    Property 1. $\log_b x + \log_b y = \log_b xy$
    Property 2. $\log_b x - \log_b y = \log_b \dfrac{x}{y}$
    Property 3. $\log_b m^n = n \log_b m$
8. **Some observations**
    (i) If $x = y$ then $\log_b x = \log_b y$
    (ii) If $\log_b x = \log_b y$ then $x = y$
    (iii) $a = \log_b b^a$ (by property 3)
    (iv) $a = b^{\log_b a}$ (by definition)

9. **Computing the value of a logarithm on a calculator** Keep an estimate of the value of the logarithm in mind. Then log N = N $\boxed{\text{LOG}}$ $\boxed{=}$ the value and Ln N = N $\boxed{\text{LNx}}$ $\boxed{=}$ the value.

10. **Base changing formula** $\log_b a = \dfrac{\log_c a}{\log_c b} = \dfrac{\log a}{\log b}$ when c = 10
$= \dfrac{\text{Lna}}{\text{Lnb}}$ when c = e.

11. **To compute the logarithm when the base is other than 10 or e** $\log_b a = \dfrac{\log a}{\log b} =$ a $\boxed{\text{LOG}}$ $\boxed{\div}$ b $\boxed{\text{LOG}}$ $\boxed{=}$ the value.

12. **To compute the anti-log N** If log N = N_1 then N = anti-log N_1 = N_1 $\boxed{\text{INV}}$ $\boxed{\text{LOG}}$ $\boxed{=}$ the value.

Review Problems

In Problems 1–7, complete the statements.

1. $\log_b x$ is defined only when b _____ and x _____ .

2. The statement $\log_b a = x$ when written in exponent form is _____ .

3. $\log_a 5 + \log_a 9 = \log_a ($ $)$.

4. $\log_4 15 - \log_4 3 = \log_4 ($ $)$.

5. $\log_5 9^7 = $ _____ $\cdot \log_5 ($ $)$.

6. $\log_b a = \dfrac{\log(\ )}{\log(\ )}$.

7. $a = b^{\log(\ )(\ )}$.

In Problems 8–11, solve for x.

Equation	Show work here	Solution
8. $2^{3x} = 16$		_____
9. $3^{2x-1} = \dfrac{1}{9}$		_____
10. $\left(\dfrac{1}{2}\right)^{2-x} = 8$		_____
11. $5^{2x} - 1 = 24$		_____

In Problems 12–15, convert the exponential statement to logarithmic form.

12. $3^4 = 81$ implies

13. $5^{-2} = \frac{1}{25}$ implies

14. $\left(\frac{3}{5}\right)^0 = 1$ implies

15. $\left(\frac{1}{2}\right)^{-3} = 8$ implies

In Problems 16–19, convert the logarithmic statement to exponential form.

16. $\log_3 9 = 2$ implies

17. $\log_2 \frac{1}{2} = -1$ implies

18. $\log_5(2x + 1) = 3$ implies

19. $\log_a(x^2 - 1) = y$ implies

In Problems 20–24, find the value.

20. $\log_5 1 =$

21. $\log_4 2 =$

22. $\log_2 \frac{1}{4} =$

23. $\log_3 \frac{1}{27} =$

24. $\log .001 =$

In Problems 25–27, convert to exponent form and solve.

    Equation   Show work here   Solution

25. $\log_3(2x - 1) = 2$

26. $\log_x 4 = 2$

27. $\log_9 3 = x$

In Problems 28–31, write the expression as a single logarithm.

28. $\log_2 25 + \log_2 5 =$

29. $\log_3 2x^2 - 2 \log_3 x =$

30. $3 \log x - 2 \log x =$ _____

31. $\frac{1}{2} \log_5 x + 2 \log_5 y - \log_5(x^2 - 1) =$ _____

In Problems 32–36, write the expression in terms of log x, log y, and log($x^2 + 1$).

32. $\log x^2 y =$

33. $\log x\sqrt{y} =$

34. $\log \dfrac{x(x^2 + 1)}{y} =$

35. $\log \dfrac{y^2(x^2 + 1)}{\sqrt{x}} =$

36. $\log[x^3 y(x^2 + 1)]^{-2} =$

In Problems 37–40, solve for the unknown.

Equation	Show work here	Solution
37. $\log_3(2x + 1) = \log_3 4$		$x =$
38. $\log 3x - \log 2 = \log 6$		$x =$
39. $\log_3(2x + 1) = \log_3 2x + 2$		$x =$
40. $\log_4 x = \log_4 14 - \log_4(x - 5)$		$x =$

In Problems 41–50, compute the logarithms on your calculator.

41. $\log 417 =$ _____   42. $\log .0075 =$ _____

43. $\log(2.793 \times 10^8) =$ _____   44. $\log(1.4 \times 10^{-6}) =$ _____

45. $\ln 1.728 =$ _____   46. $\ln .005 =$ _____

47. $\log_5 7 =$ _____ 48. $\log_{21} 25 =$ _____

49. $\log_{2.5}(.07) =$ _____ 50. $\log_{2.15} \sqrt[3]{35} =$ _____

In Problems 51–54 find N for a given log N or Ln N.

51. $\log N = -2.756$ N = _____

52. $\log N = 3.8502$ N = _____

53. $\text{Ln } N = -4.325$ N = _____

54. $\text{Ln } N = 7.324$ N = _____

In Problems 55–56, solve for x.

    Equation   Show work here   Solution

55. $3^x = 11$ 55. x = _____

56. $5^{(2x-4)} = 20$ 56. x = _____

Final Examination

1. Solve $2LM - 3N = 5LN$ for M:
   (a) $\dfrac{2L}{3N - 5NL}$ (b) $\dfrac{3N + 5NL}{2L}$ (c) $\dfrac{3N - 5NL}{2L}$ (d) $\dfrac{2L}{3N + 5NL}$ (e) None of these

2. The sum of three numbers is 16. The second number is twice the first and the third number is four less than the first. Find the numbers.
   (a) 4,11,1 (b) 5,9,2 (c) 6,8,2 (d) 5,10,1 (e) None

3. Linda is three years older than twice the age of Jane. In five years the sum of their ages will be 28. How old are they now?
   (a) 15,3 (b) 11,4 (c) 19,8 (d) 13,5 (e) None of these

4. A cashier has 21 fewer dimes than nickels, 15 more quarters than nickels, and 72 coins in total. How many nickels, dimes and quarters does he have?
   (a) 21,9,42 (b) 28,4,40 (c) 26,5,41 (d) None

5. If a car covers 420 miles in 8 hours, find the average speed of the car.
   (a) 50 mph (b) 52.5 mph (c) 60 mph (d) None

6. Perform the long division $(x^3 - 4) \div (x - 1)$
   The quotient Q and remainder R are
   (a) $Q = x^2 - x + 1, R = -3$ (b) $Q = x^2 + x + 1, R = -5$
   (c) $Q = x^2 - x - 1, R = -3$ (d) $Q = x^2 + x + 1, R = -3$
   (e) None of these

7. Consider the following statements P and Q:
   P: $(a - b)^2 = a^2 - 2ab + b^2$.
   Q: $x(3x + 4) = 3x^2 + 4x$.
   Which of the above statements are true?
   (a) P only   (b) Q only   (c) P and Q   (d) None

8. Consider the following statements P and Q:
   P: $4x(x + 1) - (x + 1) = 3x(x + 1)$.
   Q: $-3(3 - y) = 3(y - 3)$.
   Which of the above statements are true?
   (a) P only   (b) Q only   (c) P and Q   (d) None

9. One of the factors common in the two expressions $80 - 5x^2$ and $2x^2 + 17x + 36$ is
   (a) $x - 4$ (b) $x + 4$ (c) $x^2 + 4$ (d) None

10. One of the solutions of $4x^2 - 11x - 3 = 0$ is:
    (a) $\frac{1}{4}$ (b) $-\frac{1}{4}$ (c) -3 (d) $\frac{11 - \sqrt{73}}{8}$ (e) None

11. State whether the following statement is true or false. $-i^2 - 3i + 2i^2 = -1 - 3i$.
    (a) True   (b) False

12. One of the solutions of $y^2 + 2y = 24$ is:
    (a) 6   (b) $-1 - \frac{2i\sqrt{23}}{3}$ (c) 4   (d) $-1 - i\sqrt{23}$

13. The expression $\frac{x^2 + x - 6}{x^2 - 4}$ reduced completely equals:
    (a) $\frac{x + 3}{x - 2}$ (b) $\frac{x + 3}{x - 1}$ (c) $\frac{x + 3}{x + 2}$ (d) $\frac{x - 3}{x - 2}$ (e) None

14. Simplify: $\frac{x - 5}{2x^2 - x} - \frac{5}{x} =$
    (a) $\frac{3}{2x - 1}$ (b) $\frac{9}{2x - 1}$ (c) $\frac{-9}{2x - 1}$ (d) None

15. Simplify: $\frac{4x - 4}{2x^2 - x} - \frac{4}{x} =$
    (a) $\frac{2}{2x - 1}$ (b) $\frac{x - 2}{2x^2 - x}$ (c) $-\frac{4}{2x - 1}$ (d) None

16. Simplify: $\frac{4 - \frac{6}{x}}{2} =$
    (a) $\frac{2x - 3}{x}$ (b) $\frac{x - 3}{2}$ (c) $\frac{2x}{3}$ (d) $\frac{3x}{2}$

17. Which of the following pairs is a solution of $\frac{1}{x^2} + \frac{6}{x} + 8 = 0$.
    (a) $\frac{1}{4}, \frac{1}{2}$ (b) $\frac{1}{4}, -\frac{1}{2}$ (c) $-\frac{1}{4}, -\frac{1}{2}$ (d) $-\frac{1}{4}, \frac{1}{2}$ (e) None

18. Simplify and write your answer with positive exponent only: $\left[\dfrac{x^{-2/5}}{x^2 y^{-3/5}}\right]^{-5}$
    (a) $\dfrac{1}{x^8 y^2}$ (b) $\dfrac{x^8}{y^3}$ (c) $\dfrac{x^{12}}{y^3}$ (d) $x^{12} y^3$ (e) None

19. Consider the following statements P and Q:
    P: $x^{6/9} = \sqrt[3]{x^2}$.
    Q: $x^{1/4} y^{1/8} = \sqrt[8]{x^2 y}$.
    (a) P only (b) Q only (c) P and Q (d) None

20. Simplify: $\sqrt{x^5 y^4} =$
    (a) $xy^2 \sqrt{x}$ (b) $xy \sqrt{x}$ (c) $x^2 y^2 \sqrt{y}$ (d) $x^2 y^2 \sqrt{x}$ (e) None of these

21. Does $(\sqrt{3} + \sqrt{2})(\sqrt{3} - \sqrt{2}) = 1$?
    (a) Yes (b) No

22. Consider the following statements P and Q:
    P: $\dfrac{1}{5+i} = \dfrac{5-i}{24}$.
    Q: $\dfrac{1}{4+i} = \dfrac{4-i}{17}$.
    Which of the above statements are true?
    (a) P and Q (b) P only (c) Q only (d) None

23. Consider the following statements P and Q:
    P: The graph of $3y = 12$ is a line parallel to the x − axis.
    Q: The graph of $x = 6$ is a line parallel to the y − axis.
    Which of the above statements are true?
    (a) P and Q (b) P only (c) Q only (d) None

24. The two straight lines given by the equations $2x + 4y = 24$ and $4x + 2y = 26$ intersect at a point whose coordinates are:
    (a) $\left(\dfrac{11}{3}, \dfrac{14}{3}\right)$ (b) $\left(\dfrac{14}{3}, \dfrac{11}{3}\right)$ (c) $\left(\dfrac{-11}{3}, \dfrac{-14}{3}\right)$ (d) $\left(\dfrac{-14}{3}, \dfrac{-11}{3}\right)$ (e) None

25. Solve for the value of x and y.
    $3x - y = 7$.
    $x - 3y = 2$.
    (a) $\left(\dfrac{19}{8}, \dfrac{1}{8}\right)$ (b) $\left(\dfrac{20}{8}, \dfrac{10}{8}\right)$ (c) $\left(\dfrac{3}{7}, 1\right)$ (d) $\left(\dfrac{14}{20}, \dfrac{6}{2}\right)$ (e) None

26. For $x > 0$, consider the statements P and Q:
    P: $\log_x 1 = x$
    Q: $\log_2 4 = 2$
    Which of the above statements are true?
    (a) P only (b) Q only (c) P and Q (d) None

27. Consider the following statements P and Q:
    P: If $\log_3 x = 3$ then $x = 27$.
    Q: $\log_3 8 = 2$.
    Which of the above statements are true?
    (a) P only (b) Q only (c) P and Q (d) None

28. Consider the following statements P and Q:

    P: $\log_4 \sqrt[5]{4} = -\frac{1}{5}$.

    Q: $\log \frac{1}{1000} = 3$.

    Which of the above statements are true?
    (a) P and Q (b) P only (c) Q only (d) None

29. Consider the following statements P and Q:

    P: $\log \frac{A}{B} = \log A - \log B$.

    Q: $3 \log \frac{1}{3} = -3 \log 3$.

    Which of the above statements are true?
    (a) P only (b) Q only (c) P and Q (d) None

30. For $x > 0$, consider the statements P and Q:

    P: $\log_x \left(\frac{1}{x}\right) = -1$.

    Q: $\log_x (3x) = 3$.

    Which of the above statements are true?
    (a) P and Q (b) P only (c) Q only (d) None

Answers

Problem Set 1.1, Page 4

1. a
2. b
3. a
4. a
5. b
6. b
7. a
8. b
9. a
10. b
11. a
12. a
13. c
14. c
15. a

Problem Set 1.2, Page 6

1. 11
2. 13
3. 4
4. 0
5. 24
6. 11
7. 24
8. 45
9. 12
10. 18
11. 27
12. 56
13. 92
14. 45
15. 11
16. 41
17. 53
18. 26
19. 7
20. $\frac{3}{2}$
21. 1
22. 3
23. $\frac{1}{3}$
24. $\frac{1}{3}$
25. b
26. d
27. e
28. c
29. e
30. d

Problem Set 1.3, Page 11

1. <
2. <
3. >
4. <
5. >
6. >
7. >
8. <
9. 21
10. −5
11. −15
12. 3
13. −15
14. −8
15. −8
16. 6
17. −40
18. 240
19. −1
20. −15
21. 18
22. 12
23. −24
24. 4
25. 22
26. −2
27. −1
28. 1
29. −8
30. −1

Problem Set 1.4, Page 14

1. 4; 7
2. proper
3. improper
4. 1
5. 0
6. undefined
7. $2\frac{3}{5}$
8. $2\frac{2}{5}$
9. $5\frac{2}{3}$
10. $6\frac{1}{4}$
11. $2\frac{5}{8}$
12. $19\frac{1}{7}$
13. $7\frac{14}{43}$
14. $8\frac{39}{55}$
15. 3
16. 8
17. 24
18. 20
19. 63
20. 66
21. $\frac{2}{3}$
22. $\frac{3}{5}$
23. $\frac{2}{5}$
24. $\frac{6}{5}$
25. 3
26. $\frac{8}{21}$
27. 1
28. 2

Problem Set 1.5, Page 18

1. 29.5
2. 12,000,247.025
3. 8.008
4. 22.34
5. 972.69
6. 3296.91
7. 84.172
8. 60.18
9. 3175.67
10. .0958
11. 724.916
12. 863.925
13. 239.50
14. 112.362 . . .
15. 138.228
16. 42.55
17. 10.92
18. 5.79
19. 59.9
20. $68.22
21. $3267.675
23. 152.25 sq. ft.

Problem Set 1.6, Page 24

1. $\frac{2}{3}$
2. 1
3. $\frac{2}{11}$
4. $\frac{5}{12}$
5. $-\frac{1}{3}$
6. $\frac{2}{7}$
7. 3
8. 1
9. $\frac{13}{12}$
10. $\frac{3}{8}$
11. 60
12. 20
13. 8
14. 60
15. 36
16. 75
17. 30
18. 36
19. 80
20. 240
21. $\frac{61}{60}$
22. $\frac{7}{12}$
23. $\frac{1}{6}$
24. $\frac{3}{8}$
25. $\frac{7}{10}$
26. $\frac{9}{20}$
27. $\frac{9}{20}$
28. $\frac{10}{3}$
29. $\frac{16}{15}$
30. $-\frac{7}{10}$
31. $\frac{7}{3}$
32. $\frac{2}{3}$
33. $\frac{76}{75}$
34. $\frac{-269}{180}$
35. $-\frac{7}{48}$
36. $\frac{42}{5}$
37. $-\frac{2}{29}$
38. c
39. a
40. b
41. b
42. c
43. c
44. c
45. b
46. e

Problem Set 1.7, Page 28

1. 0.40, $\frac{2}{5}$
2. 0.025, $\frac{1}{40}$
3. 0.005, $\frac{1}{200}$
4. 0.01, $\frac{1}{100}$
5. 0.0004, $\frac{1}{2500}$
6. 2.15, $\frac{43}{20}$
7. 5%, $\frac{1}{20}$
8. 15%, $\frac{3}{20}$
9. 130%, $\frac{13}{10}$
10. 50%, 0.5
11. 80%, 0.8
12. 175%, 1.75
13. 1.28
14. 14.85
15. 5
16. 0.1075
17. 300
18. 20
19. 600
20. 30.24

Review Problems (1) Page 30

1. 2 · 2 · 3
2. 2 · 2 · 5
3. 2 · 3 · 5
4. 2 · 2 · 3 · 3
5. 2 · 2 · 2 · 2 · 3
6. 2 · 3 · 3 · 5
7. 2 · 2 · 17
8. 3 · 5 · 13
9. 2 · 2 · 2 · 2 · 3 · 5
10. 2 · 2 · 3 · 7 · 11
11. 17
12. 15
13. 7
14. 9
15. 12
16. 34
17. 49
18. 29
19. 7
20. 16
21. 2
22. 2
23. 1
24. 1
25. 2
26. 1
27. >
28. <
29. <
30. <
31. <
32. >
33. <
34. −7
35. −2
36. −11
37. −32
38. −8
39. −13
40. 21
41. −9
42. 19
43. 54
44. 4
45. −6
46. 16
47. −26
48. −28
49. −8
50. 1
51. −4
52. −7
53. 61
54. $2\frac{2}{5}$
55. $3\frac{2}{5}$
56. $4\frac{1}{4}$
57. $4\frac{1}{6}$
58. $2\frac{5}{11}$
59. $3\frac{10}{13}$
60. $8\frac{7}{15}$
61. $9\frac{6}{31}$
62. 8
63. 20
64. 21
65. 26
66. 48
67. 60
68. $\frac{2}{5}$
69. $\frac{1}{3}$
70. $\frac{4}{9}$
71. $\frac{3}{5}$
72. $\frac{6}{7}$

73. $\frac{4}{5}$ 74. 1 75. $\frac{4}{5}$ 76. 4.24
77. 43.59 78. 8.93 79. 227.65 80. .069
81. 22.615 82. 57 83. 3.5 84. 7.65
85. 12.15 86. .94 87. 3.163 . . . 88. 100
89. 30 90. 18 91. 60 92. 280
93. $\frac{7}{5}$ 94. $\frac{4}{7}$ 95. $\frac{17}{10}$ 96. $\frac{23}{42}$
97. $\frac{91}{60}$ 98. $\frac{3}{4}$ 99. $\frac{3}{5}$ 100. $\frac{16}{45}$
101. $\frac{43}{10}$ 102. $\frac{3}{7}$ 103. .25, $\frac{1}{4}$ 104. .035, $\frac{7}{200}$
105. 45%, $\frac{9}{20}$ 106. 105%, $\frac{21}{20}$ 107. 37.5%, .375 108. c
109. c 110. b 111. c 112. c

Chapter Test (1) Page 35

1. a 2. c 3. c 4. b 5. a 6. c 7. d 8. c 9. a 10. b
11. b 12. c 13. b 14. d 15. b 16. d 17. b 18. c 19. a 20. b

Problem Set 2.1, Page 40

1. $x \cdot x \cdot x \cdot x \cdot x$ 2. $4 \cdot 4 \cdot 4 \cdot 4 \cdot 4$ 3. $(2x)(2x)(2x)(2x)$
4. $2 \cdot x \cdot x \cdot x \cdot x$ 5. $-(4)(4)$ 6. $(-4)(-4)$
7. $-3 \cdot x \cdot x \cdot x \cdot x$ 8. $-(3x)(3x)(3x)(3x)$ 9. $(-3x)(-3x)(-3x)(-3x)$
10. $5(2x)(2x)(2x)$ 11. x,9 12. $-4,3$
13. 4,3 14. x,3 15. x,4
16. 3x,4 17. $-4x,5$ 18. $-3x,5$
19. y,3 20. ab,7 21. a
22. b 23. a 24. b
25. a 26. b 27. a
28. a 29. a 30. b

Problem Set 2.2, Page 45

1. x 2. 5x 3. x 4. x 5. $\frac{4}{x}$
6. x^2 7. y 8. 3x 9. $\frac{2x}{5y}$ 10. z
11. 1 12. 8 13. 4 14. 5 15. 4
16. 5 17. x^7 18. x^{26} 19. $4x^2$ 20. $-128x^7$
21. x^8 22. $27a^{15}$ 23. $16x^{37}$ 24. $-x^{10}$ 25. a^{bc}
26. $8x^{24}$ 27. $(2x+y)^8$ 28. x^2 29. $\frac{1}{x^2}$ 30. $\frac{1}{a^6}$
31. $8x^2$ 32. x^3 33. $\frac{8x^3}{y^3}$ 34. $\frac{4x^4y^6}{9z^2}$ 35. $\frac{1}{2x}$
36. x 37. $-3x^4y^5$ 38. $\frac{125}{16}x^5y^4$ 39. $-\frac{5}{32}yz^{15}$ 40. $\frac{243}{4} \cdot \frac{1}{x^8y}$
41. 96 42. 36 43. 15 44. 192 45. $\frac{2048}{3}$
46. 2^{29} cents = \$5,368,709.12

Problem Set 2.3, Page 49

1. 1 2. 1 3. 1 4. 2 5. 3
6. 7x 7. x^4 8. $2^9 = 512$ 9. 27 10. $\frac{1}{a^5}$

11. $\dfrac{1}{a^5}$ 12. $\dfrac{1}{3a^3}$ 13. $\dfrac{1}{a}$ 14. $\dfrac{x}{y^2}$ 15. 20
16. $\dfrac{1}{10^4}$ 17. $\dfrac{1}{x^6}$ 18. $\dfrac{1}{32x^{15}}$ 19. $\dfrac{x^{15}}{8}$ 20. $\dfrac{1}{625x^8}$
21. $\dfrac{x^8}{9}$ 22. $\dfrac{y^4}{25x^2}$ 23. $\dfrac{y^5}{2x^7}$ 24. $-\dfrac{2}{x^5}$ 25. $\dfrac{c^6}{8a^3b^6}$
26. x^{16} 27. 300 28. 100 29. $\dfrac{2}{10^7}$ 30. $\dfrac{1}{100}$
31. $\dfrac{16x^2}{9y^4}$ 32. -8 33. $\dfrac{2z^3}{x^2y^6}$ 34. $\dfrac{x^3z^6}{2y^3}$ 35. $\dfrac{y^{12}}{x^{12}}$
36. $-\dfrac{x^9}{27y^6}$ 37. d 38. c 39. e 40. c
41. d

Problem Set 2.4, Page 53

1. 7.5×10^1
2. 1.5×10^2
3. 2.97×10^1
4. 7.459×10^3
5. 1.5×10^4
6. 4.0×10^{-1}
7. 4.0×10^{-2}
8. 4.0×10^{-3}
9. 4.2975×10^2
10. 2.3×10^{-2}
11. 7.345×10^0
12. 4.3429×10^{-1}
13. 4.2×10^{-5}
14. 7.934425×10^3
15. 1.2×10^8
16. 2.7×10^9
17. 2.789×10^{12}
18. 1.3×10^{23}
19. 2.52×10^{-12}
20. 7.53×10^{-9}
21. 3.4×10^{-8}
22. 9.005×10^{-5}
23. 15000
24. 453
25. 2200
26. 90030000
27. .019
28. .000084
29. .000000752
30. .0000054
31. 6.3×10^{-4}
32. 1.1×10^{-20}
33. 3.2×10^{-4}
34. 1.05×10^0
35. 2.4×10^{-15}
36. 2×10^{17}
37. 3×10^{21}
38. 5×10^{-24}
39. 3×10^{22}
40. 8×10^{-6}
41. 9.4185×10^{12}
42. 5.2731×10^5
43. 8.49×10^{16}
44. 3.3650602×10^2
45. 1.2750254×10^3
46. 1.9805755×10^{-8}

Problem Set 2.5, Page 58

1. 5,5,25
2. 6,6,36
3. 144,12
4. 49,7
5. xy
6. $2\sqrt{5}$
7. $4\sqrt{2}$
8. $10\sqrt{2}$
9. $2\sqrt{2}$
10. $2\sqrt{3}$
11. $2\sqrt{7}$
12. $3\sqrt{2}$
13. $3\sqrt{3}$
14. $3\sqrt{5}$
15. $5\sqrt{3}$
16. $5\sqrt{5}$
17. $7\sqrt{2}$
18. $y\sqrt{y}$
19. $2x\sqrt{x}$
20. $4a\sqrt{a}$
21. $2a\sqrt{3a}$
22. $2xy\sqrt{2xy}$
23. $2xy^2\sqrt{5x}$
24. xy
25. $\dfrac{\sqrt{15}}{6}$
26. $\dfrac{7\sqrt{5}}{5}$
27. $\dfrac{10\sqrt{3}}{3}$
28. $\dfrac{2\sqrt{6}}{3}$
29. $\dfrac{5\sqrt{2}}{4}$
30. $\dfrac{7\sqrt{3}}{6}$
31. $\dfrac{5\sqrt{6}}{12}$
32. $\dfrac{5x\sqrt{y}}{y}$
33. $\dfrac{2\sqrt{11}}{11}$
34. $\dfrac{2x\sqrt{3y}}{3y}$
35. $\dfrac{5\sqrt{x}}{2x^2}$
36. $\dfrac{\sqrt{35}}{7}$
37. $\dfrac{\sqrt{5ax}}{x}$
38. $\dfrac{x\sqrt{xy}}{y}$
39. $\dfrac{2x\sqrt{3xy}}{3y}$
40. b
41. a
42. b
43. b
44. a
45. b

Problem Set 2.6, Page 61

1. 3
2. 3
3. -9
4. xy
5. x
6. $2x + 4$
7. $6x - 10$
8. $6x^2 - 3x$
9. $x^2 - 10x$
10. $x^2 + 4x$
11. $-2x^3 - x^2$
12. $-x^2 + x - 1$
13. $2a^2 - 5ab$
14. $-4T - 2$
15. $6a^4 - 9a^3 - 7a$
16. $-x^2 + 5x - 2$
17. $x^2 - 5xy - 9$
18. $A^3 + 8A^2B - 9B$
19. $x^2 - 10x - 11y + 16$
20. $x^2 - 9x + 9$
22. $3y^2 - 2y - 7$
23. $-5x$

Problem Set 2.7, Page 64

1. no, $2x + 8$
2. yes
3. no, $-x + 2$
4. yes
5. yes
6. no, $2x^2 - 8x$
7. no, $-x^3$
8. no, $-2x + 6$
9. no, $-2x^2 - 4x$
10. yes
11. $x^2 + 7x - 6$
12. $x^2 + 3x + 2$
13. $x + 6$
14. $-x - 9$
15. $2x - 7$
16. $2x^2 + 5x - 3$
17. $6x^2 - x - 1$
18. $2x^3 - 14x^2 - 10x$
19. $2x^2 - 7x - 2$
20. $-y^2 + 5y - 3$

Problem Set 2.8, Page 67

1. 44
2. 16
3. -30
4. -1
5. 26
6. -1
7. $-\dfrac{2}{5}$
8. -60
9. $\dfrac{477}{40}$
10. -8
11. $\dfrac{11}{2}$
12. 12
13. 0
14. 12.571
15. 25.143
16. 62.857
17. 113.143
18. 0
19. 25
20. 36
21. -11
22. 4
23. 97
24. -8
25. -35
26. 32
27. 68
28. 176
29. $10\dfrac{2}{5}$
30. $24\dfrac{4}{5}$
31. $15\dfrac{4}{5}$
32. 2420
33. 1650
34. 22050
35. 661.25
36. 1331

Review Problems (2) Page 70

1. x
2. 2
3. x
4. -2
5. x
6. $3x$
7. $2x$
8. x
9. x
10. $2x$
11. $(2x)^{10}$
12. $-2x^3$
13. x^4
14. $-y^3$
15. $16x^4$
16. $-27x^3$
17. x^4
18. x^{13}
19. x^7
20. $\dfrac{1}{x^4}$
21. x^6
22. $8x^9$
23. $9x^4y^6$
24. $-x^{10}$
25. $144x^{14}$
26. a^5
27. $-a^3$
28. $\dfrac{y^6}{x^4}$
29. 3
30. $\dfrac{2}{(27x^{11})}$
31. $\dfrac{27}{(8A^2B)}$
32. $\dfrac{s^7}{2t^5}$
33. $\dfrac{2y^7}{x^7z^7}$
34. $\dfrac{27x^9}{64y^{12}}$
35. 1.5×10^5
36. 2.78×10^5
37. 2.7842×10^2
38. 2.0×10^{-2}
39. 4.0×10^{-4}
40. 8.7×10^{-6}
41. 1.375×10^2
42. 3.655×10^{-2}
43. 9.0×10^8
44. 1.0×10^{-27}
45. 3
46. -7
47. y
48. 2
49. $8x^2 - 9x + 9$
50. $-4x - 6$
51. $5x^2 + 2x + 12$
52. $x^2 - 9x + 9$
53. $2x^4 - 4x^2 + 8$
54. $-x^3 - 8x^2 + 3x + 5$
55. $2\sqrt{10}$
56. 5
57. $2\sqrt{6}$
58. $3\sqrt{5}$
59. $\dfrac{7\sqrt{2}}{2}$
60. $\dfrac{\sqrt{10}}{5}$
61. $2x\sqrt{x}$
62. $\dfrac{2\sqrt{x}}{x}$
63. $\dfrac{\sqrt{xy}}{y}$
64. $8x\sqrt{x}$

Chapter Test (1-2) Page 73

1. b
2. c
3. b
4. c
5. d
6. a
7. b
8. d
9. c
10. d
11. a
12. b
13. d
14. d
15. d
16. b
17. d
18. c
19. b
20. b

Problem Set 3.1, Page 77

1. identity
2. conditional equation
3. false statement
4. identity
5. conditional equation
6. 2
7. 1
8. 1
9. 2
10. 2
11. 1
12. 2
13. 3
14. 18
15. 4
16. 2
17. 2
18. 5
19. 5
20. 5
21. 21
22. 2
23. 4
24. ±5
25. ±3
26. −1,3
27. −4
28. ±4
29. $\frac{1}{2}$
30. 2

NOTE: '±' is shorthand for '5 and −5'.

Problem Set 3.2, Page 82

1. 2
2. 4
3. 16
4. 2
5. −1
6. 3
7. 5
8. $\frac{13}{7}$
9. 9
10. 0.4
11. 14
12. $\frac{9}{8}$
13. −7
14. $-\frac{14}{3}$
15. 2
16. 2
17. 1
18. −1.1
19. $\frac{1}{10}$
20. $-\frac{19}{5}$
21. $x = .30 \times 400; 120$
22. $x = 0.20 \times 250; 50$
23. $0.15x = 45; 300$
24. $0.35x = 245; 700$
25. $x = \frac{35}{420} \times 100; 8\frac{1}{3}\%$
26. $x = \frac{250}{1250} \times 100; 20\%$
27. $x = \frac{15 - 25}{25} \times 100; -40\%$
28. $x = \frac{50 - 30}{30} \times 100; 66\frac{2}{3}\%$
29. $0.70 \times x = 40; \$57.14$
30. $1004x = 312; 300$

Problem Set 3.3, Page 87

2. $-\frac{16}{3}$
3. $\frac{9}{5}$
4. −9
5. −1
6. 2
7. 7
8. $\frac{21}{10}$
9. $\frac{29}{7}$
10. none
11. $\frac{15}{4}$
12. −9
13. $\frac{9}{2}$
14. $\frac{24}{5}$
15. $-\frac{3}{2}$
16. $\frac{5}{14}$
17. $-\frac{3}{2}$
18. $\frac{5}{2}$
19. $\frac{5}{12}$
20. 8
21. 3
22. 1
23. 9
24. 3
25. 2.4

Problem Set 3.4, Page 91

2. $\frac{7}{2}$
3. 1
4. −11
5. $\frac{17}{5}$
6. $\frac{59}{35}$
7. $\frac{18}{7}$
8. $\frac{18}{23}$
9. 1
10. −1
11. −6
12. −4
13. 5
14. 8
15. −2

Problem Set 3.5, Page 96

1. greater than or equal to
2. greater than
3. less than or equal to
4. less than
5. an infinite
6. true
7. true
8. false
9. $x \geq \frac{1}{2}$
10. $x \geq 3$
11. $x \leq 2$
12. $x \leq -2$
13. $x > -2$
14. $x \geq -\frac{5}{2}$
15. $x \geq -8$
16. $y < \frac{9}{2}$
17. $x < -2$
18. $x < 1$
19. $x > -1$
20. $x < 3$
21. $x \geq 3$
22. $x < -2$
23. $x \geq -\frac{1}{6}$
24. $x < -6$
25. $x \leq -2$

Review Problems (3) Page 98

1. identity
2. equation
3. equation
4. one
5. two
6. is
7. is
8. 3
9. 18
10. 6
11. -1
12. -14
13. $\frac{2}{3}$
14. 20
15. 50
16. $\frac{10}{3}$
17. $\frac{29}{56}$
18. 500
19. 5%
20. 20%
21. $58.33
22. 1
23. $-\frac{13}{3}$
24. $-\frac{1}{7}$
25. $-\frac{1}{2}$
26. $\frac{5}{14}$
27. $\frac{2}{7}$
28. $\frac{17}{24}$
29. $\frac{23}{4}$
30. $-\frac{34}{23}$
31. $x \leq 2$
32. $x \leq 4$
33. $x \geq 14$
34. $x > 7$
35. $x > -2$
36. $x \leq \frac{1}{7}$
37. $x > 3$
38. $x \leq \frac{51}{2}$

Chapter Test (1–3) Page 101

1. c
2. a
3. d
4. c
5. b
6. c
7. a
8. d
9. b
10. d
11. a
12. a
13. e
14. b
15. d
16. b
17. b
18. a
19. d
20. b

Problem Set 4.1, Page 105

1. $2x + 3$
2. $x - 17$
3. $x - 12$
4. $x + 7$
5. $5x$
6. $3x + 2$
7. $2x + 9$
8. $\frac{x}{7}$
9. $\frac{2}{5}x + 7$
10. $\frac{1}{3}(x + 2)$
11. $\frac{3}{8}(2x - 5)$
12. $2x + 3$
13. $x + 5; 2x + 8$
14. $1.06x$
15. $0.6x$
16. $1.04x$
17. $22x$ cents
18. $25x$ cents
19. $4x$ feet
20. $x(x + 5)$ sq. ft.
21. $5x$ miles
22. $\frac{x}{15}$ hours
23. $50t$ miles
24. $.95x$ lbs.
25. $.65(x - 4)$ gals.

Problem Set 4.2, Page 109

1. 9,21
2. 1st number: x   2nd number: $x - 5$
   Eqn: $2x - 5 = 19$; $2x = 24$; $x = 12$
   The numbers are: 12,7
3. 1st number: x   2nd number: 4x
   Eqn: $x + 4x = 35$; $5x = 35$; $x = 7$
   The numbers are: 7,28
4. 1st number: x   2nd number: $\frac{1}{2}x$
   Eqn: $x + \frac{1}{2}x = 51$; $\frac{3}{2}x = 51$; $x = 34$
   The numbers are: 34,17
5. 30,31
6. 21,23
7. 13,10,8
8. $2x - 1, x, x + 7$; $N_1 + N_2 + N_3 = 26$
   Eqn: $4x + 6 = 26$
   The numbers are: 9,5,12
9. $2x, x, 6x$; $N_1 + N_2 + N_3 = 45$
   Eqn: $9x = 45$; $x = 5$
   The numbers are: 10,5,30
10. $x, 2x, x - 4$; $N_1 + N_2 + N_3 = 16$
    Eqn: $4x - 4 = 16$; $4x = 20$; $x = 5$
    The numbers are: 5,10,1
11. Dan = 12 years, Mike = 8 years
12. now Bob is 12 years old   Rick is 16 years old
13. Sister: 14   Brother: 10
14. Linda: 13   Jane: 5
15. $P_{77} = 41$¢ per gallon
16. Actual Price = $31.00
17. Marked Price = $8.00
18. His deposit was: $2000
19. $N = 10, D = 12, Q = 5$
20. $x, 3x, x + 1; N + D + Q = 21$ $N = 4; D = 12; Q = 5$
    Eqn: $x + 3x + x + 1 = 21$; $5x = 20$; $x = 4$
21. $x, x + 2, x - 1; N + D + Q = 31$ $N = 10; D = 12; Q = 9$
    Eqn: $x + x + 2 + x - 1 = 31$; $3x = 30$; $x = 10$
22. $x + 35, x, x - 21, x - 6; P + N + D + Q = 104$ $P = 59; N = 24; D = 3; Q = 18$
    Eqn: $x + 35 + x + x - 21 + x - 6 = 104$; $4x = 96$; $x = 24$
23. 1 hr and 12 min
24. 2.55 hrs (2 hrs and 33 min)

Problem Set 4.3, Page 116

1. 50 miles per hour
2. $0.25 per lb.
3. $0.45 per lb
4. $\frac{15}{4} = \$3.75$ per lb.
5. $\frac{140}{200} = \frac{7}{10}$
6. $\frac{11}{2}$
7. 10
8. 10
9. 21
10. $\frac{45}{2}$
11. $7.70
12. 9 hrs
13. 75 miles
14. $1\frac{3}{4}$ inches
15. $5\frac{1}{2}$ hrs
16. $55.25
17. $33\frac{1}{3}$ pounds
18. $4.27
19. 9 feet
20. 7.2
21. 360
22. 60%
23. 250
24. 4%

Problem Set 4.4, Page 120

1. Beef = 3 pounds; Pork = 1 pound
2. Pink = 5 gallons; White = 10 gallons
3. Orange = 406.15 ounces; Pineapple = 73.85 ounces
4. Cashew nuts = 2 pounds; Raisins = 2 pounds
5. D = 8; N = 14; Q = 12
6. P = 15; N = 30; D = 33; Q = 11
7. at 10% = $10,000; at 8% = $20,000
8. at 20% = $4000; at −4% = $11,000
9. $3000
10. 10 gms
11. 2.5 gms
12. $\frac{8}{3}$ gms

Problem Set 4.5, Page 125

1. Distance = 207 miles
2. Speed = 52.5 miles per hour
3. Time = 12 hours
4. Time = 6 hours
5. 504 miles
6. 8 mph
7. 3 hours
8. 365 miles
9. 1 hour 20 minutes
10. 30 minutes

Problem Set 4.6, Page 128

1. Length = 28 feet; Width = 14 feet
2. Length = 27 feet; Width = 20 feet
3. Length = 16 feet; Width = 11 feet
4. Length = 10 meters; Width = 6 meters
5. Length = 6 feet; Width = 4 feet
6. Length = 13 feet; Width = 5 feet
7. Length = 20 feet; Width = 13 feet
8. Perimeter = 60 feet
9. Area = 319 sq. ft.
10. Length = 84 feet; Width = 42 feet

Review Problems (4) Page 132

1. Length = 130 feet; width = 70 feet
2. −24
3. Length = 8 cm; width = 6 cm
4. 14,15,16
5. $50.00
6. $40.00
7. $3.90
8. $5\frac{5}{6}$ hours
9. at 8% = $1000; at 9% = $3000
10. Sally = 33 years; Margie = 28 years
11. N = 7; D = 8; Q = 3
12. N = 11; D = 11; Q = 6
13. Pork = 2 pounds; Beef = 8 pounds
14. 7 hours
15. 420 miles

Chapter Test (1–4) Page 134

1. d
2. a
3. b
4. a
5. a
6. e
7. a
8. b
9. a
10. a
11. d
12. d
13. a
14. d
15. a

Problem Set 5.1, Page 139

1. three
2. two
3. one
4. -7
5. literal
6. $3x^2 - 5x + 2$
7. $-x^2 + 5x - 16$
8. $3x^2 - 10x - 2$
9. $-x^3 - 9x + 16$
10. $x^2y - 3x^3 - 5x^2$
11. $2x^2 - 7xy + 2 - 4x$
12. $-3x + 9$
13. $-2x^2 + 12x - 7$
14. $3x - 6$
15. $2x^2 - 4x$
16. $6x^2 - 15x$
17. $3x^3 - 7x^2$
18. $2x^2 - 14x + 10$
19. $2x^4 - x^3y + 3x^2y^2$
20. $15x^7 - 21x^6 - 24x^5 - 12x^4$
21. $x^2 - 4x + 6$
22. $2x^2 - 9x + 9$
23. $5x - 5$
24. $2x^2 - 17x - 28$
25. $3x^3 - 7x^2 - 17x$

Problem Set 5.2, Page 143

1. $12x^3$
2. $4x^7y^3$
3. $-6x^5y^3$
4. $6x^2y$
5. $2x^2 - 3x$
6. $8x - 4x^2$
7. $2 - x^2 + 3x$
8. $-5x + 10$
9. $2x^2 - 11x + 12$
10. $x^2 - 2x - 15$
11. $2x^2 - 3xy + y^2$
12. $-10a^2 + 7ab - b^2$
13. $8x^2 + 2xy^2 - 15y^2$
14. $6x^2 - 19xy + 15y^2$
15. $-8x^2 + 30xy - 7y^2$
16. $10x^2 - 9xy - 7y^2$
17. $2x^3 - 7x^2 - 4x + 15$
18. $8x^3 - 22x^2y + 23xy^2 - 10y^3$
19. $2a^2b - 6ab + 2b^2 - 5a^2 + 15a - 5b$
20. $x^4 - x^3 + x^2 + 3x$
21. $2y^4 - 5y^3 + 7y - 20$
22. $2x^4 + 2x^3 - 7x^2 - 5x + 5$
23. $3x^4 - 4x^3 + 7x^2 - 16x - 20$
24. $2x^4 + x^3 - 6x^2 + 28x - 16$
25. $3x^4 - 11x^3y + 11x^2y^2 - 3xy^3 - 4y^4$

Problem Set 5.3, Page 145

1. x
2. x^2
3. $\dfrac{1}{y^3}$
4. $-x$
5. $-\dfrac{1}{x}$
6. $-\dfrac{2}{x^3}$
7. $3y^2$
8. $-\dfrac{2}{x}$
9. $-b^2$
10. $\dfrac{-21}{x}$
11. $\dfrac{1}{(x-2)^2}$
12. $\dfrac{x}{4}$
13. c
14. c
15. $x + 2$
16. $4 - \dfrac{x}{3}$
17. $1 - \dfrac{y}{x}$
18. $1 - 2x$
19. $3x^2 - 2x$
20. $\dfrac{x}{3} - \dfrac{2}{3}$
21. $\dfrac{3}{2} - \dfrac{5}{2}ab^2$
22. $2 - x + 3x^2$
23. $x^2 - \dfrac{2}{5}x + \dfrac{2}{x}$
24. $\dfrac{2}{7}x - \dfrac{3}{7}y + \dfrac{4}{7}xy$
25. $-6r + 2t^3 - \dfrac{t}{r}$

Problem Set 5.4, Page 149

	Quotient	Remainder		Quotient	Remainder
1.	$x^2 + x + 1$	0	2.	$x + 1$	2
3.	$x + 2$	-3	4.	$x - 1$	0
5.	$2x + 3$	14	6.	$-x - 2$	-5
7.	$4x^2 + 3x - \dfrac{19}{3}$	$\dfrac{17}{3}$	8.	$2x^2 - 1$	0
9.	$x^4 + x^3 + x^2 + x + 1$	0	10.	$2x^2 - 8x + 35$	-141

Review Problems (5) Page 151

1. $3x^2 - 5x + 1$
2. $3x^3 - 11x^2 - 11x + 2$
3. $5x^2 - 2x + 6$
4. $x^3 - 4x^2 + 6x - 12$
5. -12
6. $-3x^2 - 2x$
7. $4x - 14$
8. $-14 + 2x$
9. $16x^2 - 24x$
10. $-6x + 2x^2$
11. $6x^4 - 8x^3 + 10x^2$
12. $4x - 11$

13. $2x^2 - 11x + 12$
14. $3x^4 - 22x^3 + 12x^2 + 4x$
15. $2x^2 + 3x + 1$
16. $10x^2 - 11x + 3$
17. $-12x^2 + 23x - 10$
18. $2x^3 - 6x^2 - 10x + 20$
19. $6x^3 + 8x^2 - 23x + 10$
20. $12x^4 - 13x^3 - 14x^2 - 20x + 35$
21. $15y^4 - 24y^3 + 59y^2 - 56y + 56$
22. $10a^4 - 35a^3 + 59a^2 - 49a + 63$
23. $4x^3 - 2x^2 - 16x + 8$
24. $4x^4 - 11x^3 + 15x^2 - 13x + 5$
25. $-5x^5 + 13x^4 - 25x^3 + 19x^2 - 16x + 6$
26. $-2x$
27. $-3x^2y$
28. $-5x^2y^2z^7$
29. $x + 2$
30. $x - 2$
31. $\dfrac{x^2}{6} + \dfrac{2}{3}x - \dfrac{2}{x}$
32. $\dfrac{1}{y} - \dfrac{3}{2}x + \dfrac{3y^2}{x}$
33. $x - 3,0$
34. $4x^2 + 6x + 17,31$
35. $8x^3 - 12x^2 + 6x - 1,0$

Chapter Test (1–5) Page 153

1. a 2. a 3. b 4. c 5. d 6. a 7. c 8. a 9. d 10. b
11. a 12. b 13. b 14. d 15. e 16. b 17. c 18. a 19. c 20. b

Problem Set 6.1, Page 157

1. trinomial
2. two, xy
3. four, -7
4. $-5, y^2$
5. x^2

Problem Set 6.2, Page 159

1. $x^2 + 4xy + 4y^2$
2. $4x^2 + 4xy + y^2$
3. $9 + 12x + 4x^2$
4. $x^4 + 2x^2 + 1$
5. $x^2 - 2x + 1$
6. $4x^2 - 4x + 1$
7. $4x^4 - 16x^2 + 16$
8. $4x^2 - 12xy + 9y^2$
9. $x^2 - 9$
10. $16x^2 - y^2$
11. $4x^2 - 9$
12. $9x^4 - y^4$

Problem Set 6.3, Page 161

1. $3x^2y(2 - 3xy)$
2. $6(x - 4)$
3. $2(y^2 + 2)$
4. $2y(x - 2y)$
5. $xy(x + y)$
6. $xy(y^2 + x^2)$
7. $7y(3y - 2x)$
8. $4xy(x + 3y)$
9. $3x^2(1 - 2y)$
10. $3x(x^2 - 4y^2 - 3x)$
11. $7x^3(2x + 3)$
12. $xy^2(2x + y)$
13. $5a^9b^4(3b + 5a^{11})$
14. $12x^5(1 - 3x^2)$
15. $7x^3(2x^2 - 3x + 5)$
16. $(x + 1)(x - 2)$
17. $(2x - 3)(4x - 5)$
18. $(x - 7)(2x - y)$
19. $2(2 - x)(3x - 5)$
20. $(x + y)^2(x + y + 3)$
21. $(x - 4)(21x - 76)$
22. $(2x^2 + y^2)(3x - 7)$
23. $5x(y + 3)(3y^2 + 4)$
24. $(3x - 5y)(x - 3y)$
25. $(2x^2 + y)(2x - 7y)$
26. $(x + 3y)(2x - y)$
27. $(4x - 3)(2 - y)$

Problem Set 6.4, Page 163

1. $2(x + 3)(x - 3)$
2. $(y - 4)(y + 4)$
3. $3(a + 2)(a - 2)$
4. $(3x + 2)(3x - 2)$
5. $(8 - x)(8 + x)$
6. $x(x + 1)(x - 1)$
7. $x^2(x + 1)(x - 1)$
8. $(2x - 3y)(2x + 3y)$
9. $(5 - 4x)(5 + 4x)$
10. $(2 - xy)(2 + xy)$
11. $(13 - y)(13 + y)$
12. $4(x - 3y)(x + 3y)$
13. $a(b - a)(b + a)$
14. $30xy(y - 2x)(y + 2x)$
15. $(x^2 + a^2)(x + a)(x - a)$
16. $(y^2 + 4x^2)(y + 2x)(y - 2x)$
17. $3(a^2 + 4)(a + 2)(a - 2)$
18. $5x(1 + 4x^2)(1 + 2x)(1 - 2x)$
19. $3x^2(9x^2 + 1)(3x + 1)(3x - 1)$
20. $(5x - y + 2)(5x + y - 2)$
21. $2(x - y - 2)(x - y + 2)$
22. $(2x - y)(2 - 2x + y)(2 + 2x - y)$
23. $(3x - y)(3x - y + 3)(3x - y - 3)$
24. $(3x - 5y + 10)(3x + 5y - 10)$

Problem Set 6.5, Page 164

1. 1,3
2. 1,5
3. −1,−5
4. −2,−3
5. 2,−3
6. −2,3
7. 2,6
8. −2,−6
9. −4,3
10. −12,1
11. −3,−8
12. −2,−12
13. −3,−5
14. −3,−16
15. −24,2

Problem Set 6.6, Page 167

1. $(x + 2)(x + 1)$
2. $(x + 1)(x + 3)$
3. $(x + 2)(x + 3)$
4. $(x − 3)(x + 2)$
5. $(x + 3)(x − 2)$
6. $(x − 4)(x + 2)$
7. $(x − 8)(x + 6)$
8. $(x + 3)(x − 5)$
9. $(x − 3)(x − 5)$
10. $(x − 6)(x − 4)$
11. $(x − 6)(x + 3)$
12. $(h − 6)(h + 4)$
13. $(m + 9)(m + 4)$
14. $(x − 9)(x − 4)$
15. $(x + 7y)(x + 2y)$
16. $(x − 7y)(x − 2y)$
17. $(c + 5d)(c − 4d)$
18. $(x − 8y)(x − 2y)$
19. $(x − 2)(x + 2)$
20. $(x − 3)(x + 3)$
21. $2(x − 2)(x + 2)$
22. $ax(x − 1)$
23. cannot
24. cannot
25. cannot
26. cannot
27. $(x − 3)(x − 2)$
28. cannot

Problem Set 6.7, Page 169

1. $(3x + 2)(2x + 1)$
2. $(5y − 4)(y − 2)$
3. $(2x + 1)(x + 3)$
4. $(4x − 3)(x − 5)$
5. $(2a + 3)(5a − 4)$
6. $(3p − 2)(2p + 5)$
7. $(3a − 2)(2a − 3)$
8. $(6x + 5)(2x − 1)$
9. $(4x + 3)(3x − 4)$
10. $(3x + 4)(2x − 1)$
11. $(6a − 5)(a + 1)$
12. $(2x − 3)(x + 5)$
13. cannot
14. $(9x + 2)(x + 1)$
15. $(4x − 1)(3x − 2)$
16. $(3x + 5)(x − 2)$
17. $(2 − 3h)(2 − 3h)$
18. cannot
19. cannot
20. $(2x + 1)(x + 3)$
21. $(4x + 1)(x + 2)$
22. $(2x − 3)(x + 4)$
23. $(4x − 1)(x + 3)$
24. $(4x + 3)(x − 3)$
25. $(2x + 1)(5x − 6)$
26. $(9x − y)(x − 3y)$
27. $x(6x − 1)(x − 3)$
28. $2(3x − 1)(3x + 2)$
29. $(3 − 2x)(1 + 6x)$
30. $(3 + 2x)(2 − 3x)$
31. $(x + y − 1)(3x + 3y + 5)$
32. $x^2(3x + 1)(x − 5)$
33. $(2x + 1)(4x^2 − 2x + 1)(x − 1)(x^2 + x + 1)$

Problem Set 6.8, Page 172

1. $3(x + 1)(x + 5)$
2. $2(x + 3)(x + 2)$
3. $5(x − 2)(x − 3)$
4. $2(x + 3)(x − 2)$
5. $a(x − 4)(x + 2)$
6. $9(x + 4)(x − 2)$
7. $4(x + 3)(x − 3)$
8. $y(x + 3)(x − 3)$
9. $(x^2 + 4)(x + 2)(x − 2)$
10. $(y + 3)(y − 3)(y^2 + 9)$
11. $2(x − 2)(x + 2)(x^2 + 4)$
12. $4(x − y)(x + y)(x^2 + y^2)$
13. $(x − 2)(x + 2)(x^2 + 2)$
14. $2(y^2 + 3)(y − \sqrt{2})(y + \sqrt{2})$
15. $3(2x + 1)(x + 3)$
16. $2(3x + 4)(2x − 1)$
17. $(x − 8y)(x + 6y)$
18. $(x − 6y)(x + 3y)$
19. $(3x^2 + 1)(x − 3)(x + 3)$
20. $(3x − 1)(3x + 1)(x^2 + 4)$

Problem Set 6.9, Page 173

1. $2x(y − 2x)(y^2 + 2xy + 4x^2)$
2. $(x − 4)(x^2 + 4x + 16)$
3. $(y − 5)(y^2 + 5y + 25)$
4. $2(x + 2)(x^2 − 2x + 4)$
5. $6(x − a)(x^2 + ax + a^2)$
6. $4x(x − 2)(x^2 + 2x + 4)$
7. $y(x − y)(x^2 + xy + y^2)$
8. $5(y − 2)(y^2 + 2y + 4)$
9. $(2x − 3)(4x^2 + 6x + 9)$
10. $y(x + y)(x^2 − xy + y^2)$
11. $2(2 − y)(4 + 2y + y^2)$
12. $2(y + 2)(y^2 − 2y + 4)$
13. $(3x − 1)(9x^2 + 3x + 1)$
14. $2x(3x − 1)(9x^2 + 3x + 1)$
15. $(x − 1)(x + 1)(x^2 + x + 1)(x^2 − x + 1)$
16. $(2 − x)(2 + x)(4 + 2x + x^2)(4 − 2x + x^2)$
17. $(x − 2y − 2)(x^2 − xy + y^2 + 2x + 2y + 4)$
18. $2(5x − 1)(25x^2 + 5x + 1)$
19. $x(x − y + 1)(x^2 + xy + y^2 − x − 2y + 1)$
20. $(3x − 2y − 1)(9x^2 + 6xy + 4y^2 + 3x + 4y + 1)$
21. $x^2(y − x)(y + x)(y^2 + xy + x^2)(y^2 − xy + x^2)$
22. $8x^3 − y^3$
23. $x^3 + y^3$
24. $8x^3 − 27$

Problem Set 6.10, Page 176

1. 9
2. 1
3. 1
4. $\frac{1}{4}$
5. 4
6. $\frac{9}{4}$
7. $\frac{9}{16}$
8. 5
9. $(x - 3 + \sqrt{13})(x - 3 - \sqrt{13})$
10. $(x - 1)(x + 3)$
11. $\left(x + \frac{3 + \sqrt{5}}{2}\right)\left(x + \frac{3 - \sqrt{5}}{2}\right)$
12. $(x + 4 + \sqrt{17})(x + 4 - \sqrt{17})$
13. $2\left(x + \frac{3}{2} - \frac{\sqrt{23}}{2}\right)\left(x + \frac{3}{2} + \frac{\sqrt{23}}{2}\right)$
14. $(3y - 2)(y + 1)$
15. $\left(y - \frac{5}{2} - \frac{\sqrt{5}}{2}\right)\left(y - \frac{5}{2} + \frac{\sqrt{5}}{2}\right)$
16. $(x + 2 - \sqrt{2})(x + 2 + \sqrt{2})$
17. $2\left(z + \frac{3}{4} - \frac{\sqrt{17}}{4}\right)\left(z + \frac{3}{4} + \frac{\sqrt{17}}{4}\right)$
18. $4\left(z + \frac{5}{8} - \frac{\sqrt{41}}{8}\right)\left(z + \frac{5}{8} + \frac{\sqrt{41}}{8}\right)$
19. $(z - 3 - \sqrt{7})(z - 3 + \sqrt{7})$
20. $2\left(b + 2 - \sqrt{\frac{7}{2}}\right)\left(b + 2 + \sqrt{\frac{7}{2}}\right)$

Review Problems (6) Page 178

1. four
2. xy
3. -3
4. 2xy
5. 2xy
6. $4x^2 - 4xy$
7. $x^2 - y^2$
8. $(x - y)y^2$
9. $(x - y)(x + y)$
10. false
11. false
12. 2,1
13. 3,2
14. 2,3
15. 6,12
16. $10(x - 3)$
17. $3(5x - 6)$
18. $a(y + b)$
19. $2(x - 1)^2$
20. $xy(x - y)(x + y)$
21. $2x^2(2x - 1)$
22. $4(x^2 + x + 2)$
23. $(2x + 5)(3x - 4)$
24. $-15,1$
25. $-5,3$
26. $-5,-5$
27. $-2,-10$
28. $-10,2$
29. $-9,3$
30. $(3x + 4)(2x - 1)$
31. $(x - 4)(x + 3)$
32. $(x - 6)(x + 3)$
33. $(x - 4)(x + 4)$
34. $(y + 5)(y - 3)$
35. $(A - 9)(A + 3)$
36. $(x + 10)(x - 2)$
37. $(2x - 1)(x - 3)$
38. $(7y + 2)(y + 1)$
39. $2(x - 1)(x + 3)$
40. $(2x - 1)^2$
41. $(4x + 1)^2$
42. $(5x - 1)(x + 3)$
43. $(4x - 7)(x + 1)$
44. $2(x - 3)(6x + 1)$
45. $(8y - 1)(2y - 3)$
46. $2(x - 2)(x + 2)(x^2 + 4)$
47. $(4x - 7y)(x + y)$
48. $3(4x - 7)(x + 1)$
49. $(x - 1)(x + 1)(x - 2)(x + 2)$
50. $(x - y)(x^2 + xy + y^2)$
51. $(x + 3)(x^2 - 3x + 9)$
52. $2x(x - 2)(x^2 + 2x + 4)$
53. $2(2x - 3y)(4x^2 + 6xy + 9y^2)$
54. $3(2 - a)(4 + 2a + a^2)$
55. a
56. d
57. a
58. b
59. b
60. d
61. c
62. b

Chapter Test (1–6) Page 182

1. b
2. e
3. c
4. d
5. d
6. e
7. a
8. d
9. b
10. a
11. a
12. d
13. c
14. c
15. b
16. d
17. b
18. c
19. b
20. a

Problem Set 7.1, Page 185

2. quadratic, two
3. quadratic, two
4. quadratic, two
5. quadratic, two
6. 2,7
7. $\frac{1}{2}, 5$
8. $-\frac{1}{3}, \frac{5}{2}$
9. $2, -\frac{4}{3}$
10. $\frac{5}{2}, -\frac{2}{3}$

Problem Set 7.2, Page 188

1. $2, 1$
2. $0, -4$
3. $0, -\frac{3}{2}$
4. $0, \frac{5}{4}$
5. $0, a$
6. $0, \frac{c}{b}$
7. $2, -2$
8. $4, -4$
9. $2, -2$
10. $4, -4$
11. $-2, -3$
12. $6, -1$
13. $2, 4$
14. $5, -3$
15. $2, 7$
16. $1, -5$
17. $2, -2$
18. $1, 6$
19. $5, \frac{1}{3}$
20. $-\frac{1}{2}, -\frac{3}{2}$
21. $\frac{1}{5}, -\frac{2}{3}$
22. $0, \frac{5}{4}$
23. $6, -3$
24. $\frac{2}{3}, \frac{3}{2}$
25. $\frac{3}{4}, -\frac{1}{3}$
26. $0, \frac{8}{3}$
27. $2, -2$
28. $1, 1$
29. $\frac{1}{2}, \frac{1}{2}$
30. $\frac{5}{3}, -\frac{3}{2}$

Problem Set 7.3, Page 192

1. b
2. a
3. c
4. d
5. b
6. c
7. b
8. c
9. b
10. c
11. $2, -1$
12. $6, -6$
13. $0, 4$
14. $-\frac{3}{2}, \frac{1}{3}$
15. $4, -4$
16. $-2, 2$
17. $-5, 1$
18. $-6, -\frac{2}{3}$
19. $\pm\sqrt{2}$
20. $-\frac{5}{2}, \frac{1}{3}$
21. $1, -3$
22. $\frac{5}{3}, -1$
23. $1, 4$
24. $5, -5$
25. $0, -\frac{1}{2}, 3$

Problem Set 7.4, Page 198

1. $a = 2, b = -3, c = 1$
2. $a = 3, b = 3, c = -5$
3. $b^2 - 4ac$
4. repeated solutions
5. distinct rational solutions
6. real irrational solutions
10. $b^2 - 4ac = 0$; repeated solutions
11. $b^2 - 4ac = 0$; repeated solutions
12. $b^2 - 4ac = 28$; irrational numbers
13. $b^2 - 4ac = 121$; rational numbers
14. $b^2 - 4ac = 36$; rational numbers
15. $b^2 - 4ac = 60$; irrational numbers
19. $3, 3$
20. $-1, -1$
21. $\frac{-1}{3} \pm \frac{\sqrt{7}}{3}$
22. $-\frac{3}{2}, \frac{1}{3}$
23. $\frac{4}{3}, -1$
24. $-\frac{1}{2} \pm \frac{\sqrt{29}}{2}$
25. $\frac{1}{3}, \frac{1}{3}$
26. $-2 \pm \sqrt{5}$
27. $4 \pm \sqrt{13}$
28. $1, -5$
29. $-1 \pm \frac{\sqrt{15}}{3}$
30. $-1 \pm \frac{\sqrt{10}}{5}$
31. c
32. c
33. d
34. c
35. b

Problem Set 7.5, Page 203

1. $3i$
2. $3i\sqrt{3}$
3. $6i$
4. $2i\sqrt{3}$
5. $2i\sqrt{7}$
6. $-5i$
7. $-4i\sqrt{2}$
8. $5i\sqrt{2}$
9. $-3 - i\sqrt{2}$
10. $2 - i\sqrt{3}$
11. $\frac{3}{2} + \frac{i}{2}$
12. $-\frac{3}{2} - \frac{(i\sqrt{5})}{2}$
13. $4, 5$
14. $-6, 11$
15. $8, 19$
16. $-2, \frac{21}{16}$
17. $-1, 1$
18. $-\frac{3}{2}, \frac{21}{16}$
19. $\frac{5}{4}, \frac{105}{64}$
20. $2, \frac{25}{16}$
22. $1 \pm i$
23. $\frac{3 \pm i\sqrt{7}}{2}$
24. $\pm 2i$
25. $\pm i\frac{\sqrt{6}}{2}$
26. $\frac{1 \pm i\sqrt{11}}{3}$
27. $\frac{1 \pm i\sqrt{3}}{2}$
28. $\frac{3 \pm i\sqrt{15}}{4}$
29. $\frac{-5 \pm \sqrt{33}}{4}$
30. $\frac{-1 \pm \sqrt{10}}{3}$
31. $\frac{1 \pm \sqrt{3}}{2}$
32. $-1 \pm \sqrt{5}$
33. $\frac{5 \pm i\sqrt{23}}{6}$
34. $\frac{5 \pm i\sqrt{15}}{2}$
35. $\frac{1 \pm \sqrt{37}}{12}$
36. $-2 \pm \sqrt{5}$
37. $\frac{21 \pm \sqrt{489}}{12}$
38. $0, -\frac{15}{4}$

Review Problems (7) Page 207

1. quadratic, two
2. $a = 3; b = -5; c = -9$
3. $-4, 3$
4. $\sqrt{-1}$
5. i
6. $2 + 3i$
7. $3, -4$
8. $-2, 2$
9. $0, \dfrac{5}{2}$
10. $-1, 8$
11. $2, -4$
12. $4, \dfrac{1}{2}$
13. $3, \dfrac{1}{2}$
14. $\dfrac{4}{3}, -\dfrac{3}{4}$
15. $\dfrac{3}{2}, -\dfrac{3}{4}$
16. $1, 2, 1; 0;$ repeated
17. $1, -5, 6; 1;$ rational
18. $1, -7, 12; 1;$ rational
19. $1, 2, -1; 8;$ irrational
20. $2, 1, 1; -7;$ complex
21. $-4, 1$
22. $\dfrac{3}{2}, 2$
23. $0, -\dfrac{5}{3}$
24. $\dfrac{6}{5}, \dfrac{4}{5}$
25. $-4, -1$
26. $3, -2$
27. $6, -4$
28. $\dfrac{1}{2}, 4$
29. $-\dfrac{1}{3}, -\dfrac{3}{2}$
30. $1, -\dfrac{3}{4}$
31. $2, -2$
32. $-1, -1$
33. $6, -3$
34. $-1 + \sqrt{3}, -1 - \sqrt{3}$
35. $-2 + \sqrt{5}, -2 - \sqrt{5}$
36. $i\sqrt{2}$
37. $2i$
38. $2i\sqrt{7}$
39. $3i\sqrt{2}$
40. $3i\sqrt{3}$
41. $\dfrac{-1 + i\sqrt{3}}{2}, \dfrac{-1 - i\sqrt{3}}{2}$
42. $\dfrac{1}{2}, -1$
43. $\dfrac{2 + 2i\sqrt{2}}{3}, \dfrac{2 - 2i\sqrt{2}}{3}$
44. c
45. c
46. d
47. a
48. b
49. e
50. a
51. d
52. b
53. c
54. b
55. b
56. d
57. c
58. b
59. b
60. b
61. b
62. a
63. d
64. a

Chapter Test (1–7) Page 212

1. c
2. c
3. d
4. a
5. d
6. b
7. b
8. c
9. c
10. c
11. d
12. c
13. b
14. c
15. c
16. b
17. b
18. c
19. a
20. d
21. a

Problem Set 8.1, Page 217

1. 0
2. 1
3. -1
4. $0, -1$
5. ± 1
6. $1, 2$
7. $\dfrac{3}{4}$
8. $-\dfrac{8}{9}$
9. $\dfrac{4}{9}$
10. $-\dfrac{x}{y}$
11. $\dfrac{(x + 2)}{x}$
12. $1 - x$
13. $\dfrac{(x + 1)}{(x - 1)}$
14. 1
15. 1
16. -1
17. $\dfrac{2}{3}$
18. $\dfrac{x}{x - 1}$
19. $\dfrac{xz}{y}$
20. $\dfrac{xy^3}{z^2}$
21. $\dfrac{6}{5} xy^2 z^3$
22. $-y^2$
23. $-2xz$
24. $-5\dfrac{y}{x}$
25. $3x$
26. $x + 1$
27. $\dfrac{2(x + 2)}{x}$
28. $\dfrac{x^2}{x - 4}$
29. $\dfrac{10x(x - 4)}{x^2 - 4}$
30. a
31. $\dfrac{xy}{x + 1}$
32. $\dfrac{b - c}{5}$
33. $\dfrac{2a - 1}{4}$
34. $\dfrac{x}{2(x - 1)}$
35. -2
36. $\dfrac{5x}{2}$
37. $\dfrac{x^2 + 3x + 9}{2}$
38. $-x$
39. $-y$
40. $\dfrac{2x + 3}{x + 2}$
41. $\dfrac{x + 2}{x + 3}$
42. $\dfrac{-2(x + 1)}{x + 2}$
43. $-15x(x + 1)$
44. $\dfrac{-(a + b)}{5}$
45. $-(x^2 + 3x + 9)$
46. $\dfrac{2x + 5y}{3x + 5y}$
47. a
48. c
49. d

Problem Set 8.2, Page 220

1. $\dfrac{y}{x}$
2. $\dfrac{ac}{b}$
3. $\dfrac{1}{x}$
4. 1
5. $2(x-1)$
6. $\dfrac{5}{x}$
7. 1
8. $\dfrac{(a+b)}{(a-b)}$
9. $\dfrac{(x+1)(x+2)}{(x+3)(x+5)}$
10. $\dfrac{x-4}{2(x+3)}$
11. $\dfrac{5(4x-1)}{4(3x-4)}$
12. $\dfrac{x+1}{3x(x-6)}$
13. $\dfrac{x+1}{x+4}$
14. $\dfrac{xy^2(x-1)}{x-2}$
15. $a+1$
16. $\dfrac{x}{x-3}$
17. $\dfrac{1}{y(x+4)}$
18. $\dfrac{x}{x+5}$
19. $\dfrac{(2b+3c)}{(b-c)}$
20. $\dfrac{5}{x}$
21. $\dfrac{1}{8}$
22. $\dfrac{3xy}{4}$
23. $\dfrac{1}{18}$
24. $\dfrac{2x^2}{(x+1)}$
25. $-\dfrac{9x}{2}$
26. $\dfrac{(x+1)}{(3x-1)}$
27. $\dfrac{(3x-1)}{(x+1)}$
28. $\dfrac{16(x-5)(x+4)}{5(x+5)}$
29. $-\dfrac{9}{[(3x+2)(x+2)]}$

Problem Set 8.3, Page 227

1. $x^2(x-1)(x-2)$
2. 45
3. $3x^2$
4. $(x-1)(x-3)$
5. $x^2(x-1)$
6. $2(x+3)(x-3)^2$
7. $(2x+5)(2x+1)(2x-1)$
8. $2x^2+x-1$
9. 20
10. $yx+y$
11. $2x^2-6x$
12. x^3+2x^2-4x-8
13. $x^4+4x^3+5x^2+2x$
14. $4x^2-4x$
16. $\dfrac{(5x+6)}{(x-2)}$
17. $\dfrac{4x}{(7x-1)}$
18. $\dfrac{2x+1}{2x-1}$
19. $\dfrac{1}{x+1}$
20. $\dfrac{2x+1}{2(x-1)}$
21. $\dfrac{7x-1}{x^2}$
22. $\dfrac{2xy+3}{3x^2}$
23. $\dfrac{4y-5x}{xy}$
24. $\dfrac{6-y}{2x}$
25. $\dfrac{9y^2+4x^2}{6xy}$
26. $\dfrac{9x-5}{x(x-1)}$
27. $\dfrac{1}{x-1}$
28. $\dfrac{-2y^2+3y-2}{y^2(y-2)(y+2)}$
29. $\dfrac{6x^2-2x-5}{x(x+1)(x-1)}$
30. $\dfrac{x^2}{(x+2)(x+1)}$
31. $\dfrac{7}{(x-1)}$
32. -1
33. $\dfrac{(2x+5)(x-1)}{x^2}$
34. $\dfrac{2x^2-3x+5}{x-1}$
35. $\dfrac{-2x^2-4x+1}{(x+1)(x+2)(x+5)}$
36. $\dfrac{11x-12}{x^2(x-3)^2}$
37. $\dfrac{-2x^3+6x^2+7x+3}{x^2(x+1)(x+2)}$
38. $\dfrac{2x^2+xy+y^2}{(x-y)(x+y)^2}$
39. 1

Problem Set 8.4, Page 233

1. $\dfrac{1}{x}$
2. $\dfrac{11}{6}$
3. $\dfrac{(x+1)}{x}$
4. $\dfrac{2}{(5-x)}$
5. $\dfrac{x^2+2x-2}{2x-1}$
6. $\dfrac{3x}{3x+1}$
7. $-x$
8. $x+2$
9. $\dfrac{x-y}{x+y}$
10. $\dfrac{x+2}{x^2+10x+24}$
11. $\dfrac{(m+3)^2(m-3)(m-2)}{m^5}$
12. $\dfrac{(3x-7)(x+1)}{3x^2-1}$
13. -1
14. $\dfrac{2}{x+4}$

Problem Set 8.5, Page 237

1. -2
2. $\dfrac{27}{14}$
3. $-\dfrac{2}{3}$
4. 6
5. $\dfrac{26}{11}$
6. 3
7. -2
8. $\dfrac{1}{2}$
9. no solution
10. $1, -\dfrac{2}{3}$
11. $\dfrac{2}{3}$
12. $1 \pm i$
13. -7
14. 13
15. no solution

Problem Set 8.6, Page 240

1. $\dfrac{C}{2\pi}$
2. $\dfrac{5-C}{2}$
3. $\dfrac{E}{c^2}$
4. $A - P$
5. $-\dfrac{b}{a}$
6. $\dfrac{d}{r}$
7. $\dfrac{y-b}{a}$
8. $\dfrac{3x-5}{4}$
9. $\dfrac{4y+5}{3}$
10. $\dfrac{9}{5}C + 32$
11. $\dfrac{A-P}{Pt}$
12. $\dfrac{M-P}{Mt}$
13. $\dfrac{2MN}{5M-2N}$
14. $\dfrac{3y-5}{y-2}$
15. $\dfrac{RR_1}{R_1 - R}$

Review Problems (8) Page 242

1. b
2. a
3. a
4. b
5. a
6. a
7. b
8. b
9. a
10. a
11. $\dfrac{(x-2)}{x}$
12. $\dfrac{3}{8}$
13. $-\dfrac{1}{3}$
14. $\dfrac{4}{7}$
15. 2
16. $2x$
17. -3
18. -4
19. $-2x$
20. $\dfrac{2}{x}$
21. $\dfrac{1}{(x+1)}$
22. $\dfrac{x}{(x+2)}$
23. $\dfrac{(x+2)}{(x+3)}$
24. $\dfrac{a(a+2)}{(a-3)}$
25. $\dfrac{-x}{(x+1)}$
26. 1
27. $\dfrac{(yx)}{2}$
28. $\dfrac{3}{2}x^2 z$
29. $\dfrac{1}{5}$
30. $5x$
31. $x - 1$
32. $\dfrac{x-1}{x-2}$
33. 1
34. $\dfrac{5x}{(x+4)}$
35. $\dfrac{1}{2}\dfrac{(y+2)}{(y-2)}$
36. $\dfrac{2a+3}{a+4}$
37. $\dfrac{8}{3}$
38. $\dfrac{128}{45}$
39. 1
40. $\dfrac{x^3}{y^3}$
41. $\dfrac{x+1}{2x}$
42. $\dfrac{5}{9y}$
43. $\dfrac{3(x-1)}{4}$
44. $\dfrac{1}{84}$
45. 2
46. $\dfrac{5}{3}$
47. $\dfrac{3}{32}$
48. $\dfrac{2(2a-3)(2a+3)}{(a-3)(a-2)}$
49. $\dfrac{22}{15}$
50. $\dfrac{5}{x}$
51. $\dfrac{3x-1}{x+1}$
52. $-\dfrac{2}{x-1}$
53. $\dfrac{2x+5}{2x-3}$
54. $\dfrac{y-2}{x}$
55. $\dfrac{5}{x-3}$
56. $\dfrac{x-2}{x+2}$
57. $-\dfrac{4}{x}$
58. $\dfrac{xy+3}{y^2}$
59. -1
60. $\dfrac{-(x+1)}{(x-1)^3}$
61. $\dfrac{3y-4x}{xy}$
62. $\dfrac{4y-xy-3}{(x-1)(y-1)}$
63. $\dfrac{5x+1}{(x-1)(x+2)}$
64. $\dfrac{-2x+5}{(x+1)(x-1)}$
65. $\dfrac{2(x+3)}{(x-1)(x+2)}$
66. $\dfrac{5x-1}{x-1}$
67. $\dfrac{-2x^2+1}{2x-1}$
68. $\dfrac{4(x^2+5x+3)}{(x-2)(x+2)^2}$
69. $-\dfrac{1}{x+3}$
70. 2
71. -1
72. 0
73. $\dfrac{1}{3}$
74. $\dfrac{2ab-a}{6a-4}$
75. $\dfrac{2y}{y-1}$
76. $\pm\sqrt{a^2-y^2}$
77. $\dfrac{-c-ax}{b}$
78. $b\left(1-\dfrac{x}{a}\right)$
79. $\dfrac{x-ax}{1+a}$
80. $\dfrac{y-b}{x}$

Chapter Test (1–8) Page 246

1. d
2. a
3. c
4. e
5. b
6. c
7. a
8. e
9. a
10. b
11. a
12. b
13. d
14. b
15. c

Problem Set 9.1, page 249

1. $x = 0$
2. x
3. 3
4. $\dfrac{2}{x^2}$
5. -4
6. 4
7. $4x^7$
8. $2x^2$
9. $\dfrac{4}{x^2}$
10. $\dfrac{2}{y^2}$
11. $-36x^6y^6$
12. $\dfrac{3x^7}{2}$
13. $16x^4$
14. $-a^3b^3$
15. $\dfrac{x^{20}}{y^6}$

Problem Set 9.2, Page 252

1. $\sqrt[n]{y}$
2. $\sqrt[n]{a}$
3. Index
4. Radicand
5. Not defined
6. Negative
7. $6x^2$
8. $3x^2$
9. x
*10. $|x|$ or x
11. -3
12. 2
13. $-3x$
*14. $2|x|y^2$ or $2xy^2$
15. -2
16. $-2xy^2$
*17. $2|xy^3|$ or $2xy^3$
18. $-3x^2$
19. $10xy^6$
20. $-5x^7$

Problem Set 9.3, Page 255

1. $\sqrt[5]{x}$
2. $x^{2/3}$
3. $m = 4, n = 5$
4. $m = 3, n = 4$
5. Negative
6. c
7. d
8. b
9. a
10. d
11. $-\dfrac{1}{2}$
12. 5
13. -5
14. $\dfrac{1}{5}$
15. $\dfrac{1}{3}$
16. 2
17. $\dfrac{1}{2}$
18. 2
19. 9
20. -2
21. $x^{7/6}y^3$
22. $x^{1/2}$
23. $x^{3/2}$
24. y^3
25. $x^{5/7}$
26. $x^{5/2}$
27. $x^{(-2/5)}$
28. $x^{(-3/7)}$
29. $\dfrac{1}{(-2)}$
30. $x^{1/4}y^{1/9}$
31. $\dfrac{1}{x^{2/3}}$
32. $x^{4/5}$
33. $x^{7/3}$
34. $x^{17/4}$
35. x^2y^{12}
36. $\dfrac{1}{x^{14/3}}$
37. x
38. $\dfrac{1}{x^6y^8z^6}$
39. $\dfrac{x^6}{y}$
40. $\dfrac{y^6}{x^5}$
41. $\dfrac{y^{1/10}}{x^{1/4}}$
42. $\dfrac{2}{y^{1/2}}$
43. $\dfrac{xz}{y}$

Problem Set 9.4, Page 260

1. $\sqrt[n]{x^2}$
2. n, m
3. 11
4. $\sqrt{7}$
5. $\sqrt[n]{x^p}$
6. b
7. b
8. d
9. b
10. d
11. 3
12. x
13. $|xy|$
14. x^2
15. x^2y^3
16. $\left(\dfrac{x}{y}\right)$
17. $\dfrac{|x|}{y^2}$
18. $2y^4$
19. $-2y^2$
20. $\dfrac{1}{2}$
21. $\sqrt{x}$
22. x^2
23. $\sqrt[3]{x}$
24. $\sqrt{x}$
25. $\sqrt[7]{x^5}$
26. $\sqrt[3]{xy^2}$
27. $x^2\sqrt{3x}$
28. $2\sqrt{5}$
29. $3\sqrt[3]{2}$
30. $x^3 \cdot \sqrt[3]{x^2}$
31. $2x^2 \cdot \sqrt[3]{3}$
32. $x^2y^2\sqrt{x}$
33. $x^4y^3\sqrt[5]{x}$
34. $3yz\sqrt[3]{3z^2}$
35. $\sqrt[3]{xy^2}$
36. $\sqrt{3xyz}$
37. $xy^2\sqrt{xy}$
38. $\dfrac{xz}{y}\sqrt[3]{xyz^2}$
39. $\dfrac{\sqrt{y}}{y}$
40. $\dfrac{x \cdot \sqrt[4]{x^2y}}{y}$
41. $\dfrac{\sqrt[4]{x}}{z}$
42. $\dfrac{\sqrt[3]{xy^2z}}{yz}$
43. $\dfrac{y\sqrt{xy}}{x^5}$
44. $\dfrac{\sqrt[3]{3xy^2}}{y}$
45. $\dfrac{\sqrt[5]{5^4}}{5}$
46. $\dfrac{-2xy^2\sqrt[3]{xz^2}}{z^3}$

Problem Set 9.5, Page 265

1. $4\sqrt{x}$
2. $\sqrt{5}$
3. 0
4. $2x\sqrt{x} + \sqrt{x}$
5. $14\sqrt{2}$
6. $4\sqrt[3]{2}$
7. $3a\sqrt[3]{a}$
8. $(3b-a)\sqrt{b}$
9. 6
10. x
11. $6x^2$
12. $\sqrt[6]{x^3y^2}$
13. $\sqrt[12]{x^8y^3}$
14. $3x\sqrt[6]{2^5x^5y^4}$
15. $\sqrt[12]{a^6b^8c^3}$
16. $-12 - 2\sqrt{2} + 8\sqrt{3} + \sqrt{6}$
17. $-12 + \sqrt{6}$
18. -1
19. $8 - 2\sqrt{15}$
20. $-1 - \sqrt{6}$
21. $\sqrt{3}$
22. $\dfrac{\sqrt{3y}}{y}$
23. $\dfrac{\sqrt[3]{4xy^2}}{2y}$
24. $\dfrac{\sqrt[6]{200}}{2}$
25. $\dfrac{\sqrt{2xy}}{y}$
26. $\dfrac{9 - 3\sqrt{2}}{7}$
27. $\sqrt{5} - 1$
28. $\dfrac{x(\sqrt{y} + 1)}{y - 1}$
29. $7 - 4\sqrt{3}$
30. $\dfrac{7 - 3\sqrt{6}}{5}$

Problem Set 9.6, Page 270

1. $3,4$
2. real part
3. 25
4. $-1,-4$
5. $2,-2$
6. c
7. b
8. b
9. $-10 + 11i$
10. $-3i$
11. $3 + 4i$
12. 6
13. $2\sqrt{3}$
14. $2 + 8i$
15. $-15 - 6i$
16. 10
17. 5
18. $11 + 13i$
19. $-5 + 15i$
20. $3 - 2i$
21. $\dfrac{-1}{2} + \dfrac{1}{2}i$
22. $\dfrac{4 - 2i}{5}$
23. $-i$
24. $\dfrac{4}{13} - \dfrac{7}{13}i$
25. $\dfrac{-2}{5} - \dfrac{11}{5}i$

Review Problems (9) Page 274

1. $\dfrac{3}{x^2}$
2. $\dfrac{3y^3}{x^4}$
3. $-\dfrac{1}{27}$
4. $8x^8$
5. $\dfrac{y^{15}}{x^{10}z^{10}}$
6. $32x$
7. $\dfrac{x^9}{y^4}$
8. $2|x|$
9. x^2
10. $2|x|$
11. -3
12. $-2x$
13. $3x^3$
14. -2
15. $\dfrac{x^{1/2}}{y}$
16. 3
17. 512
18. x
19. $\dfrac{x}{y^2}$
20. $\dfrac{y^2}{2xz^5}$
21. $\dfrac{y^3}{8^{1/4}z^6}$
22. $\dfrac{3^{1/6}x^{1/3}z^{1/2}}{y^{1/3}}$
23. $x^{4/3}$
24. $x^{5/7}$
25. y^2
26. $y^{1/2}x^{1/3}$
27. $x^{7/2}$
28. $x^{7/2}$
29. $x^{(-2/3)}$
30. $xy^{7/2}$
31. x^3
32. x^2
33. $\sqrt{x}$
34. $\sqrt[3]{x^2}$
35. $x\sqrt{x}$
36. $-x^2\sqrt[5]{x}$
37. $\dfrac{x}{y}$
38. $\dfrac{\sqrt{2}}{2}$
39. $\dfrac{\sqrt{y}}{y}$
40. $\dfrac{x\sqrt[3]{x^2}}{y}$
41. $xy^3\sqrt{xy}$
42. $-3xy^2z^3\sqrt[3]{xz}$
43. $\dfrac{\sqrt[3]{z^2y}}{z^2}$
44. $\dfrac{y\sqrt{z}}{z^2}$
45. $-4\sqrt{3}$
46. $4\sqrt{3}$
47. $-2\sqrt{x}$
48. x^2
49. $2x^2$
50. $\sqrt[12]{x^8y^3}$
51. $\dfrac{\sqrt{3}}{3}$
52. $\dfrac{3\sqrt{2}}{2}$
53. $-2 - \sqrt{3}$
54. $\sqrt{5} + \sqrt{2}$
55. $-5 - 2\sqrt{6}$
56. $-3 + 2i$
57. $4 - 4i$
58. $1 - i$
59. $\dfrac{2}{5} - \dfrac{1}{5}i$
60. $\dfrac{3}{4} - i$

Chapter Test (1–9) page 277

1. c
2. a
3. c
4. c
5. b
6. b
7. c
8. b
9. d
10. d
11. b
12. a
13. a
14. c

Problem Set 10.3, Page 299

1. $(3,0)$
2. $(4,7)$
3. $(1,2)$
4. $(-1,-8)$
5. $(2,-6)$
6. $(1,1)$
7. $(5,2)$
8. no solution (parallel)
9. infinitely many solutions (same line)
10. $\left(2, -\dfrac{2}{3}\right)$

Problem Set 10.4, Page 303

1. yes
2. yes
3. yes
4. no
5. no
6. yes
7. $(3,7)$
8. $(4,9)$
9. $\left(\frac{8}{3}, \frac{7}{3}\right)$
10. $\left(\frac{1}{3}, \frac{11}{3}\right)$
11. $\left(\frac{7}{3}, -\frac{2}{3}\right)$
12. $\left(\frac{31}{7}, \frac{-2}{7}\right)$
13. $\left(-\frac{1}{4}, \frac{11}{8}\right)$
14. no solution
15. $\left(\frac{7}{4}, \frac{5}{8}\right)$
16. $\left(2, \frac{1}{4}\right)$
17. $\left(-\frac{1}{4}, -\frac{3}{4}\right)$
18. $\left(\frac{31}{5}, \frac{17}{3}\right)$

Problem Set 10.5, Page 307

1. $(5,-6)$
2. $(2,5)$
3. $(1,3)$
4. $(1,-1)$
5. $\left(\frac{13}{11}, \frac{19}{11}\right)$
6. $\left(\frac{12}{5}, \frac{3}{5}\right)$
7. $\left(\frac{19}{8}, \frac{1}{16}\right)$
8. $(13,9)$
9. $\left(-\frac{7}{26}, \frac{29}{26}\right)$
10. $\left(-6, \frac{1}{2}\right)$
11. no solution
12. $\left(\frac{37}{22}, -\frac{9}{22}\right)$
13. $\left(-\frac{13}{6}, -\frac{16}{9}\right)$
14. $\left(\frac{130}{89}, -\frac{110}{89}\right)$
15. $\left(\frac{37}{21}, \frac{244}{21}\right)$
16. $\left(\frac{57}{26}, \frac{46}{39}\right)$
17. no solution
18. $(-1,1)$
19. infinitely many
20. $\left(\frac{57}{55}, -\frac{10}{11}\right)$

Review Problems (10) Page 309

8. $\left(\frac{1}{2}, \frac{3}{2}\right)$
9. $(3,1)$
10. $(2,1)$
11. $(7,11)$
12. $\left(3, \frac{2}{3}\right)$
13. $\left(\frac{28}{5}, \frac{6}{5}\right)$
14. $(2,1)$
15. $\left(-\frac{44}{7}, -\frac{48}{7}\right)$
16. $\left(-\frac{38}{31}, -\frac{2}{31}\right)$
17. $(-2,-2)$
18. $\left(-\frac{1}{8}, \frac{3}{8}\right)$
19. $\left(\frac{60}{17}, \frac{12}{17}\right)$
20. $\left(\frac{21}{5}, \frac{14}{5}\right)$
21. no solution
22. no solution
23. infinitely many
24. $\left(-\frac{9}{5}, -\frac{2}{5}\right)$
25. $(2,-1)$

Chapter Test (1–10) Page 313

1. d
2. c
3. a
4. e
5. c
6. b
7. a
8. b
9. c
10. a
11. d
12. c
13. d
14. b
15. e

Problem Set 11.1, Page 317

1. positive, real
2. 2
3. 2
4. 3
5. 5
6. 2
7. 6
8. 5
9. $\frac{1}{2}$
10. $\frac{1}{4}$
11. $\frac{1}{2}$
12. $\frac{1}{4}$
13. $\frac{1}{4}$
14. $\frac{1}{9}$
15. $\frac{1}{4}$
16. 8
17. $\frac{1}{8}$
18. 2
19. $\frac{1}{9}$
20. $\frac{1}{9}$
21. 4.711113
22. 4.728805
23. 18.2955
24. 2.394826
25. 15.67551
26. 1669.854
27. 35.01564
28. 26.575997
29. -26.01082
30. 27.1206
31. $\frac{3}{2}$
32. 3
33. 2
34. $-\frac{3}{2}$
35. 2
36. $-\frac{1}{4}$
37. 3
38. $\frac{3}{2}$
39. $\frac{1}{2}$
40. $\frac{9}{2}$

Problem Set 11.2, Page 321

1. 2; 16
2. positive and not equal to; positive
3. 25
4. 0
5. b
6. b
7. e
8. b
9. c
10. a
11. $\log_2 32 = 5$
12. $\log_5 125 = 3$
13. $\log_8 8 = 1$
14. $\log_5 1 = 0$
15. $\log_4\left(\dfrac{1}{16}\right) = -2$
16. $\log_8(2) = \dfrac{1}{3}$
17. $\log_4 \dfrac{1}{2} = -\dfrac{1}{2}$
18. $\log_{4/5} 1 = 0$
19. $\log_{1000} \dfrac{1}{10} = -\dfrac{1}{3}$
20. $\log_{64} 32 = \dfrac{5}{6}$
21. -2
22. 2
23. 3
24. 1
25. -2
26. 4
27. -1
28. $5^2 = 25$
29. $2^2 = 4$
30. $3^4 = x$
31. $25^{1/2} = 5$
32. $2^{-3} = \dfrac{1}{8}$
33. $10^2 = 100$
34. $x^2 = x^2$
35. $36^{1/2} = 6$
36. $5^{-1} = \dfrac{1}{5}$
37. $29^0 = 1$
38. $64^{-1/2} = \dfrac{1}{8}$
39. $\dfrac{17}{2}$
40. $\dfrac{1}{4}$
41. 5
42. 4
43. 1000
44. $\dfrac{1}{4}$
45. 4
46. 0
47. 10
48. 1
49. -2
50. $\dfrac{1}{8}$

Problem Set 11.3, Page 328

1. b
2. b
3. a
4. a
5. a
6. 1
7. 0
8. 4
9. $\dfrac{9}{6}$
10. 2
11. 3
12. $\dfrac{1}{2}$
13. 1
14. $\dfrac{1}{2}$
15. $\dfrac{1}{5}$
16. -2
17. $\dfrac{2}{3}$
18. -8
19. $\dfrac{10}{3}$
20. -2
21. $\dfrac{11}{2}$
22. 10
23. 100
24. 9
25. 10
26. x
27. $\log_5\left(\dfrac{20}{9}\right)$
28. $\log \dfrac{22}{7}$
29. $\log_2 x$
30. $\log(xy^2)$
31. $\log_3(x^4 y^2 z^{-1/2})$
32. $\log_5(x^{13/3} y^2)$
33. $\log_{15} x^{2/3}$
34. $\text{Ln}(y^x x^y)$
35. $\log 1000 = 3$
36. $3 \log x + 2 \log y$
37. $5 \log y + \dfrac{1}{2} \log x$
38. $4 \log x - \log(2x - 1)$
39. $2 \log x + 3 \log y + 4 \log(2x - 1)$
40. $\dfrac{1}{3} \log x + 5 \log y - 2 \log(2x - 1)$
41. $8 \log x + 4 \log y - 4 \log(2x - 1)$
42. $-4 \log x - 8 \log y - 8 \log(2x - 1)$
43. $\dfrac{1}{2}[\log y + \log(2x - 1)]$
44. $3 \log x + \dfrac{1}{2} \log y - \dfrac{1}{2} \log(2x - 1)$

Problem Set 11.4, Page 332

1. $\dfrac{9}{8}$
2. 2
3. $\dfrac{4}{5}$
4. 3
5. $\dfrac{3}{2}$
6. $\dfrac{5}{4}$
7. 20
8. 10
9. 12
10. $\dfrac{4}{3}$
11. $\dfrac{16}{5}$
12. $\dfrac{9}{5}$
13. $\dfrac{125}{2}$
14. $\dfrac{1}{2}$
15. $\dfrac{1}{12}$
16. $4; -3$
17. no solution; -4
18. no solution; -1
19. $2; -\dfrac{1}{2}$
20. $e\sqrt{\dfrac{e}{5}}$

Problem Set 11.5, Page 338

1. 2,3
2. 2,3
3. 3,4
4. 0,1
5. −1,0
6. 1,2
7. 3,4
8. −2,−1
9. −4,−3
10. 0,1
11. 5
12. 2
13. 3
14. 6 or 7
15. −4 or −5
16. 2.4361626
17. 3.678791434
18. 0.445604203
19. −0.319664487
20. 0.138239524
21. −7.6143937
22. 6.862012051
23. 4.304065093
24. 5.645446898
25. −2.617295838
26. 1.025790668
27. 22.130947
28. 286.4177970
29. 0.051286138
30. 56754.46054
31. 0.000331894
32. 3.475362×10^{-9}
33. 11.370247
34. 2.718281828
35. 4.218353×10^3
36. 6.437692×10^{-2}
37. 1.747305×10^5
38. 7.5840636
39. 4.954196310
40. 1.134239374
41. 1.716734736
42. .1054868

Problem Set 11.6, Page 342

1. 2.2297158
2. 1.796631835
3. 0.424495717
4. −7.57938892
5. −6.628291111
6. 104.0516619
7. 1.197802×10^5
8. 0.263148808
9. 2.166434456
10. 19.12262405

Review Problems (11) Page 344

1. $b > 0, b \neq 1, x$ real, $x > 0$
2. $b^x = a$
3. 45
4. 5
5. 7; 9
6. $\dfrac{\log a}{\log b}$
7. $a = b^{\log_b a}$
8. $\dfrac{4}{3}$
9. $-\dfrac{1}{2}$
10. 5
11. 1
12. $4 = \log_3 81$
13. $-2 = \log_5 \dfrac{1}{25}$
14. $0 = \log_{3/5}(1)$
15. $-3 = \log_{1/2}(8)$
16. $3^2 = 9$
17. $2^{-1} = \dfrac{1}{2}$
18. $2x + 1 = 5^3$
19. $x^2 - 1 = a^y$
20. 0
21. $\dfrac{1}{2}$
22. −2
23. −3
24. −3
25. 5
26. 2
27. $\dfrac{1}{2}$
28. $\log_2 125$
29. $\log_3 2$
30. $\log x$
31. $\log_5 \left[\dfrac{x^{1/2} y^2}{(x^2 - 1)} \right]$
32. $2 \log x + \log y$
33. $\log x + \dfrac{1}{2} \log y$
34. $\log x + \log(x^2 + 1) - \log y$
35. $2 \log y + \log(x^2 + 1) - \dfrac{1}{2} \log x$
36. $-6 \log x - 2 \log y - 2 \log(x^2 + 1)$
37. $\dfrac{3}{2}$
38. 4
39. $\dfrac{1}{16}$
40. 7
41. 2.620136055
42. −2.124938737
43. 8.446070936
44. −5.853871964
45. 0.546964670
46. −5.298317367
47. 1.209061955
48. 1.057267894
49. −2.902201173
50. 1.548224439
51. 1.753880×10^{-3}
52. 7.082719×10^3
53. 1.323355×10^{-2}
54. 1.516257×10^3
55. 2.182658338
56. 2.930676559

Final Examination, Page 347

1. b
2. d
3. d
4. c
5. b
6. d
7. c
8. b
9. b
10. b
11. a
12. c
13. c
14. c
15. c
16. a
17. c
18. c
19. c
20. d
21. a
22. c
23. a
24. b
25. a
26. b
27. a
28. d
29. c
30. b